Urn Models and Their Applications in Finance

Masato Hisakado

Urn Models and Their Applications in Finance

Springer

Masato Hisakado
Nomura Hodings, Inc.
Chiyoda-ku, Tokyo, Japan

ISBN 978-981-96-3824-6 ISBN 978-981-96-3825-3 (eBook)
https://doi.org/10.1007/978-981-96-3825-3

This Springer imprint is published by the registered company Springer Nature Singapore Pte Ltd.
The registered company address is: 152 Beach Road, #21-01/04 Gateway East, Singapore 189721, Singapore

Competing Interests The author has no competing interests to declare that are relevant to the content of this manuscript.

Contents

List of Figures

List of Tables

Chapter 1
Introduction

We cannot mathematize the specific evolution of the biosphere. We can, at best, seek statistical laws about the distributions of aspect of this evolution.
Kauffman S A, A World Beyond Physics: The Emergence and Evolution of Life (Kauffman 2019)

1.1 Positive Feedback and Urn Process

According to Kauffman, we have not obtained mathematics that describes complex systems. However, the urn model can be the first step to represent complex systems. In fact, the Pólya urn model is a simple model with a positive feedback system, and the distribution does not converge to the delta function. Hence, it is difficult to predict the future. The positive feedback system has following four properties (Arthur 1994)

1. Unpredictability
2. Rigidity
3. Non-ergodic
4. Potential path inefficiencies

We can see these properties in Pólya urn and nonlinear Pólya urn models, which are urn models. These properties are related to the non-self-averaging process of these urn models. When the law of large numbers (LLN) can be applied, the ensemble average is the same as the sample average for a large time limit. However, in the non-self-averaging process, the ensemble average differs from the sample average. The origin of the non-self-averaging process of Pólya and nonlinear Pólya urn models is the conditional probability.

Here, we consider red balls and white balls in the urn. By changing in the probabilities by positive feedback, the average of the ratio of red balls depends on the sample path. The sample path is the history of the balls taken out. In other words, the impact affects the sample path and changes the average.

M. Hisakado, *Urn Models and Their Applications in Finance*,
https://doi.org/10.1007/978-981-96-3825-3_1

Abut "unpredictability", for example, the Pólya urn process does not converge to the delta function. This is a continuous distribution function with variance. We cannot predict the convergence point which is called the non-self-averaging process. According to Kauffman, it corresponds to the word "unprestatable".

In the case of the nonlinear Pólya urn process, multiple equilibria exist. It is difficult to predict which equilibrium is selected in the process. In nonlinear Pólya process, two phases are observed. One is the one-peak phase and the other is the two-peak phase. The one-peak phase includes only the absorbing sate. The two-peak phase has two equilibria. This is an absorption transition that belongs to the non-equilibrium phase transition. In other words, it is the phase transition between the self-averaging process and non self-averaging process.

"Rigidity" is the difficulty of switching the convergence point as time passes. In the case of nonlinear Pólya urn process, two equilibria exist. It is difficult to change one equilibrium to another equilibrium even if the change is to the lowest potential point. The difficulty of changing one equilibrium increases as time passes. In the equilibrium process, the process converges to the equilibrium, the lowest potential point. Rigidity is also called "rock in" of the process. The cost to change from the first selected equilibrium (dominant design) to another one is switching cost (Shapiro et al. 1998).

"Ergodic" means equilibrium and the process converges to the lowest potential point. In the case of "non-ergodic", the process does not converge to the lowest potential point. This corresponds to the non-self-averaging process which converges to several points which included a worse equilibrium. In contrast, in the equilibrium model, the solution oscillates between the two states of the nonlinear Pólya process. If we use the simulated annealing, the process converges to a good equilibrium. This is the Ising model, which is a toy model for the phase transition. In this book we apply annealing to several algorithms.

The fourth property, the process has path dependence. In other words, the process depends on its history of the process, and there is a probability that the process will converge to the worse equilibrium. In the equilibrium process, the process converges to a good equilibrium only. Therefore, the history is important for the non-equilibrium process. This corresponds to the non-self-averaging process.

In this book, we consider these properties using the urn process (Jhonson 1977; Mahmoud 2008). We find that the urn processes are simple but have a rich potential. We believe it is important to study complex systems.

1.2 Applications of Urn Models to Finance

A positive feedback system is useful to represent the correlation as the Pólya urn process which is a non-self-averaging process. Correlation is important for financial models. In contrast, in the non-self-averaging process, the ensemble average differs from the sample average. This indicates the difficulty of the parameter estimation because the observed average (sample average) differs from the ensemble average.

To estimate the parameters correctly, we need many samples or the assumption of independence. In the last half of the book, we discuss how to apply urn models to the time series of default data, including the temporal correlation.

One method to include the temporal correlation is the multi-term urn model. The effects of the previous time decay as time goes by. In this book, we use the exponential and power decays. When the decay is sufficient, the process becomes a self-averaging process and the parameters can be estimated correctly.

Another method is to change from the Gaussian process to the Poisson process by applying the double scaling limit. In this limit, we obtain the Negative Binomial distribution (NBD) that does not use the positive feedback. These processes are self-averaging processes.

Moreover, when we add positive feedback to NBD and Poisson processes by applying the multi-term, we obtain self-excited NBD and Hawkes processes (SE-NBD). Using this feedback, the model can represent the clustering of defaults. In both models, there is a phase transition between the steady state and non-steady state, and we can estimate the parameter in the steady state. These processes are applied in several fields including finance.

1.3 Overview of the Book

The remainder of this book is organized as follows. In Chap. 2, we introduce the correlated binomial model (CBM), which includes Pólya urn process. We discuss a general method to construct correlated binomial distributions by imposing several consistent relations on the joint probability function. We obtain self-consistency relations for conditional correlations and conditional probabilities. The beta-binomial distribution (BBD) is derived by a strong symmetric assumption on conditional correlations and it corresponds to Pólya urn process. Correlations are important for the application of mathematical models. We discuss the correlations of the financial applications in Chap. 10 and afterwards. Our derivation clarifies the correlation structure of the BBD. It is also possible to study the correlation structures of other solutions.

In Chap. 3, we consider the mixed process of Pólya urn and simple urn processes. The process converges to the delta function by adding the simple urn process, and we can see the phase transition of the convergence speed. We introduce a voting model that is similar to a Keynesian beauty contest and analyze it from a mathematical perspective. There are two types of voters—herders and independents—and two candidates. The herder's response function is linear. Herders correspond to the Pólya urn and the independents correspond to the simple urn process. This simple model contains three phases. If the herders constitute a majority or even half of the total voters, the voting rate converges more slowly than in a binomial distribution. If independents constitute the majority of voters, the voting rate converges at the same rate in a binomial distribution. We call Super-normal phase transition. This is the phase transition between strong self-averaging process and weak self-averaging process.

In Chap. 4, we discuss the nonlinear Pólya process. In the voting model, the herder's response function is nonlinear. In this case, the herding is stronger than the linear response function. We can identify two types of phase transitions. One is like the phase transition in the Ising model. However, this is a non-equilibrium phase transition. The other is the transition of super and normal diffusions, which was discussed in Chap. 3. These phase transitions co-exist for nonlinear Pólya process.

In Chap. 5, we discuss the extreme case of the nonlinear Pólya process, discussed in the previous section. Herders always select the majority of the previous votes, which is visible to them. We call them digital herders. As the fraction of herders increases, the model features a phase transition beyond which a state in which most voters make the correct choice coexists with one where most of them are incorrect. In this case, we obtain the exact solution of the model. Using the exact solution, we can discuss the critical exponents of this phase transition.

In Chap. 6, the herders obtain information from a finite number, r, of previous voters. The total number of balls in the urn is a constant case. It is the equilibrium process or ergodic process. The solution oscillates between the two states. A good (bad) equilibrium is where a majority of r select the correct (incorrect) candidate. I We show that there is no phase transition in this model. We demonstrate that if the annealing schedule is sufficiently slow from finite r to infinite r, the voting rate converges only to a good equilibrium. This simulated annealing process can be applied in several algorithms in Chap. 9. A phase transition can be observed in the infinite-size limit of r when the response function is nonlinear. This equilibrium process becomes the Ising model.

In Chap. 7, we extend the model in Chap. 6 to a multi-color balls case. This is the equilibrium process. When the herder is analog, we can calculate the distribution exactly. We obtain the Pitman sampling formula as the equilibrium distribution.

After Chap. 8, we describe some application of the urn models. In Chap. 8, we consider the relation between the urn process and the network. The number of links at the node increases as the urn process. In fact Barabási-Albert (BA) model, which is a scale-free network, is related to the Pólya urn process. We extend the urn process by introducing a parameter, ω that adjusts the strength of the feedback system. Note that the sample average of the distribution of links converges to the ensemble average for the large network limit. Hence, the distribution of links in the network does not depend on the sample, and it is the self-averaging process. It covers not only the positive feedback and negative feedback. The parameter is related to the preferential attachment of nodes in the network and has different weights for incoming and outgoing links. Using this parameter we can discuss the random graph, Barabási-Albert (BA) model, and lattice networks from a unified viewpoint, feedback, with the parameter ω with values $1, 0, -1$ characterizing these networks. ω is the number of added balls in the urn model, and negative ω corresponds to negative feedback.

In Chap. 6, we discussed the annealing of the equilibrium urn process. The equilibrium process is an ergodic process that converges to the optimized solution. In Chap. 9, we apply these properties to ant colony optimization (ACO) and the generic algorithm (GA). Here, we consider the phase transition of these algorithms and how to converge to good equilibrium.

After Chap. 10, we apply urn models to finance. Probability of default (PD) estimation is an important process for financial institutions. The difficulty of estimation depends on the correlations between borrowers. In Chap. 10, we consider a multi-year case with a temporal correlation using the multi-term urn process. A phase transition occurs when the temporal correlation decays by the power decay. When the power index is less than one, the PD estimator does not converge. It is difficult to estimate the PD using the limited historical data. Conversely, when the power index is greater than one, the convergence is the same as that of the binomial distribution.

In Chap. 11, we adopt the Merton model, which uses asset correlation instead of default correlation, and find that a phase transition occurs when the temporal correlation decays according to the power law. The phase transition is the phase transition between the slow and normal convergence speeds. It differs from the phase transition discussed in Chap. 10. The Merton model is a standard tool in financial institutions. In this chapter, we compare it to the urn process in Chap. 10. When the power index is less than one, the PD estimator converges slowly. Thus, estimating the PD is difficult with limited historical data. Conversely, when the power index is larger than one, the convergence speed is inversely proportional to the number of samples.

In Chap. 12, we consider the limit of a multi-term urn process with a correlation in the same term and temporal correlation. Here we obtain the Poisson process. The objective is to clarify the relationship between the urn model and the Hawkes process. In the double-scaling limit of this multi-term urn process, the self-exciting negative binomial distribution (SE-NBD) process, which is a marked Hawkes process, is obtained. In the standard continuous limit, this process becomes the Hawkes process, which has no correlation in the same term. The difference is the variance of the intensity function in that the phase transition from the steady state to the non-steady state can be observed. The critical point, at which the power law distribution is obtained, is the same for the Hawkes and urn processes.

We extend the SE-NBD process to a multidimensional process in Chap. 13. We apply a multidimensional self-exciting negative binomial distribution (SE-NBD) process to the default portfolios, including several sectors. Using the multidimensional SE-NBD process (MD-SE-NBD), we can estimate the interactions between these sectors as a network. By applying impact analysis to the data, we can classify the upstream and downstream sectors. Upstream sectors are real-estate and financial institution (FI) sectors. The shock spreads from the upstream to the downstream sectors. This result resembles an amplifier of the shock. This is consistent with the analysis of bubble bursts.

In Chap. 14, we consider the Poisson process limit of the Merton model, which is discussed in Chap. 11. We show that the Merton model becomes a Poisson process with the log-normal distributed intensity function. We discuss the relation between this model and the Hawkes process. In addition, we demonstrate why the beta distribution and Merton model have similar properties.

In Appendix, we consider a voting model on networks. It is the combination of models in Chaps. 5 and 8. We discuss the differences between the phases in which the networks depend. In $\omega \neq -1$, the following two types of phase transitions can be observed. The first is the transition between a state where most voters make correct choices and a state where most of them are incorrect. This is an absorption transition that belongs to the non-equilibrium transition. In the asymmetric case, a discontinuous phase transition, where the gap depends on the network. The super-normal transition is the transition of the convergence speed, and the critical point of the convergence speed transition depends on ω. At $\omega = 1$, in the BA model, this transition disappears. Both phase transitions disappear at $\omega = -1$ in the lattice case. In conclusion, as the performance near the lattice case, $\omega \sim -1$ exhibits the best performance of the voting in all networks. As the hub size decreases, the performance improves.

1.4 Taxonomy of the Chapters

In Table 1.1, we show the taxonomy of the models from the viewpoints of self-averaging and non-self-averaging process. Basically, the self-averaging process is the equilibrium and the non-self-averaging process is in non-equilibrium. The transitions discussed in this book are followings,

Table 1.1 The taxonomy of chapters

Self-averaging	Strong self-averaging	Weak self-averaging	Non-self-averaging	Phase transition
	Equilibrium		Non-equilibrium	
Chapter 2			○	
Chapter 3	○	○		○
Chapter 4	○	○	○	○
Chapter 5	○		○	○
Chapter 6	○			
Chapter 7	○			
Chapter 8	○ Distribution			
Chapter 9	○	○	○	○
Chapter 10	○		○	○
Chapter 11	○	○		○
Chapter 12	○		○	○
Chapter 13	○		○	○
Chapter 14	○	○		○
Appendix	○	○	○	○

1. Between weak and strong self-averaging processes.
2. Between non-self-averaging and the self-averaging processes.
3. Combination of 1 and 2.

References

Arthur B W, Increasing returns and path dependence in the economy. University of Michigan Press, (1994).
Johnson N L, and Kotz S, Urn models and their application; an approach to modern discrete probability theory (1977).
Kauffman S A, A world beyond physics: the emergence and evolution of life Oxford University Press, (2019).
Mahmoud H, Pólya urn models. Chapman and Hall/CRC (2008).
Shapiro C, Varian H R, Information Rules, Harvard business review press, (1998)

Chapter 2
Correlated Binomial Models and Pólya Urn Model

2.1 Introduction

Incorporation of correlation ρ into Bernoulli random variables $X(i)(i = 1, 2, \ldots N)$ taking the value 1 with probability p and taking the value 0 with probability $1 - p$ has a long history and has been widely discussed in a variety of fields such as science, mathematics, and engineering. They are also called binary variables. The expectation value of the random variable A as $\mathrm{E}(A)$, the correlation ρ between $X(i)$ and $X(j)$ is defined as follows

$$\rho = \mathrm{Corr}(X(i), X(j)) = \frac{\mathrm{E}(X(i)X(j)) - \mathrm{E}(X(i))\mathrm{E}(X(j))}{\sqrt{\mathrm{E}(X(i))(1 - \mathrm{E}(X(i)))\mathrm{E}(X(j))(1 - \mathrm{E}(X(j)))}}. \quad (2.1)$$

If there is no correlation between the random variables, the number n of the variables taking the value 1 obeys the binomial probability distribution $\mathrm{b}(N, p)$. Necessity of the correlation ρ comes from the fact that there are many phenomena in which dependency structures in random events are crucial or necessary for explaining experimental data.

In this chapter, we propose a general method for constructing correlated binomial models (CBM) based on consistent conditional probabilities and correlations. The proposed method makes it possible to study the correlation structure for any probability distribution function for exchangeable correlated Bernoulli random variables. The CBM represents the correlation using conditional probabilities, which corresponds to the non-self-averaging process.

The organization of this chapter is as follows. In Sect. 2.2, we introduce the conditional probabilities p_{ij} and conditional correlations ρ_{ij} and show how to construct CBMs. Section 2.3 presents solutions to the self-consistent relations. We obtain a beta binomial distribution (BBD) under strong symmetric conditional correlations assumptions. For other probability distribution functions, including the Moody's correlated binomial model (MCB) and distributions constructed by the superposition of binomial distributions (Bernoulli mixture model), we calculate p_{ij} and ρ_{ij}. We study

M. Hisakado, *Urn Models and Their Applications in Finance*,
https://doi.org/10.1007/978-981-96-3825-3_2

the probability distribution functions of these solutions from the viewpoint of their correlation structures ρ_{ij}. In Sect. 2.4, we discuss correlation structures of the solutions using examples. We summarize this chapter in Sect. 2.5. This chapter is based on Hisakado et al. (2006).

2.2 Correlated Binomial Models and Their Constructions

In this section, we construct the joint probabilities and distribution functions of the CBMs.

2.2.1 Construction of CBMs

We introduce the following definitions. The first one is the products of $X(i)$ and $1 - X(j)$, and they include all observables of the model,

$$\Pi_{ij} = \prod_{i'=1}^{i} X(i') \prod_{j'=i+1}^{i+j} (1 - X(j')). \tag{2.2}$$

The following definitions are their unconditional and **conditional expectation values** (see Fig. 2.1). **The conditional expectation is important to represent correlation.**

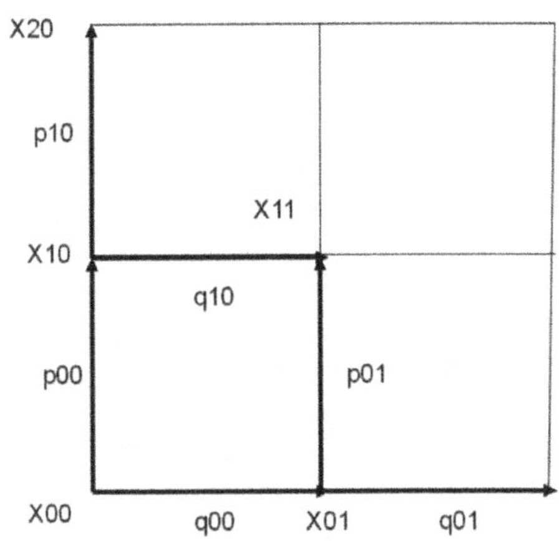

Fig. 2.1 Pascal's triangle like representation of X_{ij} and p_{ij}, q_{ij} up to $i + j \le 2$. $X_{00} = 1$, $X_{10} = \mathrm{E}(X_1) = p$, $X_{01} = \mathrm{E}(1 - X_1) = 1 - p = q$ etc

$$X_{ij} = \mathrm{E}(\Pi_{ij}), \tag{2.3}$$

$$p_{ij} = \mathrm{E}(X(i+j+1)|\Pi_{ij}=1) = \frac{X_{i+1j}}{X_{ij}}, \tag{2.4}$$

$$q_{ij} = \mathrm{E}(1-X(i+j+1)|\Pi_{ij}=1) = \frac{X_{ij+1}}{X_{ij}}, \tag{2.5}$$

where $X_{00}=1$, $X_{10}=p$ and $X_{01}=1-p=q$. Furthermore, the relation $p_{ij}+q_{ij}=1$ should hold for any i,j given the conservation of probabilities. All information is contained in X_{ij}. The joint probability $P(X(1)=x(1), X(2)=x(2), \ldots, X(N)=x(N))$ with $\sum_{i'=1}^{N} X(i')=n$ is given by X_{nN-n} and the distribution function $P_N(n)$ is also calculated as

$$P_N(n) = {}_N C_n \cdot X_{nN-n}. \tag{2.6}$$

To estimate X_{ij}, we must calculate the products of p_{kl} and q_{kl} from $(0,0)$ to (i,j). As at path, we can choose anyone, and the product must not depend on the choice. This property is guaranteed by the next condition on p_{ij} and q_{ij} as

$$q_{i+1j}\cdot p_{ij} = p_{ij+1}\cdot q_{ij} = \frac{X_{i+1j+1}}{X_{ij}}. \tag{2.7}$$

To satisfy these conditions, we introduce the following conditional correlations

$$\mathrm{Corr}(X(i+j+1), X(i+j+2)|\Pi_{ij}=1) = \rho_{ij}. \tag{2.8}$$

Here, we set $\rho_{00}=\rho$. $(1-X(i+j+1))$ and $(1-X(i+j+2))$ are also correlated with the same strength, and the following relation holds,

$$\mathrm{Corr}((1-X(i+j+1)), (1-X(i+j+2))|\Pi_{ij}=1) = \rho_{ij}. \tag{2.9}$$

From these relations, we obtain the following recursive relations for p_{ij} and q_{ij} as

$$\begin{aligned} p_{i+1j} &= p_{ij} + (1-p_{ij})\rho_{ij} \\ q_{ij+1} &= q_{ij} + (1-q_{ij})\rho_{ij}. \end{aligned} \tag{2.10}$$

If we assume the identity $p_{ij}+q_{ij}=1$, we obtain $q_{ij}=1-p_{ij}$, $q_{i+1j}=1-p_{i+1j}=(1-p_{ij})(1-\rho_{ij})$ and $p_{ij+1}=1-q_{ij+1}=p_{ij}(1-\rho_{ij})$. Then $q_{i+1j}\cdot p_{ij}=p_{ij+1}\cdot q_{ij}=p_{ij}(1-p_{ij})(1-\rho_{ij})$ holds and we see that the above consistency relation Eq. (2.7) does hold. We show the image of the communication relation in Fig. 2.2.

The remaining consistency relations or the probability conservation identity is $p_{ij}+q_{ij}=1$. We prove the identity using the inductive method. For $i=j=0$, the identity holds trivially as $p_{00}+q_{00}=p+q=1$. For $j=0$ or $i=0$, q_{i0} and p_{0j} are calculated as $q_{i0}=1-p_{i0}$ and $p_{0j}=1-q_{0j}$ and the identity also holds trivially.

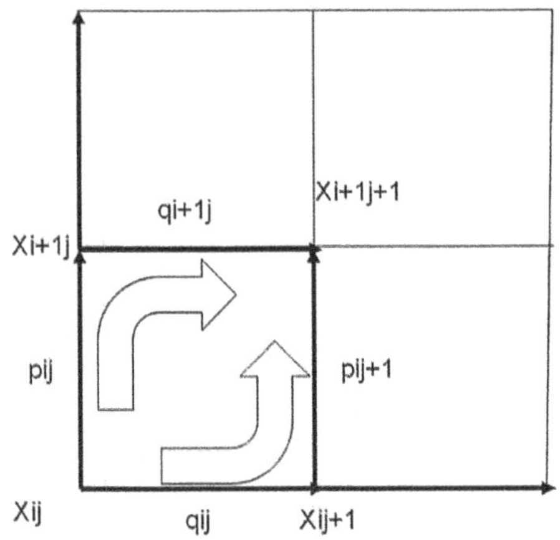

Fig. 2.2 Proof of the commutation relation $q_{i+1j} \cdot p_{ij} = p_{ij+1} \cdot q_{ij}$

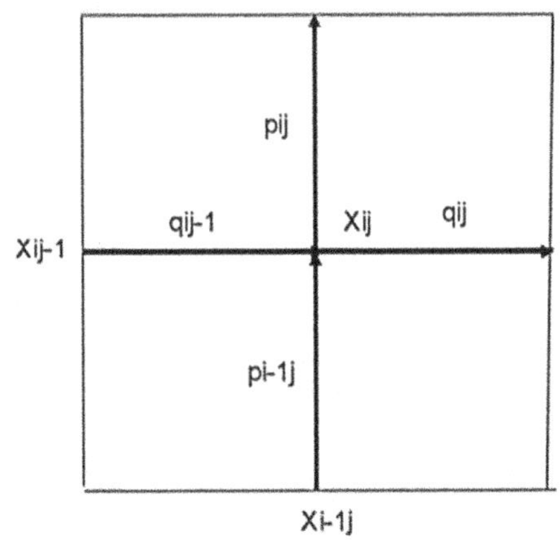

Fig. 2.3 Picture for the $p_{ij} + q_{ij} = 1$ condition

Then, we assume $p_{ij-1} + q_{ij-1} = 1$ and prove the identity $p_{ij} + q_{ij} = 1$ (see Fig. 2.3). From the recursive equations, we obtain Eq. (2.10) on p_{ij} and q_{ij}, we have the following relations:

$$
\begin{aligned}
1 &= p_{ij} + q_{ij} \\
&= p_{i-1j} + (1 - p_{i-1j})\rho_{i-1j} + (1 - p_{ij-1}) + p_{ij-1}\rho_{ij-1}.
\end{aligned} \tag{2.11}
$$

To satisfy the identity, the conditional correlation ρ_{ij-1} and ρ_{i-1j} must satisfy the following relations:

$$
p_{i-1j} - p_{ij-1} = -(1 - p_{i-1j})\rho_{i-1j} - p_{ij-1}\rho_{ij-1}. \tag{2.12}
$$

If the conditional correlations ρ_{ij} are fixed to satisfy the relations, the model becomes self-consistent. In other words, it guarantees the normalization of the resulting probability distribution.

2.2.2 Moments of CBMs

We estimate the moments of the CBM. For this purpose, we introduce the following operators H and D_k. The first linear operator H maps the polynomial in p, q to the joint probabilities. By its linearity, we only need to fix its action on the monomial $p^i q^j$ as

$$H[p^i q^j] = p_{00} p_{10} \cdots p_{i-10} q_{i0} q_{i1} \cdots q_{ij-1}. \tag{2.13}$$

The joint probability X_{nN-n} is expressed as $X_{nN-n} = H[p^n q^{N-n}]$. Here, we choose the far left path from $(0, 0)$ to $(n, N-n)$ in Pascal's triangle (see Fig. 2.1). The action of H on the binomial expansion $(p+q)^N = 1^N$ can be interpreted as the probability distribution and its normalization condition as follows,

$$1 = H[1^N] = H[(p+q)^N] = \sum_{n=0}^{N} {}_N C_n \cdot H[p^n q^{N-n}] = \sum_{n=0}^{N} {}_N C_n \cdot X_{nN-n}. \tag{2.14}$$

In order to calculate the moments of CBM, it is necessary to put n^k in the above summation. Instead, we will put $n(n-1)(n-2)\cdots(n-k+1)$ and introduce the following differential operators D_k.

$$D_k = \sum_{0 \le i_1, i_2, \ldots \le N-1}^{i_1 \ne i_2, i_1 \ne i_3, \ldots, i_{k-1} \ne i_k} p_{i_1 0} p_{i_2 0} \cdots p_{i_k 0} \frac{\partial^k}{\partial p_{i_1 0} \partial p_{i_2 0} \cdots \partial p_{i_k 0}} \tag{2.15}$$

The action of D_k on X_{nN-n} for $n \ge k$ is

$$D_k X_{nN-n} = n(n-1)(n-2)\cdots(n-k+1) X_{nN-n}. \tag{2.16}$$

In contrast, the same expression can be obtained as follows,

$$\begin{aligned} H[p^k \frac{d^k}{dp^k} p^n q^{N-n}] &= H[n(n-1)(n-2)\cdots(n-k+1) p^n q^{N-n}] \\ &= n(n-1)(n-2)\cdots(n-k+1) X_{nN-n}. \end{aligned} \tag{2.17}$$

This relation defines the action of D_k on the operator H with any polynomial $f(p, q)$ as

$$D_k H[f(p, q)] = H[p^k \frac{d^k}{dp^k} f(p, q)]. \tag{2.18}$$

Calculation of the expectation value of $n(n-1)\cdots(n-k+1)$ is performed by the action of operator D_k on the binomial expansion of $H[1^N] = H[(p+q)^N]$.

$$D_k H[(p+q)^N] = \sum_{n=0}^{N} {}_N C_n \cdot D_k X_{nN-n} \tag{2.19}$$

RHS is nothing but the expectation value $\mathrm{E}[n(n-1)(n-2)\cdots(n-k+1)]$. LHS is calculated using Eq. (2.18) as

$$\begin{aligned} D_k H[(p+q)^N] &= H[p^k \frac{d^k}{dp^k}(p+q)^N] \\ &= N(N-1)(N-2)\cdots(N-k+1)H[p^k(p+q)^{N-k}] \\ &= N(N-1)(N-2)\cdots(N-k+1)H[p^k] \\ &= N(N-1)(N-2)\cdots(N-k+1)p_{00}p_{10}p_{20}\cdots p_{k-10}. \end{aligned} \tag{2.20}$$

We obtain the relation,

$$\mathrm{E}[n(n-1)(n-2)\cdots(n-k+1)] = N(N-1)(N-2)\cdots(N-k+1)p_{00}p_{10}p_{20}\cdots p_{k-10} \tag{2.21}$$

From this relation, we can estimate the moments of the CBM. We obtain the average and variance,

$$E(S(N)) = Np, \tag{2.22}$$

and

$$V(S(N)) = Npq + N(N-1)pq\rho = \frac{Npq(1+N\theta)}{1+\theta}, \tag{2.23}$$

where $S(N) = \sum_i^N X(i)$ and $\theta = \rho/(1-\rho)$[1]. The first term of Eq. (2.23) represents for the variance of the binomial distribution, and the second term represents the variance of the correlations. Hence, we can consider that the CBM is the binomial distribution including the correlation.

2.2.3 *Self-Averaging and Non-Self-Averaging Processes*

Here, we consider self-averaging and non-self-averaging processes (Lifshitz 1968). The $\bar{X}(N)$ depending on the system size N is defined by the average,

$$\bar{X}(N) = \frac{S(N)}{N} = \frac{\sum_{i=1}^{N} X(i)}{N}. \tag{2.24}$$

[1] θ is the parameter for the correlation. We use it in Chap. 7.

It is the sample average. In contrast, the ensemble average of $X(N)$ is denoted $\mathrm{E}(X(N))$.

When

$$|\bar{X}(N) - \mathrm{E}(X(N))| \to 0, \tag{2.25}$$

in the limit $N \to \infty$, the sequence $\{X(i)\}$ is called self-averaging. The sample averages converges to the ensemble average. Otherwise, this sequence is non-self-averaging. Here, consider the case of $\mathrm{E}(X(N))$. In general, we can consider the ensemble and sample averages of the statistics. In Chap. 8, we consider the self-averaging of the distribution.

The law of large numbers (LLN) holds when $X(i)$ are independent of each other and LHS converges to 0. However, in non-self-averaging process, LLN does not hold because of the correlation.

In Sect. 2.3.1, we discuss the impact process. When there is a difference between the no impact and impact process in the large N limit,

$$|\bar{X}(N) - \bar{X}(N)^{+}| \to\neq 0, \tag{2.26}$$

where $X(N)^{+}$ is the average of the impact process, the process is equivalent to the non-self-averaging process. This definition means that the average depends on the path. This is the path dependency discussed in Chap. 1.

In some studies, non-self-averaging was defined differently,

$$CV(X(N))^2 = \frac{\mathrm{E}(X(N)^2) - \mathrm{E}(X(N))^2}{\mathrm{E}(X(N))^2} = \frac{\mathrm{V}(X(N))}{\mathrm{E}(X(N))^2} \to 0, \tag{2.27}$$

where $CV(X(N))$ is the coefficients of the variables (CV) and $\mathrm{V}(X(N)) = \mathrm{E}(X(N)^2) - \mathrm{E}(X(N))^2$ is the variance.

CV can be calculated using Eqs. (2.22) and (2.23) for the CBM,

$$CV(X(N))^2 = \frac{\mathrm{V}(X(N))}{\mathrm{E}(X(N))^2} = \frac{q(1+N\theta)}{p(1+\theta)N}. \tag{2.28}$$

In the limit, we obtain

$$\lim_{N\to\infty} CV(X(N))^2 = \frac{\theta q}{(1+\theta)p} = \frac{q}{p}\rho. \tag{2.29}$$

When $\lim_{N\to\infty} CV(X(N))^2 = 0$, the process is self-averaging. In fact, in the absence of correlation, $\rho = 0$, we obtain $CV(X(N)) = 0$. **We can conclude that CBM is non-self-averaging process.** Hence, Pólya process that corresponds to the BBD is also a non-self-averaging process because it is included in CBMs. We discuss the BBD in the next section.

2.3 Beta Binomial Distribution and Other Solutions in the CBM

In the previous section, we derived self-consistent equations for p_{ij} and ρ_{ij}. They are summarized as follows,

$$p_{i+1j} = p_{ij} + (1 - p_{ij})\rho_{ij}, \tag{2.30}$$

$$p_{ij+1} = p_{ij} - p_{ij}\rho_{ij}, \tag{2.31}$$

$$p_{i-1j} - p_{ij-1} = -(1 - p_{i-1j})\rho_{i-1j} - p_{ij-1}\rho_{ij-1}. \tag{2.32}$$

In this section, we present several solutions to these equations. We note that, if one knows the joint probabilities X_{ij}, from the definitions for p_{ij} and q_{ij}, we can estimate p_{ij}. Then ρ_{ij} are estimated from the recursive equation Eq. (2.30). In addition, we interpret the behaviors of the solutions from the viewpoint of the correlation structures. We present several solutions for CBMs.

2.3.1 *Beta Binomial Distribution*

Reduction to Beta Binomial Distribution

To solve the above relations on ρ_{ij} and p_{ij}, we use the symmetry viewpoint. For $p = \frac{1}{2}$ case, the model should have particle-hole duality between $X(j)$ and $1 - X(j)$ or Z_2 symmetry. Then $\rho_{ij} = \rho_{ji}$ should hold. **We make a stronger assumption that for any p, the system has the symmetric and ρ_{ij} depends on i, j only through the combination** $n = i + j$.[2] With a suitable choice of indexes $i \to i + 1$ and $j = n - i$, Eq. (2.32) reduces to

$$p_{in-i} - p_{i+1n-i-1} = \rho_n(-1 + p_{in-i} - p_{i+1n-i-1}). \tag{2.33}$$

From this relation, we see that p_{ij} with the same $n = i + j$ consists of an arithmetic sequence with the common difference Δ_n.

$$p_{i+1n-i-1} - p_{in-i} = \Delta_n. \tag{2.34}$$

Δ_n satisfy the following equation

$$\Delta_n = \rho_n(1 + \Delta_n). \tag{2.35}$$

ρ_n can be solved with Δ_n as

$$\rho_n = \frac{\Delta_n}{1 + \Delta_n}. \tag{2.36}$$

[2] This property is an important property of the BBD.

From the relation (2.30) for p_{ij}, we obtain the following recursive relation for ρ_n as,

$$\rho_n = \frac{\Delta_n}{1+\Delta_n} = \frac{\Delta_{n-1}(1-\rho_{n-1})}{1+\Delta_{n-1}(1-\rho_{n-1})} = \frac{\rho_{n-1}}{1+\rho_{n-1}}. \tag{2.37}$$

The explicit forms for ρ_n, and Δ_n are

$$\rho_n = \frac{\rho}{1+n\rho} \quad \text{and} \quad \Delta_n = \rho_{n-1}. \tag{2.38}$$

Then p_{ij} and q_{ij} can be obtained explicitly and the results are

$$p_{ij} = p_{i+j0} - j\Delta_{i+j} = \frac{p(1-\rho)+i\rho}{1+(i+j-1)\rho}, \tag{2.39}$$

$$q_{ij} = 1 - p_{ij} = \frac{q(1-\rho)+j\rho}{1+(i+j-1)\rho}. \tag{2.40}$$

X_{nN-n} are then obtained by taking the products of producing these conditional probabilities from $(0,0)$ to $(n, N-n)$,

$$X_{nN-n} = \prod_{i=0}^{n-1} p_{i0} \prod_{j=0}^{N-n-1} q_{nj}. \tag{2.41}$$

Putting the above results for p_{ij} and q_{ij} into them, we obtain

$$X_{nN-n} = \frac{\prod_{i=0}^{n-1}(p(1-\rho)+i\rho)\prod_{j=0}^{N-n-1}(q(1-\rho)+j\rho)}{\prod_{k=0}^{N-1}(1+(k-1)\rho)}. \tag{2.42}$$

Here $q = 1-p$. By multiplying the binomial coefficients ${}_N C_n$, we obtain the distribution function $P_N(n)$ as

$$P_N(n) = {}_N C_n \cdot X_{nN-n}. \tag{2.43}$$

This distribution is nothing but the beta binomial distribution function (BBD).

The BBD is also represented using the parameters α and β,

$$P(n) = {}_N C_n \cdot \frac{\mathrm{B}(\alpha+n, N+\beta-n)}{\mathrm{B}(\alpha,\beta)}. \tag{2.44}$$

where

$$p = \frac{\alpha}{\alpha+\beta}, \quad q = 1-p = \frac{\beta}{\alpha+\beta} \quad \text{and} \quad \rho = \frac{1}{\alpha+\beta+1}, \tag{2.45}$$

and $\mathrm{B}(\alpha, \beta)$ is the beta function. The beta function is defined as follows,

$$\mathrm{B}(\alpha, \beta) = \int_0^1 t^{\alpha-1}(1-t)^{\beta-1}dt. \tag{2.46}$$

Pólya Urn Problem and BBD

In this subsection, we introduce the Pólya urn problem and discuss its relation to BBD (Pólya 1931; Eggenberger et al. 1923). We now consider the urn process. At time 0, the urn contains θ_0 red balls and $n_0 - \theta_0$ white balls. It is the initial condition of the urn process. A ball is taken out at time t, and the balls are taken out sequentially from the urn. For example, a single ball is taken out, after which we return the ball taken out to the urn and add ω balls of the same color to the urn. Thus, the total number of balls increases by ω a process, which is a correlation parameter. In fact, as we take out more red balls, the probability that a red ball is taken out increases, because the number of red balls in the urn increases. This is a correlation created by the positive feedback. We repeat the process N times. The end of the process set to $t = N$. Hence, the total number of balls is $n_0 + N\omega$ at $t = N$.

Here, we introduce the variable N_1[3] which is the number of red balls taken out in this process. At time t, the number of red (white) balls is c_0 (c_1) in the urn.

$$\begin{aligned} c_0(t) &= \theta_0 + \omega k \\ c_1(t) &= n_0 - \theta_0(t-k) \end{aligned} \tag{2.47}$$

The probability that a red (white) ball is taken out is $c_0(t)/(c_0(t) + c_1(t))$ $(c_1(t)/(c_0(t) + c_1(t)))$. Because c_0 is proportional to k, we call this the linear Pólya process. When c_0 is not proportional to k, we will study this in the following chapters. When k red balls are taken out at time $t = N$, $N_1 = k$. the probability $P(N_1 = k)$ can be calculated as follows,

$$\begin{aligned} P(N_1 = k) &= \frac{N!}{k!(N-k)!}[\frac{\theta_0}{n_0} \cdots \frac{\theta_0 + (k-1)\omega}{n_0 + (k-1)\omega}][\frac{n_0 - \theta_0}{n_0 + k\omega} \cdots \frac{n_0 - \theta_0 + (N-k-1)\omega}{n_0 + (N-1)\omega}] \\ &= \frac{N!}{k!(N-k)!} \frac{\Pi_{i=0}^{k-1}(\theta_0 + i\omega)\Pi_{j=0}^{N-k-1}(n_0 - \theta_0 + j\omega)}{\Pi_{l=0}^{N-1}(n_0 + l\omega)} \\ &= \frac{N!}{k!(N-k)!} \frac{\Pi_{i=0}^{k-1}(p + i\rho/(1-\rho))\Pi_{j=0}^{N-k-1}(q + j\rho/(1-\rho))}{\Pi_{l=0}^{N-1}(1 + l\rho/(1-\rho))}, \\ &= \frac{N!}{k!(N-k)!} \frac{\Pi_{i=0}^{k-1}(p(1-\rho)/\rho + i)\Pi_{j=0}^{N-k-1}(q(1-\rho)/\rho + j)}{\Pi_{l=0}^{N-1}((1-\rho)/\rho + l)}, \\ &= \frac{N!}{k!(N-k)!} \frac{\Pi_{i=0}^{k-1}(\alpha + i)\Pi_{j=0}^{N-k-1}(\beta + j)}{\Pi_{l=0}^{N-1}(\alpha + \beta + l)}, \end{aligned} \tag{2.48}$$

[3] "1" means the first term. This step is preparation for the extension to the multi-term urn process.

Table 2.1 Summary of parameters for the urn process

Parameter and definition	
θ_0	Number of red balls at the start of the process
n_0	Number of total balls in the initial condition for each term
N	Number of balls taken out
ω	Increase of balls when we take out a ball. It is related to correlation.
k	Number of red balls taken out in the process

where $p = \theta_0/n_0$, $q = 1 - \theta_0/n_0$, and $\omega/n_0 = \rho/(1-\rho)$.[4] This is known as the beta binomial distribution BBD(α, β,N) where $p = \alpha/(\alpha+\beta)$ and $q = \beta/(\alpha+\beta)$. Using the parameters of urn, $\alpha = \theta_0/\omega$ and $\beta = (n_0 - \theta_0)/\omega$.

Here, α and β correspond to the parameters of the beta distribution in the continuous limit of the BBD. Note that $\rho = 1/(1+\alpha+\beta)$ plays the role of a correlation. The mean of k is $\mu = Np$ and the variance of k is given by $Npq + N(N-1)pq\rho$ and the second term corresponds to the correlation. Hence, ω which is the number of added balls, is the correlation parameter. In the limit $\omega \to 0$, $P(N_1 = k)$ becomes a binomial distribution that corresponds to the no correlation case. The parameters of the urn process are summarized in Table 2.1. We show an example of the urn process under the conditions $n_0 = 2$, $\theta_0 = 1$, and $\omega = 1$ in Fig. 2.4.

It is customary to represent the urn process using a square all addition matrix,

$$\mathrm{A} = \begin{pmatrix} A_{1,1} & A_{1,2} \\ A_{2,1} & A_{2,2} \end{pmatrix}. \tag{2.49}$$

The rows are indexed by the color of the ball picked, and the columns are indexed by the color of the balls added. In the case of the above urn process, the matrix is,

$$\mathrm{A} = \begin{pmatrix} \omega & 0 \\ 0 & \omega \end{pmatrix}. \tag{2.50}$$

In this book, mainly, we discuss positive feedback. When we set the matrix,

$$\mathrm{A} = \begin{pmatrix} 0 & 1 \\ 1 & 0 \end{pmatrix}, \tag{2.51}$$

the urn process is negative feedback and Bernard Friedman's urn (Freedman 1965; Friedman 1949). We discuss the negative feedback in Chaps. 3 and 8.

[4] We use the parameter $\theta = (1-\rho)/\rho$ instead of the correlation ρ in Chap. 7.

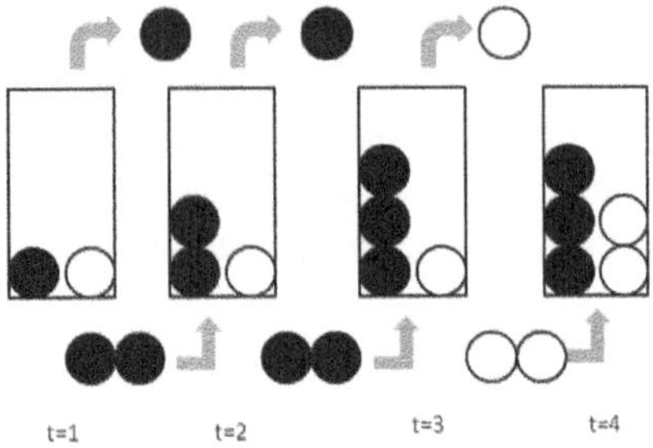

Fig. 2.4 Pólya urn process under conditions $n_0 = 2$, $\theta_0 = 1$, and $\omega = 1$

Impact Analysis of the BBD

We consider the impact

$$\delta(t_0) = \delta(t - t_0) = 1 \qquad t = t_0$$
$$= 0 \qquad t \neq t_0. \tag{2.52}$$

The response function $\phi(t - t_0)$ is defined by the difference $\delta S(t)$ between $S(t)$ and $S(t)^+$,

$$\delta S(t) = S(t)^+ - S(t) = \sum_{t_0=1}^{t} \phi(t, t - t_0)\delta(t_0), \tag{2.53}$$

where $S(t)^+$ is an impact process, $S(t)$ is a non-impact process and $S(t) = \sum_i^t X(i)$ and $S^+(t) = \sum_i^{t_0-1} X(i) + 1 + \sum_{i=t_0}^t X^+(i)$. X^+ is the process after the impact. The response function generally depends on t and t_0. We obtain

$$\mathrm{E}(\phi(t, t - t_0)) = \mathrm{E}(S(t)^+) - \mathrm{E}(S(t)) = \sum_k P(k|t_0)\mathrm{E}(S^+(t)|k, t_0) - \sum_k P(k|t_0)\mathrm{E}(S(t)|k, t_0)$$
$$= \sum_k P(k|t_0)\frac{\rho}{1 + t_0\rho}(t - t_0) = \frac{\rho}{1 + t_0\rho}(t - t_0), \tag{2.54}$$

where $P(k|t_0)$ is the distribution of the BBD at $t = t_0$ and $\sum_k P(k|t_0) = 1$. Here, we use the relation,

$$\mathrm{E}(S^+(t)|k, t_0) - \mathrm{E}(S(t)|k, t_0) = \frac{\rho}{1 + t_0\rho}, \tag{2.55}$$

which does not depend on k. The reduced response function is given by

$$\frac{\mathrm{E}(\phi(t, t-t_0))}{N} \to \frac{\rho}{1+t_0\rho}, \tag{2.56}$$

in the large $t = N$ limit. **The average of the reduced function does not decay to 0. This means that the effect of the impact continues. In other words, the impact changes the history of the process, which corresponds to the path dependency which we introduced in Chap.** 1.

Limit of BBD

Here, we consider the limit, $N \to \infty$. Using Eq. (2.48), we obtain in this limit,

$$\begin{aligned}
P(N_1 = k) &= \frac{N!}{k!(N-k)!}\frac{\Pi_{i=0}^{k-1}(\alpha+i)\Pi_{j=0}^{N-k-1}(\beta+j)}{\Pi_{l=0}^{N-1}(\alpha+\beta+l)}, \\
&= \frac{N!}{k!(N-k)!}\frac{(\alpha+k-1)!(\beta+N-k-1)!}{(\alpha+\beta+N-1)!}\frac{(\alpha-1)!(\beta-1)!}{(\alpha+\beta-1)!} \\
&= \frac{(\alpha+k-1)\cdots(k+1)(\beta+N-k-1)\cdots(N-k+1)}{(\alpha+\beta+N-1)\cdots(N+1)}\frac{1}{B(\alpha,\beta)}\frac{1}{B(\alpha,\beta)} \\
&\approx \frac{1}{N}(\frac{k}{N})^{\alpha-1}(1-\frac{k}{N})^{\beta-1}
\end{aligned} \tag{2.57}$$

where $B(\alpha, \beta) = \Gamma(\alpha)\Gamma(\beta)/\Gamma(\alpha+\beta)$. In the continuous limit,

$$P(t) = t^{\alpha-1}(1-t)^{\beta-1}dt, \tag{2.58}$$

where $t = k/N$. **It is the beta distribution**, $B(\alpha, \beta)$, $\alpha = p(1-\rho)/\rho$ and $\beta = q(1-\rho)/\rho$. The parameters of the beta distribution are related to the average and correlation. It is called the Balding-Nichols model, which has the mean p and the variance $\rho p(1-p)$ (Balding et al 1995). In Chap. 12, we obtain another limit of the BBD.

2.3.2 *Moody's Correlated Binomial Model*

When $\rho_{i,0} = \rho$ for all i (Witt 2004), we call this model Moody's Correlated Binomial (MCB) model. Comparing Eq. (2.36). MCB has a higher correlation ρ_n than BBC. The above consistent equations are difficult to solve, and the available analytic expressions are those for p_{i0} as $p_{i0} = 1-(1-p)(1-\rho)^i$. With this result, we have the following formal expression for X_{ij} as

$$X_{ij} = \mathrm{E}(\Pi_{ij}) = \mathrm{E}(\prod_{i'=1}^{i} X(i') \prod_{j'=i+1}^{i+j} (1 - X(j')))$$

$$= \sum_{k=0}^{j} (-1)^k {}_jC_k \mathrm{E}(\prod_{i'=1}^{i+k} X(i')) = \sum_{k=0}^{j} (-1)^k {}_jC_k \cdot p_{i+k0}. \tag{2.59}$$

With this expression, it is possible to estimate p_{ij}, q_{ij} and ρ_{ij} from their definitions. We can obtain the distribution as

$$P(S(N) = k) = \frac{N!}{k!(N-k)!} \sum_{j=0}^{N-k} [(-1)^j \frac{(N-k)!}{j!(N-k-j)!} \Pi_{i=0}^{j+k-1} p_{i0}]. \tag{2.60}$$

However, Eqs. (2.59) and (2.60) contain ${}_jC_k(-1)^k$ and as N becomes large, it becomes difficult to calculate them. We discuss the limit of MCB in Sect. 12.7 Appendix A.

2.3.3 Bernoulli Mixture Models

For Bernoulli mixture model with a mixing probability distribution function $f(p)$, the expression for the joint probability function X_{ij} is calculated with

$$X_{ij} = \mathrm{E}(\Pi_{ij}) = \int_0^1 dp f(p) p^i (1-p)^j. \tag{2.61}$$

If we use the beta distribution for $f(p)$, we obtain the BBD, Eq. (2.42). Consistent equations completely determine any correlated binomial distribution for exchangeable Bernoulli random variables. Each correlated binomial distribution follows these relations. **With the assumption $\rho_{ij} = \rho_{i+j}$, we are automatically led to the BBD. In other words, the probability distribution with this symmetry $\rho_{ij} = \rho_{i+j}$, we can prove that it is the BBD.** No other probability distribution has this symmetry as discussed in Sect. 2.3.1.

Here, we consider the relation between the CBM and the Bernoulli mixture model. According to De Finetti's theorem, the probability distribution of any infinite exchangeable Bernoulli random variables can be expressed using a mixture of the binomial distributions (De Finetti 1974). The CBM in the $N \to \infty$ limit should be expressed by this mixture. From Eq. (2.61), we obtain the relation $P(X(1) = 1, X(2) = 1, \ldots, X(k) = 1) = X_{k0} = \int f(p) p^k dp$. X_{k0} is expressed as $X_{k0} = p_{00} p_{10} \cdots p_{k-10}$, we have a correspondence between the moments of $f(p)$ and the CBM. That is, if one knows p_{i0} for any i, we know the mixing function $f(p)$ and vice versa. This correspondence shows the equivalence of the CBM and the Bernoulli mixture model at the large N limit.

2.3.4 Ising-Type Models

As with other mixing functions $f(p)$, we consider the cases that correspond to the Ising model with a strength of magnitude of correlation ρ[5] (Kitsukawa et al. 2006). If the system size N is sufficiently large, the distribution can be approximated by the superposition of two binomial distributions. If we take $N \to \infty$, the system loses its ergodicity, and the phase space breaks into two peaks (Goldenfeld 1992). When there appear two peaks, only one of them represents the real equilibrium state. This is a non-ergodic process and we can confirm in the nonlinear Pólya model in Chaps. 4 and 5.

We consider the cases that correspond to $p = 1/2$ (symmetric case) and $p \simeq 0$ (asymmetric case). For the symmetric case, $f(p)$ has two peaks at p and $q = 1 - p$ with the same height. For the assymetric case, $f(p)$ has a high peak at $p' = p'' \simeq 0$ and a short peak at $p' \simeq 1$. The following mixing functions are called them Ising-type models.

Symmetric Case

Here, we set

$$f(p') = \frac{1}{2}\delta(p' - p) + \frac{1}{2}\delta(p' - q), \tag{2.62}$$

where $q = 1 - p$. This mixing function corresponds to the symmetric Ising model. X_{ij} are given as

$$X_{ij} = \frac{1}{2}(p^i q^j + p^j q^i). \tag{2.63}$$

p_{ij} and ρ_{ij} are calculated easily as

$$p_{ij} = \frac{p^{i+1}q^j + p^j q^{i+1}}{p^i q^j + p^j q^i} \tag{2.64}$$

$$\rho_{ij} = \frac{p^{i+j}q^{i+j}(p-q)^2}{p^{i+j}q^{i+j}(p^2+q^2) + qp(p^{2i}q^{2j} + q^{2i}p^{2j})}. \tag{2.65}$$

This solution has the symmetry, $\rho_{ij} = \rho_{ji}$. We call this model the Two-Binomial model.

Asymmetric Case

Next, we set

$$f(p') = (1-\alpha)\delta(p' - p'') + \alpha\delta(p' - 1). \tag{2.66}$$

[5] See Chap. 6 for Ising model

This mixing function corresponds to the asymmetric Ising model. We call this model Binomial plus (B+) model, because it is a binomial distribution plus one small peak at $n = N$. Between p, ρ and p'', α, we have the relations

$$p = \alpha + (1-\alpha)p'', \quad \text{and} \quad \rho = \frac{\alpha(1-p'')}{\alpha + (1-\alpha)p''}, \tag{2.67}$$

and

$$\alpha = \frac{\rho p}{1 - p + \rho p}. \tag{2.68}$$

X_{ij} are given as

$$X_{ij} = (1-\alpha)p''^{i}(1-p'')^{j} + \alpha\delta_{j,0} \tag{2.69}$$

p_{ij} and ρ_{ij} are calculated easily as

$$p_{i0} = \frac{\alpha + (1-\alpha)p''^{i+1}}{\alpha + (1-\alpha)p''^{i}} \quad \text{and} \quad p_{ij} = p'' \quad \text{for} \quad j \neq 0 \tag{2.70}$$

and

$$\rho_{i0} = \frac{\alpha(1-p'')}{\alpha + (1-\alpha)p''^{i+1}} \quad \text{and} \quad \rho_{ij} = 0 \quad \text{for} \quad j \neq 0. \tag{2.71}$$

2.4 Correlation Structures of Solutions

In this section, we study the relations between probability distribution and the correlation structure.

2.4.1 *Symmetric Case*

Figure 2.5 shows the probability distribution profiles of three correlated models, MCB, BBD, and Two-Binomial models. We set $p = 0.5$, $\rho = 0.3$ and $N = 30$. We also show the binomial distribution for comparison. The first three curves have the same p and ρ, however, their profiles are drastically different. The symmetric Two-Binomial model has two peaks, and their overlap decreases as N increases. In the limit $N \to \infty$, the overlap disappears, and the system loses its ergodicity. In contrast, the BBD's profile is broad, and even if we set $N \to \infty$, we obtain the beta distribution as discussed in Sect. 2.3, and the shape is nearly unchanged.

The profile of the MCB model is peculiar. It is not symmetric and shows singular rippling. The origin of the ripping can be understood by inspecting of its correlation structure. Figure 2.6 shows the correlation structures for the above three models. The parameters are equal and we show ρ_{i30-i}. In contrast to the BBD's correlation,

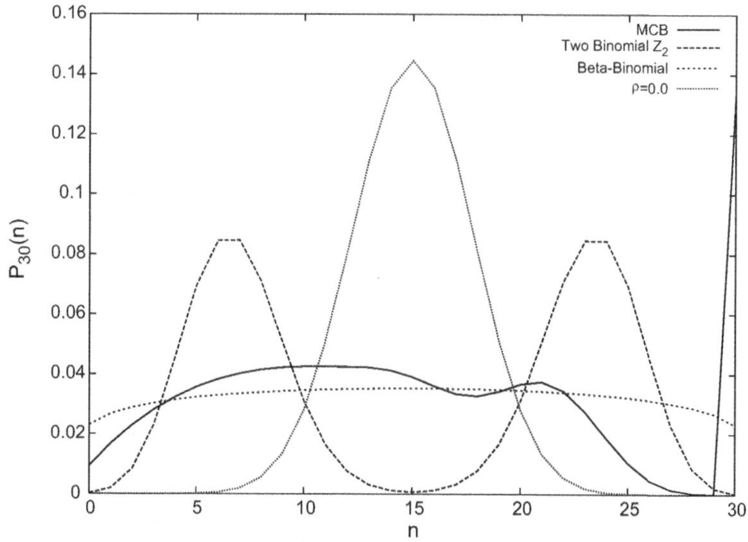

Fig. 2.5 Probability distribution $P_{30}(n)$ for $p = 0.5$, $\rho = 0.3$, and $N = 30$. The three distributions, MCB (solid line), BBD (dotted line) and Two-Binomial (thin dotted line). These distributions have the same variance, which is determined by ρ. We also show a binomial distribution ($\rho = 0.0$) as the comparison

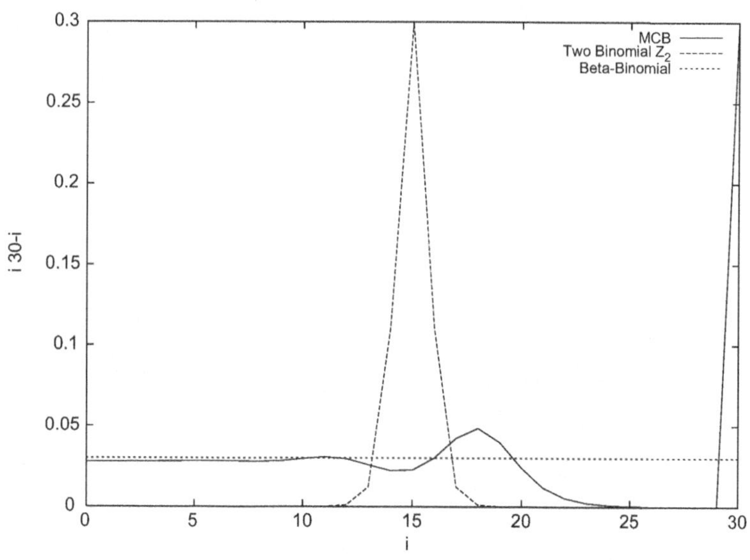

Fig. 2.6 Correlation $\rho_{i\,30-i}$ for the MCB (solid line), BBD (thin dotted line), and Two-Binomial (dotted line) models. We set $\rho = 0.3$ and $p = 0.5$ as in the previous figure

which is constant with $i + j$ fixed, the correlation for the MCB has a sharp peak at $i = 30$ and shows a strong rippling structure. The curve is not symmetric and the distortion is reflected in the shape of the probability distribution. On the other hand, the correlation curve for a Two-Binomial distribution has a strong peak at $i = \frac{N}{2}$ and is much different from the BBD's correlation curve. This strong peak and rapid decay may be reflected in the decomposition of the probability distribution.

2.4.2 Asymmetric Case

Figure 2.7 shows the probability distribution of MCB, BBD, and B+ models. We set $p = 0.1$, $\rho = 0.3$, and $N = 30$. We also show the pure binomial distribution for comparison. MCB and BBD have almost the same bulk shape, however, MCB has a small peak at $n = 30$. B+ has a stronger peak at $n = 30$, and its bulk shape can be obtained from a small left shift of the pure binomial distribution $p = 0.1$. These profile differences are reflected in their correlation structures in Fig. 2.8. The result shows the correlation structures for the above three models. The parameters are equal to those in the previous figure. In contrast to the constant BBD structure, MCB and B+ models have a peak at $i = 30$.

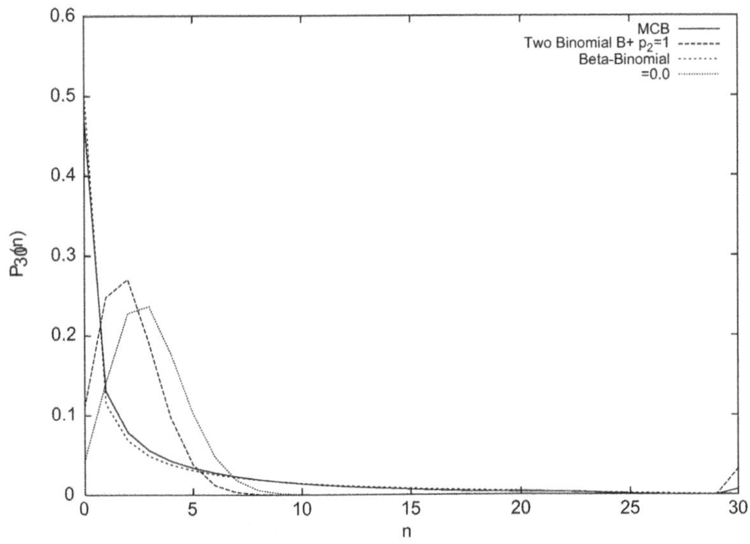

Fig. 2.7 Probability distribution $P_{30}(n)$ for $p = 0.1$, $\rho = 0.3$, and $N = 30$. We show three distributions, MCB (solid line), BBD (dotted line), and B+ (thin dotted line). These distributions have the same variance, which is decided by ρ. We also show a binomial distribution ($\rho = 0.0$) for comparison

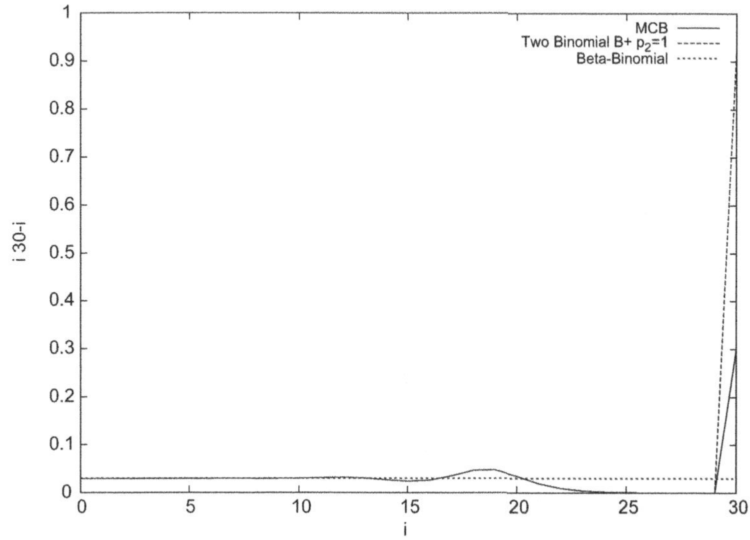

Fig. 2.8 Correlation ρ_{i30-i} for MCB (solid line), BBD (thin dotted line), and B+ (dotted line) models. Here, we set $\rho = 0.3$ and $p = 0.1$

2.5 Summary

Correlation is important but it is difficult to introduce in the model. Pólya urn model is one of the methods to introduce the correlation using the positive feedback. In this chapter, we presented a general method to constructing correlated binomial models (CBM) and calculated their moments. The CBM includes Moody's correlated binomial model (MCB) and the beta binomial distribution (BBD), which corresponds to Pólya urn model.

References

Balding DJ, Nichols RA A method for quantifying differentiation between populations at multi-allelic loci and its implications for investigating identity and paternity Genetica **96 (1–2)** 3–12 (1995).

de Finetti B, Theory of Probability: A Critical Introductory Treatment, Wiley (1974-5).

Eggenberger F, and Pólya G, Über die Statistik Verketteter Vorgänge, Z. Angew. Math. Mech. 1, 279–289 (1923).

Freedman D A, Bernard Friedman's urn, Ann. Math. Statist **36** 956–970 (1965).

Friedman B, A simple urn model, Comm. Pure Appl. Math. **2**, 59–70 (1949).

Goldenfeld N, *Lectures on Phase Transitions and the Renormalization Group*, Addison-Wesley Publishing Company (1992).

Hisakado M, Kitsukawa K, and Mori S, Correlated binomial models and correlation structures, Journal of Physics A: Mathematical and General, **39(50)** 15365 (2006).

Kitsukawa K, Mori S and Hisakado M, Evaluation of Tranche in Securitization and Long-range Ising Model, Physica A, **368(1)** 191–206 (2006).

Lifshitz I M, Theory if fulcutuation level in disordered systems, Zh. Eksp. Teor. Fiz. **53**, 743–758 (1968).

Pólya G, Sur quelques points de la théorie des probabilités, Ann. Inst. Henri Poincaré, **1**, 117–161 (1931).

Witt G, Moody's Correlated Binomial Default Distribution, (Moody's Investors Service) **August 10** (2004).

Chapter 3
Phase Transition of Super and Normal Diffusion Phases

3.1 Introduction

The Keynesian beauty contest is a popular concept that is used to explain price fluctuations in equity markets (Keynes 1936). Keynes described the actions of rational agents in a market using an analogy based on a fictional newspaper contest.

It is not a case of choosing those that, to the best of one's judgment, are really the prettiest, nor even those that average opinion genuinely thinks the prettiest.

J.M Keyens The General Theory of Employment, Interest and Money (1936)

In the contest, entrants were asked to choose a set of the six most beautiful faces from photographs of different women. Entrants who select the most popular face will be awarded a prize. A naive strategy would be to choose the most beautiful face according to the opinion of the entrant. Entrants are known for employing such a strategy as independent voters. A more sophisticated entrant, aiming to maximize his/her chances of winning a prize, would try to deduce the majority's perception of beauty. This implies that the entrant would make a selection based on some inference from his/her knowledge of public perception. Such voters are known as herders. To estimate public perception, people observe the actions of other individuals; then, they make a choice similar to those of others. Because it is usually sensible to do what other people are doing, this phenomenon is assumed to be the result of a rational choice. Nevertheless, this approach can sometimes lead to arbitrary or even erroneous decisions. This phenomenon is known as an information cascade (Bikhchandani et al. 1992). In this chapter, we use the Pólya urn model and the simple urn for the voting model.

We introduce a voting model with two candidates C_1 and C_2. As mentioned above, we set two types of voters—independents and herders. Independents' voting is based on their fundamental values; on the other hand, herders' voting is based on the number of votes. The herders' making decision process is the same as the Pólya urn model. We call these herders analog herders. In this chapter, we consider the case of analog

M. Hisakado, *Urn Models and Their Applications in Finance*,
https://doi.org/10.1007/978-981-96-3825-3_3

herder. On the other hand, the process of independents are the simple urn model, the binomial model. Hence, in the case where all the voters are herders, the process is the Pólya process, and the voting rate converges to a beta distribution in a large time limit.[1] However, when there are independents as voters, the process converges to the delta function. Independents play the role of noise. Although this model is very simple, it contains a super-normal phase transition. It is the phase transition of the speed of convergence.

The super-diffusive behavior in the sense that variance $V(T)$ grows asymptotically faster than T (where T is the time or the length) in several fields (Eisenstein et al. 1995; Schenkel et al. 1993; Hod et al. 2004; Usatenko et al. 2003; Huillet 2008). It is characterized by variance $V \sim T^{\alpha}$ when $\alpha > 1$. When $\alpha = 1$, the diffusion of variance becomes a Brownian motion. For example, in the case of daily financial data, T represents the time series of the data. The past price affects the present price, and diffusion becomes faster than the Brownian motion. Such phenomena can be attributed to long-range positive correlations. We may observe a dynamical phase transition (from normal to super-diffusive behavior) (Hod et al. 2004). In such a phase transition, the correlation plays an important role. In this chapter, we analyze super-normal phase transition.

The organization of this chapter is as follows. In Sect. 3.2, we introduce our voting model and mathematically define the two types of voters—independents and herders. In Sect. 3.3, we show the relation between the voting model and the urn process. In the urn model, this voting model is a mixture of the two types of urns. In Sect. 3.4, we calculate the distribution functions strictly for the special cases—independent voters always vote for either of the two candidates; their behavior is not probabilistic. Then, we obtain the solution of the master equation. In this case, there is no phase transition. In Sect. 3.4, we discuss more general cases. We use a stochastic differential equation, the Fokker-Planck equation, and a numerical simulation. In this model, we can observe the phase transition at the ratio of herders to independents through the variance of the distributions. This is the super-normal transition. If the herders constitute a majority or half of the total number of voters, the voting rate converges more slowly than in a binomial distribution. If independents constitute the majority of voters, the voting converges at the same rate as it would in binomial distribution. In Sect. 3.5, we confirm the phase transition by numerical simulations. The last section presents a summary of this chapter. This chapter is based on Hisakado et al. (2010).

3.2 Voting Model and Master Equation

We model the voting of two candidates, C_0 and C_1. At time t, each candidate has $c_0(t)$ and $c_1(t)$ votes. Here, we consider the Bernoulli random variables $X(t)$, which corresponds to the vote at time t. $X(t) = 1(0)$ is the vote to the candidate $1(0)$ at t where $t > 1$. Then, we obtain

[1] See Chap. 2.

$$c_1(t) = c_1 + \sum_{i=2}^{t} X(i), \tag{3.1}$$

and

$$c_0(t) = c_0 + \sum_{i=2}^{t} (1 - X(i)). \tag{3.2}$$

Initially ($t = 1$), the two candidates, C_0 and C_1, have $c_0(1)$ and $c_1(1)$ votes, respectively. Hereafter, we omit the time for the initial votes ($c_0 \equiv c_0(1)$ and $c_1 \equiv c_1(1)$) and define $\theta \equiv c_0 + c_1$. At each time step, one voter votes for one candidate. Voters are allowed to see the number of votes for each candidate when they vote so that they have knowledge of public perception.

There are two types of voters—independents and herders. Independents vote for C_0 and C_1 with probabilities $1 - q$ and q, respectively. Their votes are independent of others' votes and depend on what they believe is their fundamental value. Therefore, its opinion decision process is a simple urn model. The herders vote for each candidate with the probabilities proportional to the candidates' votes. If the numbers of votes are $c_0(t)$ and $c_1(t)$ at time t, a herder votes with probability $c_0(t)/(c_0(t) + c_1(t))$ for C_0 and $c_1(t)/(c_0(t) + c_1(t))$ for C_1. Herders' votes are determined by the number of votes.

Here, we set the ratio of independents to herders as $1 - p$ and p, respectively. If we set $p = 1$, this system becomes a Pólya model with c_0 starting elements of type C_0 and c_1 starting elements of type C_1. In this case, the distribution of the voting rate is a beta distribution. As such, this system is a mixture of the Pólya process and the binomial process.

The evolution equation of candidate C_0 is given by

$$\begin{aligned} P(k,t) = {} & p\frac{k-1}{\theta+t-2}P(k-1,t-1) + p(1 - \frac{k}{\theta+t-2})P(k,t-1) \\ & +(1-p)qP(k,t-1) + (1-p)(1-q)P(k-1,t-1). \end{aligned} \tag{3.3}$$

$P(k,t)$ is the distribution of the number of votes k at time t for candidate C_0. The first and second terms of Eq. (3.3) denote the votes of the herders; the third and fourth terms denote the votes of the independents. **Eq.** (3.3) **is the master equation.**

If we set $Q(k,t)$ as the distribution of the number of votes k at time t for candidate C_1, we have the following relation

$$Q(k,t) = 1 - P(\theta + t - 1 - k, t). \tag{3.4}$$

The initial condition is $P(c_0, 1) = 1$.

3.3 The Relation Between the Urn Models and the Voting Model

We model an urn model with two color balls, C_0 and C_1. At time t, $c_0(t)$ and $c_1(t)$ balls have been taken out. They are the sum of the balls that were taken out from both urns. In the voting model, we set $\omega = 1$ in the urn model, which we explain in Chap. 2. Therefore, the number of total balls increases one by one.

Initially ($t = 1$), the two colors of balls, C_0 and C_1, have c_0 and c_1, respectively.[2] At each time step, one ball is taken out.

There are two types of urns—simple and Polya urn. In the simple urn, a ball that is C_0 or C_1 is taken out with probabilities $1 - q$ and q, respectively. Therefore, it is a sequential process. In the Pólya urn, a ball which is C_0 or C_1 is taken out with probabilities that are proportional to the records. After that, we return the ball taken out to the urn and add one ball of the same color to the urn for the Pólya urn.

If the number of balls that have been taken out are $c_0(t)$ and $c_1(t)$ at time t, with probability $c_0(t)/(c_0(t) + c_1(t))$ for C_0 and $c_1(t)/(c_0(t) + c_1(t))$ for C_1. Pólya urn is based on the number of balls that have been taken out. Then, in the voting model, the herder is called an analog herder.

Here, we set the ratio of the urn from which we take out the balls, the simple urn to Pólya urn as $1 - p$ and p, respectively. We show an example of the process in Fig. 3.1. If we set $p = 1$, this system becomes a Pólya model with c_0 starting elements of type C_0 and c_1 starting elements of type C_1. In this case, the distribution of the $C_0(C_1)$ ball is a beta distribution. As such, this system is a mixture of Pólya process and binomial distribution. We show an image of the process in Fig. 3.2.

3.4 Master Equation and Exact Solutions for $q = 1$ (or $q = 0$)

In this section, we study the exact solution of Eq. (3.3) for a special case. For $q = 1$, we obtain the following master equation:

$$\begin{aligned} P(k,t) &= p\frac{k-1}{\theta+t-2}P(k-1,t-1) + p\left(1-\frac{k}{\theta+t-2}\right)P(k,t-1) \\ &\quad +(1-p)P(k,t-1). \end{aligned} \tag{3.5}$$

At this limit, the independents always vote for only one candidate, C_1 (if we set $q = 0$, independent voters vote only for C_0). There is one color in the simple Urn case. The master equation has a simpler form:

$$P(k,t) = P(k,t-1) - \frac{kp}{\theta+t-2}P(k,t-1) + \frac{p}{\theta+t-2}(k-1)P(k-1,t-1). \tag{3.6}$$

[2] Note that $c_1 = \theta_0$ and $\theta = c_1 + c_0 = n_0$. Using α and β, $c_1 = \alpha$ and $c_0 = \beta$, because we set $\omega = 1$ in the voting model.

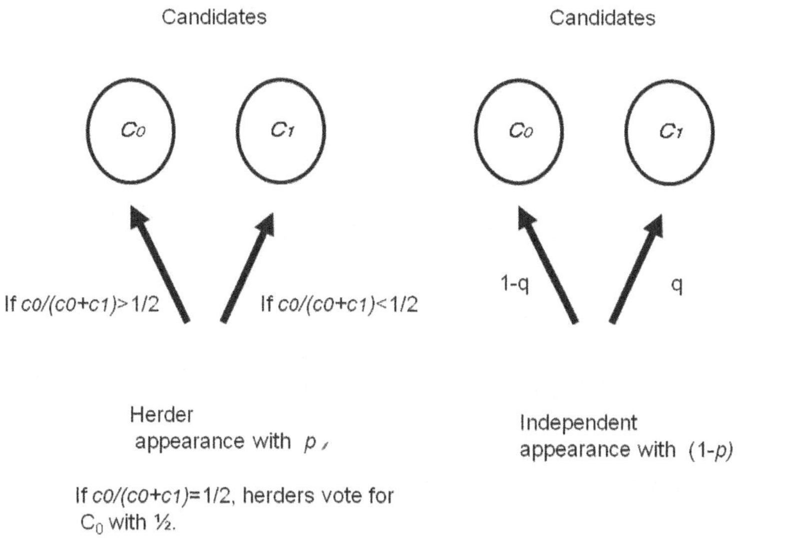

Fig. 3.1 Demonstration of model

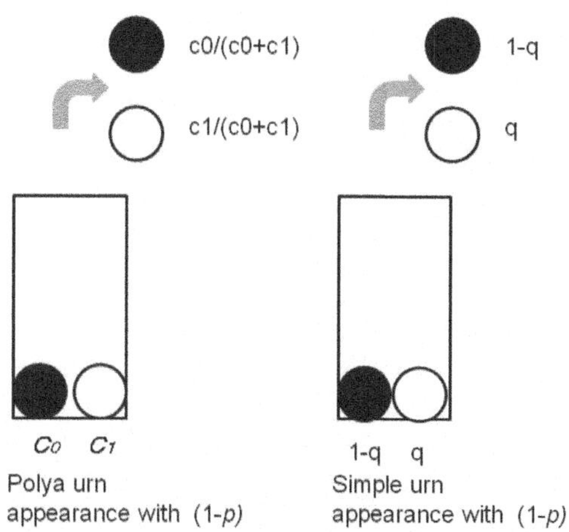

Fig. 3.2 This is an example of the urn process. The balls are taken out sequentially. The type of urn is selected randomly. The Pólya urn depends on the history of the taken out balls. The simple urn does not depend on the history of the balls taken out. The Pólya urn appears randomly at the probability p

When we substitute $k = c_0$ in the above equation, the last term of the RHS vanishes, and thus, the probability $P(k, t)$ can be calculated easily:

$$P(c_0, t) = (1 - \frac{c_0 p}{\theta})(1 - \frac{c_0 p}{\theta + 1}) \cdots (1 - \frac{c_0 p}{\theta + t - 2}), \tag{3.7}$$

where $(z)_i = z \cdot (z + 1) \cdots (z + i - 1)$. For $k > C_0$, we can prove (see Sect. 3.7 Appendix A) that the following equality holds:

$$P(k',t) = \sum_{l=1}^{k'-c_0+1} (-1)^{l-1} \frac{(c_0)_{k'-c_0}}{(1)_{k'-c_0-l+1}(1)_{l-1}} \frac{(\theta - (l+c_0-1)p)_{t-1}}{(\theta)_{t-1}}. \tag{3.8}$$

This is the distribution of votes for the special case wherein the independent voters always vote for only one candidate, C_1.

If we set $p = 1$, all voters are herders, and we obtain the following reduction:

$$\begin{aligned} P(k',t) &= \frac{(c_0)_{k'-c_0}}{(\theta)_{t-1}(1)_{k'-c_0}} \sum_{l=1}^{k'-c_0+1} (-1)^{l-1} \binom{k'-c_0}{l-1} (c_1 - l + 1)_{t-1}, \\ &= \binom{t-1}{k'-c_0} \frac{(c_0)_{k'-c_0}(c_1)_{t-1-k'+c_0}}{(\theta)_{t-1}}. \end{aligned} \tag{3.9}$$

This is a beta binomial distribution. At the limit $t \to \infty$, the above equation becomes a beta distribution. Note that to obtain Eq. (3.9), we use the identity

$$\sum_{l=1}^{k'-c_0+1} (-1)^{l-1} \binom{k'-c_0}{l-1} (c_1 - l + 1)_{t-1} = \frac{(1)_{t-1}}{(1)_{t-1-k'+c_0}} (c_1)_{t-1-k'+c_0}.$$

In the limit $t \to \infty$ and $c_0, c_1 \to 0$ with $\alpha = c_1/c_0$ fixed, the scale invariance holds over the entire range.

Here, we can calculate the moment of these distributions to analyze them. The moment is given by

$$\mu_i(t) = \sum_{k=c_0}^{c_0+t-1} k^i P(k,t). \tag{3.10}$$

We introduce the quasi-moment as

$$\hat{\mu}_i(t) \equiv \sum_{k=c_0}^{c_0+t-1} k(k+1)\cdots(k+i-1)P(k,t). \tag{3.11}$$

We can prove (see Sect. 3.9 Appendix B) that the quasi-moment can take the following form:

$$\hat{\mu}_i(t) = \sum_{l=1}^{t} (-1)^{l-1} \frac{(c_0)_{t+i}}{(1)_{l-1}(1)_{t-l}(l+c_0+i-1)} \frac{(\theta - (l+c_0-1)p)_{t-1}}{(\theta)_{t-1}}. \tag{3.12}$$

If we set $i = 1$, we obtain the average vote

$$\hat{\mu}_1(t) = \sum_{l=1}^{t} (-1)^{l-1} \frac{(c_0)_{t+1}}{(1)_{l-1}(1)_{t-l}(l+c_0)} \frac{(\theta - (l+c_0-1)p)_{t-1}}{(\theta)_{t-1}}. \tag{3.13}$$

We study $t \to \infty$. The coefficients of master equation Eq. (3.5) do not contain the initial votes for C_0; given by c_0. If we set $t >> \theta = c_0 + c_1$, the master equation does not depend on the initial conditions. Therefore, for a large t limit, the behavior of the moment does not depend on the initial conditions c_0 and c_1. We can also observe this in Eqs. (3.23) and (3.24) in the next section. Here, we set $c_0 = c_1 = 1$ for the representative case. Direct calculation using Eq. (3.12) is difficult. Hence, we study $P(k', t)$ as $t \to \infty$. In the above case, distribution Eq. (3.8) becomes a constant distribution with cut-off k^*. The cut-off implies a fast decay for larger values ($k' > k^*$). If we set $p = 1$, we obtain a constant distribution. Using Eq. (3.8), we obtain

$$\lim_{t\to\infty} P(k', t) = \sum_{l=1}^{k'} (-1)^{l-1} \frac{(1)_{k'-1}}{(1)_{k'-l}(1)_{l-1}} t^{-p}. \tag{3.14}$$

For large time values, the only time dependent term is t^{-lp}. In the case of $t >> k'$, we can assume that only the first term of the summation is non-negligible. Therefore,

$$\lim_{t\to\infty} P(k', t) \sim t^{-p}. \tag{3.15}$$

Th cut-off k^* is the inflection point of $P(k', t)$. Using Eq. (3.14), we can obtain $k^* = t^p + 2$. Then, the moment is given by

$$\mu_i(t) \sim \sum_{k=1}^{k^*} k^i P(k, t) \sim t^{ip}. \tag{3.16}$$

Equation (3.16) is a continuous function of p. Hence, there is no phase transition throughout. In the next section, we discuss the general case for the continuous limit.

3.5 Super-Normal Phase Transition

3.5.1 Fokker-Plank Equation

To investigate long-range correlations, we analyze in the limit $t \to \infty$. We can rewrite Eq. (3.5) as

$$\begin{aligned} c_0(t) = k \to k+1 : P(k, t) &= \frac{kp}{\theta + t - 1} + (1-p)(1-q) \\ &= \frac{p}{2}\left(1 + \frac{2k - (\theta + t - 1)}{\theta + t - 1}\right) + (1-p)(1-q). \end{aligned} \tag{3.17}$$

We define $\Delta_t = 2c_0(t) - (\theta + t - 1)$ with the initial condition $\Delta_1 = c_0 - c_1 = 2c_0 - \theta$. We change the notation from k to Δ_t for convenience. Then, we have $|\Delta_t| = |2k - (\theta + t - 1)| < c + t - 1$. The support for the law of Δ_t is thus $\{\Delta_1 -$

$(t-1), \Delta_1 + (t-1)\}$. Given $\Delta_t = s$, we obtain a random walk model

$$\Delta_t = s \to s+1 : P_{\frac{s+\theta+t-1}{2},t} = \frac{p(s+\theta+t-1)}{2(\theta+t-1)} + (1-p)(1-q),$$
$$\Delta_n = s \to s-1 : Q_{\frac{s+c+t-1}{2},t} = 1 - P_{\frac{s+c+t-1}{2},t}. \tag{3.18}$$

Let $\epsilon = 1/c \to 0$. We now consider

$$X_\tau = \epsilon \Delta_{[t/\epsilon]},$$
$$P(x,\tau) = \epsilon P(\Delta_t/\epsilon, t/\epsilon), \tag{3.19}$$

where $\tau = t/\epsilon$ and $x = \Delta_t/\epsilon$.

We use $\delta X_\tau = X_{\tau+\epsilon} - X_\tau$ and ζ_τ, a standard i.i.d. Gaussian sequence; our objective is to identify the drift f_τ and variance g_τ^2 such that

$$\delta X_\tau = f_\tau(X_\tau)\epsilon + \sqrt{\epsilon} g_\tau(X_\tau)\zeta_{\tau+\epsilon}. \tag{3.20}$$

Given $X_\tau = x$, by using the transition probabilities of Δ_n, we obtain

$$\mathrm{E}(\delta X_\tau) = \epsilon \mathrm{E}(\Delta_{[\tau/\epsilon]+1} - \Delta_{[\tau/\epsilon]}) = \epsilon(2p_{[\frac{1/\epsilon+c+\tau/\epsilon-1}{2}],\tau/\epsilon} - 1)$$
$$= \epsilon[(1-p)(1-2q) + \frac{px}{\tau+1}]. \tag{3.21}$$

The drift term is $f_\tau(x) = (1-p)(1-2q) + px/(\tau+1)$. Moreover,

$$\sigma^2(\delta X_\tau) = \epsilon^2[1^2 p_{[\frac{1/\epsilon+c+\tau/\epsilon-1}{2}],\tau/\epsilon} + (-1)^2(1 - p_{[\frac{1/\epsilon+c+\tau/\epsilon-1}{2}],\tau/\epsilon})] = \epsilon^2, \tag{3.22}$$

such that $g_{\epsilon,\tau}(x) = \sqrt{\epsilon}$. In the continuous limit, we can obtain the Fokker-Plank diffusion equation for this process:

$$\frac{\partial P}{\partial \tau} = \frac{1}{2}\frac{\partial^2 P}{\partial x^2} - \frac{p}{\tau+1}\frac{\partial (xP)}{\partial x} - (1-p)(1-2q)\frac{\partial P}{\partial x}. \tag{3.23}$$

3.5.2 Stochastic Difference Equation and Super-Normal Phase Transition

We can also obtain X_τ such that it obeys a diffusion equation with small additive noise:

$$\mathrm{d}X_\tau = [(1-p)(1-2q) + \frac{pX_\tau}{\tau+1}]\mathrm{d}\tau + \sqrt{\epsilon}\mathrm{d}B_\tau, \qquad X_0 = \frac{c_0 - c_1}{c}, \tag{3.24}$$

where dB_τ is the Brownian motion. It is the stochastic differential equation or Langevin equation. The parameters, $q = 1/2$ and $p = -1$, corresponds to the Bernard Friedman's urn (Friedman 1949; Freedman et al. 1965). Though Eqs. (3.23) and (3.24) are equivalent, hereafter, we only deal with Eq. (3.24) for simplicity. Assume c_0 is random or deterministic. Let

$$\sigma_0^2 \equiv \sigma^2(Y_0) = 4\epsilon^2\sigma^2(s) \tag{3.25}$$

be the variance of X_0. If X_0 is a Gaussian ($X_0 \sim \mathrm{c}(y_0, \sigma_0^2)$) or deterministic ($X_0 \sim \delta_{x0}$), the law of X_τ ensures that the Gaussian is in accordance with the density

$$p_\tau(x) \sim \frac{1}{\sqrt{2\pi}\sigma_\tau} \mathrm{e}^{-(x-x_\tau)^2/2\sigma_\tau^2}, \tag{3.26}$$

where $x_\tau = \mathrm{E}(X_\tau)$ is the expected value of X_τ and $\sigma_\tau^2 \equiv v_\tau$ is its variance. If $\Phi_\tau(\lambda) = \log(\mathrm{e}^{\mathrm{i}\lambda X_\tau})$ is the logarithm of the characteristic function of the law of X_τ, we have

$$\partial_\tau \Phi_\tau(\lambda) = \frac{p}{1+\tau}\lambda\partial_\lambda \Phi_\tau(\lambda) + \mathrm{i}(1-p)(1-2q)\lambda - \frac{\epsilon}{2}\lambda^2 \tag{3.27}$$

and

$$\Phi_\tau(\lambda) = \mathrm{i}\lambda x_\tau - \frac{\lambda^2}{2} v_\tau. \tag{3.28}$$

By identifying the real and imaginary parts of $\Phi_\tau(\lambda)$, we obtain the dynamics of the mean of X_τ as

$$\dot{x}_\tau = \frac{p}{1+\tau} x_\tau + (1-p)(1-2q). \tag{3.29}$$

The solution for x_τ is

$$x_\tau = (x_0 + 2q - 1)(1+\tau)^p + (1-2q)(1+\tau) \sim (x_0 + 2q - 1)\tau^p + (1-2q)\tau. \tag{3.30}$$

Since we are interested in the voting rate obtained, we introduce a new scaled variable:

$$\tilde{x}_\tau \equiv \frac{x_\tau}{\tau}.$$

The solution for $\tilde{x}_\tau$ is

$$\tilde{x}_\tau \sim (x_0 + 2q - 1)\tau^{(p-1)} + (1-2q). \tag{3.31}$$

When $p \neq 1$, $\tilde{x}_\tau \sim 1 - 2q$. This implies that the average ratio of C_0's votes to the total votes is $1 - q$. When $p = 1$, $\tilde{x}_\tau \sim x_0$. This agrees with the assertion that the scaled distribution of votes becomes a beta distribution when τ is large. In this case, the mean value does not change.

From the above discussion, we can infer that the distribution is similar to a delta function. The question of how this distribution converges to a delta function constitutes the next problem. To investigate this, we analyze the dynamics of variance. The dynamics of v_τ are given by the Riccati equation

$$\dot{v}_\tau = \frac{2p}{1+\tau}v_\tau + \epsilon. \tag{3.32}$$

If $p \neq 1/2$, we obtain

$$\begin{aligned} v_\tau &= v_0 + \int_0^\tau (\frac{1+\tau}{1+r})^{2p}(\frac{2p}{1+r}v_0 + \epsilon)dr = v_0(1+\tau)^{2p} + \epsilon \int_0^\tau (\frac{1+\tau}{1+r})^2 p dr \\ &= v_0(1+\tau)^{2p} + \frac{\epsilon}{1-2p}(1+\tau)^{2p}((1+\tau)^{1-2p} - 1). \end{aligned} \tag{3.33}$$

If $p = 1/2$, we have

$$v_\tau = v_0 + \int_0^\tau (\frac{1+\tau}{1+r})(\frac{1}{1+r}v_0 + \epsilon)dr = v_0(1+\tau) + \epsilon(1+\tau)\log(1+\tau). \tag{3.34}$$

Now, we can summarize the temporal behavior of the variance as follows,

$$v_\tau \sim \frac{\epsilon}{1-2p}\tau \qquad \text{if} \quad p < \frac{1}{2}, \tag{3.35}$$

$$v_\tau \sim (v_0 + \frac{\epsilon}{2p-1})\tau^{2p} \qquad \text{if} \quad p > \frac{1}{2}, \tag{3.36}$$

$$v_\tau \sim \epsilon\tau\log(\tau) \qquad \text{if} \quad p = \frac{1}{2}. \tag{3.37}$$

Here, we introduce the rescaled variables

$$\tilde{v}_\tau \equiv \frac{v_\tau}{\tau^2}.$$

The solution for $\tilde{v}_\tau$ is

$$\tilde{v}_\tau \sim \frac{\epsilon}{1-2p}\tau^{-1} \qquad \text{if} \quad p < \frac{1}{2}, \tag{3.38}$$

$$\tilde{v}_\tau \sim (v_0 + \frac{\epsilon}{2p-1})\tau^{2p-2} \qquad \text{if} \quad p > \frac{1}{2}, \tag{3.39}$$

$$\tilde{v}_\tau \sim \epsilon\frac{\log(\tau)}{\tau} \qquad \text{if} \quad p = \frac{1}{2}. \tag{3.40}$$

If $p = 1$, $\tilde{v}_\tau$ becomes v_0. This agrees with our assertion that the distribution of votes becomes a beta distribution. If $p > 1/2$ or $p = 1/2$, candidate C_0 gathers $1 - q$ of all votes in the scaled distributions, however, the voting rate converges more slowly than that in a binomial distribution. If $0 < p < 1/2$, the voting rate becomes $1 - q$, and the distribution converges as in a binomial distribution. Hence, if independent voters form a majority, the distribution of votes becomes similar to a delta function and the convergence is at the same rate as that in a binomial distribution. If the herders form the majority, the distribution remains the same, but the convergence is slower than that in a binomial distribution. We call this the super-normal phase transition Similar phenomena can be observed in several fields. In daily financial data, the motion of the price does not represent a Markov process, and it is difficult to forecast the future price (Usatenko et al. 2003). In fact, it has been pointed out that the motion of price is a super-diffusive behavior and the stochastic differential equation for price is similar to Eq. (3.24) (Hod et al. 2004; Schütz at al. 2004). When all voters are herders, the distribution becomes a beta distribution and does not converge as the Pólya urn model.

We now consider the correlation. For a beta binomial distribution, we can define the parameter $\rho \equiv 1/(c + 1)$ (Hisakado et al. 2006). This parameter represents the strength of following a decision. If we set $\rho = 1$, everyone votes for the candidate who received the first voter's vote. In contrast, when $\rho = 0$, the herders become independent. Note that our conclusion does not depend on ρ except when $\rho = 0$. For large t, convergence is not related to ρ, but is related to p, the appearance probability of independent voters.

Here, we discuss the solution presented in the previous section. If we set $q = 0$ ($q = 1$) is the same as the relation (3.4), Eq. (3.17) becomes 0 for large t. This is because independent voters' votes become deterministic. Hence, in Eq. (3.23), the diffusion term disappears. In Eq. (3.32), the noise term ϵ disappears, and the dynamics of v_τ are given by

$$\dot{v}_\tau = \frac{2p}{1 + \tau} v_\tau. \tag{3.41}$$

Then, the phase transition disappears, and the behavior of v_0 is continuous

$$\tilde{v}_\tau \sim (v_0)\tau^{2p-2} \qquad \text{for all} \quad p. \tag{3.42}$$

This result is acceptable, following the discussion in the previous section and that Eq. (3.12) is continuous with respect to p. (See Eq. (3.16).) If there is a consensus about the fundamental value, the herders affect the convergence in proportion to their ratio. In addition, in this case, convergence does not depend on the correlation ρ.

3.5.3 Self-Averaging Process

We consider the coefficients of variables (CV). $CV(X_\tau)$ can be calculated as,

$$
\begin{aligned}
CV(X_\tau)^2 = \frac{\nu_\tau}{x_\tau^2} &\sim \frac{\epsilon}{(1-2p)(1-2q)^2}\tau^{-1} \quad \text{if} \quad p < \frac{1}{2}, \\
&\sim (v_0 + \frac{\epsilon}{2p-1})\frac{1}{(1-2q)^2}\tau^{2p-2} \quad \text{if} \quad p > \frac{1}{2}, \\
&\sim \epsilon \frac{1}{(1-2q)^2}\frac{\log(\tau)}{\tau} \quad \text{if} \quad p = \frac{1}{2}.
\end{aligned}
\tag{3.43}
$$

If $0 < p < 1/2$, $CV(X_\tau)^2$ converges to 0, and the behavior is $CV(X_\tau)^2 \sim \tau^{-1}$. It is a strong self-averaging process. If $p > 1/2$ or $p = 1/2$, $CV(X_\tau)^2$ converges to 0 and slower than the strong self-averaging process. This is called weak self-averaging process. **We can confirm this is the phase transition between the weak self-averaging process and the strong self-averaging process.**

3.5.4 Mean Field Approximation

Next we solve Eq. (3.24) using the mean field approximation. We can assume a stationary solution as

$$X_\infty = \bar{v}\tau + (1-p)(1-2q)\tau, \tag{3.44}$$

where $\bar{v}$ is constant. Substituting Eq. (3.44) into Eq. (3.24), we obtain the self-consistent equation,

$$\bar{v} = p\bar{v} + (1-p)(1-2q). \tag{3.45}$$

We can obtain the solution,

$$\bar{v} = 1 - 2q. \tag{3.46}$$

It is the mean field equation (Fig. 3.3a). This is consistent with the conclusion of Eq. (3.46).

We show Eq. (3.45) in Fig. 3.3(a) is the case $p \neq 0$ and (b) is the case $p = 0$. In the case of (a), there is the solution, $\bar{v} = 1 - 2q$. On the other hand, in the case of (b), the solutions are on the line. In this case, all voters are herders and this corresponds to the Pólya urn process. The solution is spread across the range. It is the beta binomial distribution.

We expand X_τ around the solution $\bar{v}\tau + (1-p)(2q-1)\tau$.

$$X_\tau = \bar{v}\tau + (1-p)(1-2q)\tau + W_\tau. \tag{3.47}$$

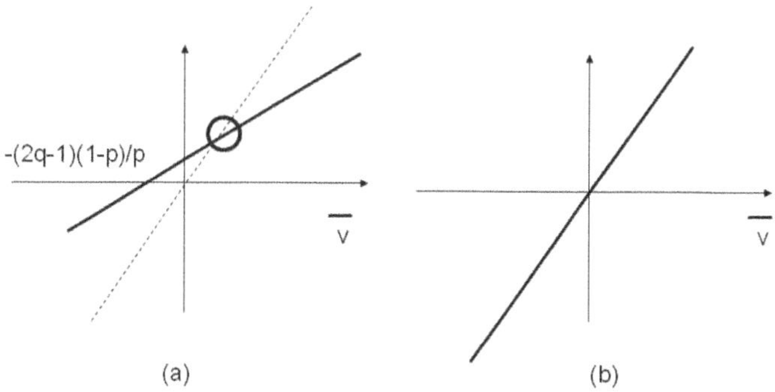

Fig. 3.3 Solutions of the self-consistent equation (3.45). **a** $p \neq 0$ and **b** $p = 0$. In the case of **a**, there is a solution, $\bar{v} = 1 - 2q$. In contrast, in the case of **b**, the solutions are on the line

Here, we set $X_\tau \gg W_\tau$, that is $\tau \gg 1$. We rewrite Eq. (3.24) using Eq. (3.47) and obtain as follows:

$$dW_\tau = p\frac{W_\tau}{\tau + 1}d\tau + \sqrt{\epsilon}dB_\tau. \tag{3.48}$$

The critical point $p_{vc} = 1/2$ and it does not depend on q. This is the super-normal phase transition (Fig. 3.3).

3.5.5 Impact Analysis

Here, we study the discrete process, Eq. (3.18) when $q = 1/2$ using the property of martingale. The definition of a martingale is

$$\mathrm{E}[f(t+1)|F_t] = f(t), \tag{3.49}$$

where F_t is the historical record. We obtain using Eq. (3.18),

$$\mathrm{E}[\Delta_{t+1}|F_t] = \frac{p}{t}\hat{S}(t), \tag{3.50}$$

where $\hat{S}(t) = \sum_i^t \Delta_i$. Hence, we obtain

$$\mathrm{E}[S_{t+1}|F_t] = S_t + \mathrm{E}[\Delta_{t+1}] = (1 + \frac{p}{t})\hat{S}(t), \tag{3.51}$$

Here, we set

$$\gamma_i = 1 + \frac{p}{i}, \tag{3.52}$$

and

$$a_t = \prod_{i=1}^{t-1} \gamma_i. \tag{3.53}$$

We consider the process,

$$\hat{M}_t = \frac{\hat{S}(t)}{a_t}. \tag{3.54}$$

We can calculate

$$\mathrm{E}[\hat{M}_{t+1}|F_t] = \frac{1}{a_{t+1}}\mathrm{E}[\hat{S}(t+1)|F_t] = \frac{1}{a_{t+1}}\gamma_t \hat{S}(t) = \frac{\hat{S}(t)}{a_t} = \hat{M}_t. \tag{3.55}$$

Hence, the process $\{\hat{M}(t)\}$ is Martingale.

Here, we consider the impact process $\hat{S}(t)^+ = \sum_i^t \Delta_i$ with the initial condition,

$$\Delta(0) = 1, \tag{3.56}$$

and the non-impact process $\hat{S}(t) = \sum^t \Delta_i$ with this condition,

$$\begin{aligned} \Delta(0) &= 1 \qquad \text{the probability} \quad 1/2, \\ \Delta(0) &= -1 \qquad \text{the probability} \quad 1/2. \end{aligned} \tag{3.57}$$

Using Eq. (3.55) we can obtain,

$$\mathrm{E}[\hat{S}(t)^+] = a_t = \frac{\Gamma(t+p)}{\Gamma(t)\Gamma(p+1)} \sim \frac{t^p}{\Gamma(p+1)} \tag{3.58}$$

in the limit $t \to \infty$ and $\mathrm{E}[\hat{S}(t)] = 0$ because of the symmetry. Here, we use the relation

$$\frac{\Gamma(t+p}{\Gamma(t)} \sim t^p. \tag{3.59}$$

Then, we obtain the scaled impact

$$\frac{\mathrm{E}[\hat{S}(t)^+] - \mathrm{E}[\hat{S}(t)]}{t} = \frac{t^{p-1}}{\Gamma(p+1)}, \tag{3.60}$$

in the limit $t \to \infty$. **Then, the impact affects the process as the power curve and the difference converges to 0 in the limit without** $p = 1$. In the limit $p = 1$ the impact does not converge to 0 which corresponds to Pólya urn process discussed in Chap. 2.

For the Brownian motion, the difference is the diffusion, $\mathrm{E}[\hat{S}(t)^{+}] - \mathrm{E}[\hat{S}(t)] \sim \sqrt{t}$. Hence, we can estimate the super-normal phase transition at $p_{vc} = 1/2$ as the previous discussions.

3.6 Numerical Simulations

To confirm the analytical results of the asymptotic behavior, we perform numerical simulations. We use the master equation Eq. (3.3) directly.

Figure 3.4 displays the deformation of the distribution of votes for C_0 over time t. Figure 3.4(a) shows the case in which independent voters vote for C_0 with the probability $1 - q$, and Fig. 3.4(b) shows the case in which independent voters always vote for C_0. We can see that the distribution converges to a delta function for the votes of the independent voters. If all voters are herders ($p = 1$), the distribution becomes a beta binomial distribution. Because of the independent voters, the distribution is deformed. For the case $q = 0$, we can obtain the exact solution in the previous section.

Figure 3.5 displays the resulting scaled variance for different p. The distribution converges to a delta function over time. However, there are differences between the left and the right figures at $p \geq 0.5$. The difference is in the speed of convergence, which is characterized by the scaled variance $\tilde{v}$. In the general case ($q \neq 0, 1$), we can recognize the phase transition as Fig. 3.5(a). If $p > 0.5$, the variance converges at the same rate as that in a binomial distribution (slope $= -1$). If $p < 0.5$ or $p = 0.5$, super-diffusive behavior is exhibited, and convergence is slower than that in a binomial distribution (slope > -1). The cases $p < 0.5$ and $p = 0.5$ represent two different phases. In contrast, for other cases ($q = 0, 1$), the slope changes continuously as Fig. 3.5(b). There is no phase transition.

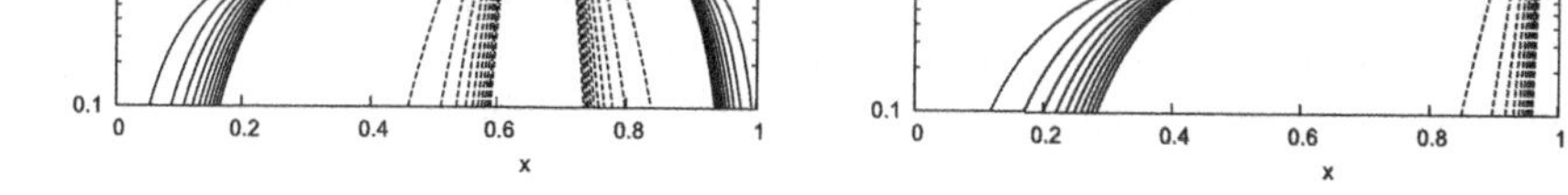

(a) (b)

Fig. 3.4 Asymptotic behavior of the scaled variance $\tilde{v}_\tau$ for $q = 0.5$ (Left) and $q = 0$ (Right), $p = 0.9, 0.7, 0.5, 0.3, 0.1$. Asymptotic behavior of the distribution of votes for C_0 at $t = 100 \to 10000$ for $q = 1/3$ (Left) and $q = 0$ (Right) in $p = 0.9, 0.3$

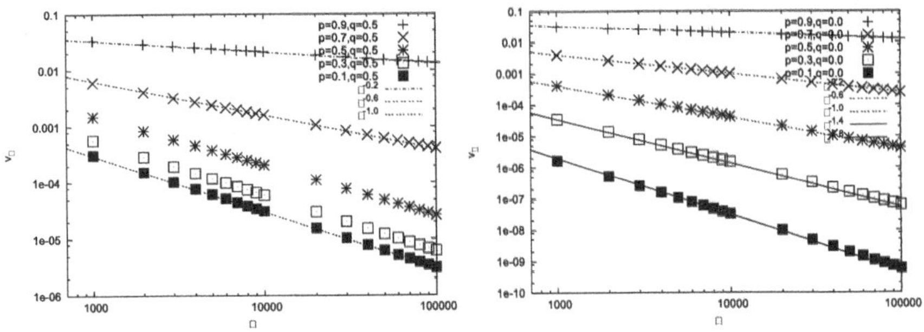

(a) (b)

Fig. 3.5 Asymptotic behavior of the scaled variance $\tilde{v}_\tau$ for $q = 0.5$ (left) and $q = 0$ (right), $p = 0.9, 0.7, 0.5, 0.3, 0.1$

3.7 Summary

We investigated a voting model similar to a Keynesian beauty contest. Mathematically, our model is a mixture of the binomial process (independents) and the Pólya urn process (herders). We calculated the exact solution for special cases of the master equation and analyzed it using a stochastic differential equation for the general case.

In the special cases, $q = 0.1$, no phase transition occurs because the independents vote deterministically. In general q, the correlation structure, exhibits a dramatic change at a critical value. If herders constitute a majority or half of the total number of voters, the variance converges slower than it would in binomial distributions. It is the super-normal phase transition or the phase transition between the weak and strong self-averaging processes. We discussed this phase transition using solutions of stochastic differential equation, mean field approximation and impact analysis. In the impact analysis, we used the martingale property.

3.8 Appendix A Proof of Eq. (3.8)

We prove the assumption of Eq. (3.8). Multiplying Eq. (3.5) by $(-1)^{k-c_0}(l-1)!/\{(k-1)(k-2)\cdots c_0(l-k+c_0-1)!\} = (-1)^{k-c_0}(1)_{l-1}/(c_0)_{k-c_0}(1)_{l-k+c_0-1}$ and summing over $k = c_0, c_0+1, \ldots, l+c_0-1$, we obtain

$$
\begin{aligned}
&\sum_{k=c_0}^{l+c_0-1}(-1)^{k-c_0}\frac{(1)_{l-1}}{(c_0)_{k-c_0}(1)_{l-k+c_0-1}}P(k,t)\\
&=\sum_{k=c_0}^{l+c_0-1}(-1)^{k-c_0}\frac{(1)_{l-1}}{(c_0)_{k-c_0}(1)_{l-k+c_0-1}}P(k,t-1)-\frac{p}{c+t-2}\\
&\times\sum_{k=c_0}^{l+c_0-1}(-1)^{k-c_0}\frac{(1)_{l-1}}{(c_0)_{k-c_0}(1)_{l-k+c_0-1}}[kP(k,t-1)-(k-1)P(k-1,t)]],
\end{aligned}
\tag{3.61}
$$

where $(z)_i = z\cdot(z+1)\cdots(z+i-1)$. We call the second and third terms of the RHS without the coefficient $p/(\theta+t-2)$ A and B, respectively. We can rewrite A as

$$
\begin{aligned}
A=&\sum_{k=c_0}^{l+c_0-2}(-1)^{k-c_0}\frac{(1)_{l-1}}{(c_0)_{k-c_0}(1)_{l-k+c_0-1}}kP(k,t-1)\\
&+(-1)^{l-1}\frac{(1)_{l-1}(l+c_0-1)}{(c_0)_{l-1}}P(l+c_0-1,t-1).
\end{aligned}
\tag{3.62}
$$

We can rewrite B as

$$
\begin{aligned}
B=&\sum_{k=c_0+1}^{l+c_0-1}(-1)^{k-c_0}\frac{(1)_{l-1}}{(c_0)_{k-c_0-1}(1)_{l-k+c_0-1}}P(k-1,t-1)\\
=&-\sum_{k=c_0}^{l+c_0-2}(-1)^{k-c_0}\frac{(1)_{l-1}(l-k+c_0-1)}{(c_0)_{k-c_0}(1)_{l-k+c_0-1}}P(k,t-1).
\end{aligned}
\tag{3.63}
$$

Then, $A-B$ is given by

$$
A-B=(l+c_0-1)\sum_{k=c_0}^{l-1}(-1)^{k-c_0}\frac{(1)_{l-1}}{(c_0)_{k-c_0}(1)_{l-k+c_0-1}}P(k,t-1). \tag{3.64}
$$

Substituting Eq. (3.64) in Eq. (3.61), we can obtain the time evolution of the summation:

$$
\begin{aligned}
&\sum_{k=c_0}^{l-k-1}(-1)^{k-c_0}\frac{(1)_{l-1}}{(c_0)_{k-c_0}(1)_{l-k+c_0-1}}P(k,t)\\
&=\frac{\theta+t-2-(l+c_0-1)p}{\theta+t-2}\sum_{k=c_0}^{l-k-1}(-1)^{k-c_0}\frac{(1)_{l-1}}{(c_0)_{k-c_0}(1)_{l-k+c_0-1}}P(k,t-1).
\end{aligned}
\tag{3.65}
$$

For $k > C_0$, we can prove that the following equality holds:

$$\sum_{k=c_0}^{l+c_0-1} (-1)^{k-c_0} \frac{(1)_{l-1}}{(c_0)_{k-c_0}(1)_{l-k+c_0-1}} P(k,t) = \frac{(\theta-(l+c_0-1)p)_{t-1}}{(\theta)_{t-1}}. \quad (3.66)$$

The analytic form can be obtained by multiplying both sides with $(-1)^{l-1}(k'-1)(k'-2)\cdots s/[(k'-l-c_0+1)!(l-1)!] = (-1)^{l-1}(c_0)_{k'-c_0-1}/[(1)_{k'-l-c-0+1}(1)_{l-1}]$ and summing over $l = 1, \ldots, k'-c_0+1$

$$P(k',t) = \sum_{l=1}^{k'-c_0+1} (-1)^{l-1} \frac{(c_0)_{k'-c_0}}{(1)_{k'-c_0-l+1}(1)_{l-1}} \frac{(c-(l+c_0-1)p)_{t-1}}{(c)_{t-1}}. \quad (3.67)$$

3.9 Appendix B A Proof of Eq. (3.12)

By replacing the analytical form Eq. (3.11), we obtain

$$\hat{\mu}_i(t) = \sum_{k'=c_0}^{c_0+t-1} (k')_i \sum_{l=1}^{k'-c_0+1} (-1)^{l-1} \frac{(c_0)_{k'-c_0}}{(1)_{k'-l-c_0+1}(1)_{l-1}} \frac{(\theta-(l+c_0-1)p)_{t-1}}{(\theta)_{t-1}}. \quad (3.68)$$

Further,

$$\sum_{k'=c_0}^{t+c_0-1} \sum_{l=1}^{k'-c_0+1} = \sum_{l=1}^{t} \sum_{k'=l+c_0-1}^{t+c_0-1} \quad (3.69)$$

and

$$\sum_{k'=l+c_0-1}^{t+c_0-1} (k)_i \frac{(c_0)_{k-c_0}}{(1)_{k'-l-c_0+1}(1)_{l-1}} = \frac{(c_0)_{i+t}}{(1)_{l-1}(1)_{t-l}(l+c_0+i-1)}. \quad (3.70)$$

the quasi-moment can have the following form:

$$\hat{\mu}_i(t) = \sum_{l=1}^{t} (-1)^{l-1} \frac{(c_0)_{t+i}}{(1)_{l-1}(1)_{t-l}(l+c_0+i-1)} \frac{(\theta-(l+c_0-1)p)_{t-1}}{(\theta)_{t-1}}. \quad (3.71)$$

References

Bikhchandani S, Hirshleifer D and Welch I, A Theory of Fads, Fashion, Custom, and Cultural Change as Informational Cascades, *Journal of Political Economy* **100** 992–1026 (1992).

Eisenstein E, Kanter I and Kessler D A and Kinzel W, Generation and Prediction of Time Series by a Neural Network, *Phys. Rev. Lett.* 74 4559 (1995).

Freedman D A, Bernard Friedman's urn, Ann. Math. Statist **36** 956–970 (1965).

Friedman B, A simple urn model, Comm. Pure Appl. Math. **2**, 59–70 (1949).

Hisakado M and Mori S, Phase transition and information cascade in a voting model *J. Phys. A* **43(31)** 3152 7 (2010).

Hisakado M, Kitsukawa K and Mori S, Correlated binomial models and correlation structures, *J. Phys. A.* **39** 15365 (2006).

Hod S and Keshet U, Phase transition in random walks with long-range correlations, *Phys. Rev. E.* **70** 015104 (2004).

Huillet T, On Pólya-Friedman random walks, *J. Phys. A* **41** 505005 (2008).

Keynes J M, *General Theory of Employment Interest and Money* (1936).

Schenkel A, Zhang J and Zhang Y C, Long Range Correlation in Human Writings, *Fractals* **1** 47–57 (1993).

Schütz G.M. and Timper S, Elephants can always remember: Exact long-range memory effects in a non-Markovian random walk, *Phys. Rev. E* **70** 045101 (2004).

Usatenko O V and Yampol'skii, Binary N-step Markov chains and long-range correlated systems, *Phys. Rev. Lett* **90** 110601 (2003).

Chapter 4
Nonlinear Pólya Urn Model and Phase Transitions

4.1 Introduction

In Chap. 3, we introduced a sequential voting model that is similar to a Keynesian beauty contest and considered the Pólya urn model as an analog herder. Herders vote for each candidate with probabilities that are proportional to the candidates' votes. We can consider the extreme herders who always choose the candidate with majority of the previous votes, which is visible to them. We refer to these herders as digital herders. In the next chapter, we discuss the digital herders. In the experiments, human beings exhibit behaviors between that of digital and analog herders (Mori et al. 2011, 2019a, b). The probability can be fitted by a tanh function. In this chapter, we discuss the rich phases of the models in which herders exhibit behavior that can be fitted to tanh-type function. We identify two types of phase transitions. **In the urn model, the** tanh**-type herder is that the number of added red (white) balls depends on nonlinear function the ratio of the red (white) balls in the urn.** In general, it is called the nonlinear Pólya model.

The remainder of this chapter is organized as follows. In Sect. 4.2, we introduce our voting model and mathematically define the two types of voters—independents and herders. In Sect. 4.3, we derive a stochastic differential equation and discuss the phase transition using it. In Sect. 4.4, we discuss the phase diagram and demonstrate the coexistence of these phase transitions. In Sect. 4.5, we verify these transitions through numerical simulations for the symmetric case. Finally, the conclusions are presented in Sect. 4.6. This chapter is based on Hisakado et al. (2012).

M. Hisakado, *Urn Models and Their Applications in Finance*,
https://doi.org/10.1007/978-981-96-3825-3_4

4.2 Voting Model

We model the voting of two candidates, C_0 and C_1; at time t, C_0 and C_1 have $c_0(t)$ and $c_1(t)$ votes, respectively. In each time step, one voter votes for one candidate; the voting is sequential. Hence, at time t, the tth voter votes, after which the total number of votes is t. Voters are allowed to see all previous votes for each candidate; thus, they are aware of public perception.[1]

Here, we consider the Bernoulli random variables, $X(t)$ which corresponds to the vote at time t. $X(t) = 1(0)$ denotes the vote to candidate 1(0) at t. Then, we obtain

$$c_1(t) = \sum_{i=1}^{t} X(i), \tag{4.1}$$

and

$$c_0(t) = \sum_{i=1}^{t} (1 - X(i)). \tag{4.2}$$

There are two types of voters—independents and herders; we assume an infinite number of voters. The independents vote for C_0 and C_1 with probabilities $1 - q$ and q, respectively. Their votes are independent of those of others, i.e., their votes are based on fundamental values.

Here, we set C_0 as the incorrect candidate and C_1 as the correct candidate to validate the performance of the votes. We can set $q \geq 0.5$ because we believe that independents vote for the correct candidate C_1 rather than the incorrect candidate C_0. In other words, we assume that the intelligence of the independents is virtually accurate.

In contrast, the herders' votes are based on the number of previous votes. The following functions as used. If the number of votes is $c_0(t)$ and $c_1(t)$ at time t, herder votes for C_1 at time $t + 1$ with the following probability:

$$f(z) = \frac{1}{2}[\tanh \lambda\{\frac{c_1(t)}{(c_0(t) + c_1(t))} - \frac{1}{2}\} + 1], \tag{4.3}$$

where $z(t) = c_1(t)/(c_1(t) + c_0(t))$. We call $f(z)$ the response function. If C_1 receives majority votes, i.e., $z(t) = c_1(t)/(c_0(t) + c_1(t)) > 1/2$, the ratio of votes for C_1, denoted by $f(z)$, increases exponentially to 1. λ reflects how previous answers affect a voter's choice. If λ is positive and large, the voter has high confidence in the previous votes. We use function Eq. (4.3) for our social experiments, and it fits well with human behaviors in the experiments (Mori et al. 2011). In our experiments, we can estimate $\lambda = 4.69, 5.3$, respectively. We call the herder tanh-type herder. If $c_0(t) = c_1(t)$, the herders vote for C_0 and C_1 with the same probability, i.e., 1/2. This model can also be derived from Bayes' theorem (see Sect. 4.6 Appendix A). In this case, λ is not

[1] The tth voter can see $c_0(t-1)$ and $c_1(t-1)$ votes at time t.

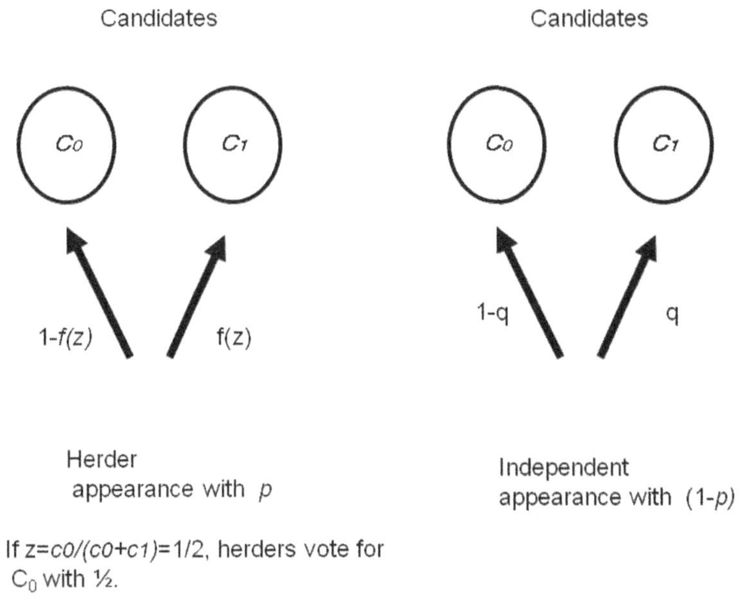

Fig. 4.1 Representation of our model. Independents vote for C_0 and C_1 with probabilities $1-q$ and q, respectively. Herders vote for C_0 and C_1 with probabilities $1 - f(z)$ and $f(z)$, respectively. We set the ratio of independents to herders as $(1-p)/p$. In the limit $\lambda \to \infty$, a herder is a digital herder. In the limit $\lambda = 2$, a herder is similar to an analog herder. In general, the herder is a hybrid of an analog herder and a digital herder with the parameter λ

constant and increases as the number of votes increases, according to the central limit theorem. The case is based on the assumption that all voters are independents. Hereafter, we treat λ as a parameter (see Fig. 4.1).

In the upper limit of λ, herders will vote for the candidate who receives majority votes. If $c_0(t) > c_1(t)$, the herders vote for candidate C_0, whereas if $c_0(t) < c_1(t)$, they vote for candidate C_1. These herders are known as digital herders which we discuss in Chap. 5. We expand Eq. (4.3) as follows:

$$f(z(t)) \sim \frac{1}{2} + \frac{\lambda}{2}\left(\frac{c_1(t)}{(c_0(t)+c_1(t))} - \frac{1}{2}\right) - \frac{\lambda^3}{6}\left(\frac{c_1(t)}{(c_0(t)+c_1(t))} - \frac{1}{2}\right)^3 + \cdots . \tag{4.4}$$

If $z(t) = c_1(t)/(c_0(t)+c_1(t)) \sim 1/2$ and $\lambda = 2$, a herder votes for C_1 with a probability of $z(t) = c_1(t)/(c_0(t)+c_1(t))$. These herders are analog herders with $q = 1/2$. Therefore, the herders who vote with probability Eq. (4.3) are hybrids of analog and digital herders.

This is an extension of the Pólya urn process. We call this model is the nonlinear Pólya model. In this chapter, we explain the case, $f(z) = \tanh(z)$.

Independents and herders appear randomly and vote. We set the ratio of independents to herders as $(1-p)/p$. In this study, we mainly focus on the upper limit of t. This refers to the voting of infinite voters.

The evolution (master) equation of candidate C_1 is given by

$$\begin{aligned}P(k,t) &= pf(z)P(k-1,t-1) + p(1-f(z))P(k,t-1)\\ &\quad +(1-p)qP(k-1,t-1) + (1-p)(1-q)P(k,t-1), \quad (4.5)\end{aligned}$$

where $z = k/(t-1)$. Here, $P(k,t)$ denotes the distribution of the number of votes k at time t for candidate C_1. The first and second terms of Eq. (4.5) denote the votes of the herders; the third and fourth terms denote the votes of the independents.

4.3 Mean Field Approximation

4.3.1 Stochatastic Differential Equation

To investigate long-range correlations, we analyze in the limit $t \to \infty$. We can rewrite Eq. (4.5) as

$$c_1(t) = k \to k+1 : P_{k,t} = \frac{p}{2}[\tanh\lambda(\frac{k}{t-1} - \frac{1}{2}) + 1] + (1-p)q. \quad (4.6)$$

In the scaling limit $t = c_0(t) + c_1(t) \to \infty$, we define

$$\frac{c_1(t)}{t} \Longrightarrow Z. \quad (4.7)$$

Z is the ratio of voters who vote for C_1.

We define a new variable Δ_t such that

$$\Delta_t = 2c_1(t) - t = c_1(t) - c_0(t). \quad (4.8)$$

We change the notation from k to Δ_t for convenience. Then, we have $|\Delta_t| = |2k - t| < t$. Thus, Δ_t holds within $\{-t, t\}$. Given $\Delta_t = u$, we obtain the following random walk model:

$$\begin{aligned}\Delta_t &= u \to u+1 : P_{\frac{u+t}{2},t} = \frac{p}{2}[\tanh\{\frac{\lambda u}{2(t-1)}\} + 1] + (1-p)q,\\ \Delta_t &= u \to u-1 : Q_{\frac{u+t}{2},t} = 1 - P_{\frac{u+t}{2},t}.\end{aligned}$$

We now consider the continuous limit $\epsilon \to 0$,

$$\begin{aligned} X_\tau &= \epsilon\Delta_{[t/\epsilon]}, \\ P(x,\tau) &= \epsilon P(\Delta_t/\epsilon, t/\epsilon), \end{aligned} \tag{4.9}$$

where $\tau = t/\epsilon$ and $x = \Delta_t/\epsilon$. We use $\delta X_\tau = X_{\tau+\epsilon} - X_\tau$ and ζ_τ, a standard i.i.d. Gaussian sequence; our objective is to identify the drift f_τ and the variance g_τ^2 such that

$$\delta X_\tau = f_\tau(X_\tau)\epsilon + \sqrt{\epsilon} g_\tau(X_\tau)\zeta_{\tau+\epsilon}. \tag{4.10}$$

Given $X_\tau = x$, by using the transition probabilities of Δ_n, we obtain

$$\begin{aligned} \mathrm{E}(\delta X_\tau) &= \epsilon \mathrm{E}(\Delta_{[\tau/\epsilon]+1} - \Delta_{[\tau/\epsilon]}) = \epsilon(2p_{[\frac{l/\epsilon+\tau/\epsilon}{2}],\tau/\epsilon} - 1) \\ &= \epsilon[(1-p)(2q-1) + p\tanh(\frac{\lambda x}{2\tau})]. \end{aligned} \tag{4.11}$$

Then, the drift term is $f_\tau(x) = (1-p)(2q-1) + p\tanh(\lambda x/2\tau)$. Moreover,

$$\sigma^2(\delta X_\tau) = \epsilon^2[1^2 p_{[\frac{l/\epsilon+\tau/\epsilon}{2}],\tau/\epsilon} + (-1)^2(1 - p_{[\frac{l/\epsilon+\tau/\epsilon}{2}],\tau/\epsilon})] = \epsilon^2, \tag{4.12}$$

such that $g_{\epsilon,\tau}(x) = \sqrt{\epsilon}$. We can obtain X_τ such that it obeys a diffusion equation with small additive noise:

$$\mathrm{d}X_\tau = [(1-p)(2q-1) + p\tanh(\frac{\lambda X_\tau}{2\tau})]\mathrm{d}\tau + \sqrt{\epsilon}\mathrm{d}B_\tau. \tag{4.13}$$

Here, we change the variable X_τ to Y_τ by using the expression,

$$pY_\tau = X_\tau - (1-p)(2q-1)\tau. \tag{4.14}$$

We rewrite Eq. (4.13) by using Y_τ:

$$\mathrm{d}Y_\tau = \tanh\frac{p\lambda}{2\tau}(Y_\tau + \frac{(2q-1)(1-p)}{p}\tau)\mathrm{d}\tau + \sqrt{\epsilon}\mathrm{d}B_\tau. \tag{4.15}$$

Using Eqs. (4.7), (4.8), and (4.14), we obtain the relations of the variables

$$2Z - 1 = \frac{X_\infty}{\tau} = \frac{pY_\infty}{\tau} + (1-p)(2q-1). \tag{4.16}$$

4.3.2 Mean Field Approximation and Phase Transtion

In this section, we discuss the phase transition using the mean field approximation. We are interested in the behavior in the limit $\tau \to \infty$. Here, we consider the solution $Y_\infty \sim \tau^\alpha$, where $\alpha \le 1$, since $\tanh x \le 1$. The slow solution is $Y\infty \sim \tau^\alpha$, where

$\alpha < 1$ is hidden by the fast solution $\alpha = 1$ in the upper limit of τ. Hence, we can assume a stationary solution as

$$Y_\infty = \bar{\upsilon}\tau, \tag{4.17}$$

where $\bar{\upsilon}$ is the constant. Substituting Eq. (4.17) into Eq. (4.15), we obtain

$$\Delta Y_\infty = \tanh \frac{p\lambda}{2}(\bar{\upsilon} + \frac{(2q-1)(1-p)}{p})\Delta\tau = \bar{\upsilon}\tau. \tag{4.18}$$

The second equality can be obtained from Eq. (4.17). **Then, we can obtain the self-consistent equation,**

$$\bar{\upsilon} = \tanh \frac{p\lambda}{2}(\bar{\upsilon} + \frac{(2q-1)(1-p)}{p}). \tag{4.19}$$

The second term on the RHS of Eq. (4.19) corresponds to the external field. For $q = 1/2$ and $p = 1$, the external field disappears. If $\lambda > 2$, a phase transition occurs in the range $0 \leq p \leq 1$. This is a phase transition in the non-equilibrium process.

Symmetric Case

In the symmetric case, the phase transition is the continuous one. In the critical case, the solution is a single tangent point and the process converges to the point. In the case Fig. 4.2(a) and (b), there is only one solution. It is one-peak phase. In the case of (c), there are three solutions. Two of them are stable and mid one is unstable and one of them is chosen for the non-equilibrium process. It is two-peak phase. **The phase transition is continuous.**

Asymmetric Case

In contrast, when asymmetric case, the phase transition is the discontinuous phase transition. In Fig. 4.3(b), in the critical case, the process converges at the tangent point. Hence, there are two points where the process converges when the critical case. **The probability corresponds to the gap, and the phase transition is discontinuous.**[2]

Equation (4.19) admits one solution below the critical point $p \leq p_c$ (see Fig. 4.3(a)) and three solutions for $p > p_c$ (see Fig. 4.3(b)). When $p \leq p_c$, we refer to the phase as the single peak phase. When $p > p_c$, the upper and lower solutions are stable; on the other hand, the intermediate solution is unstable. Then, the two stable solutions attain good and bad equilibrium, and their distribution becomes the sum of the two delta functions. We refer to this phase as the two-peak phase.

[2] We discuss the gap on Appendix.

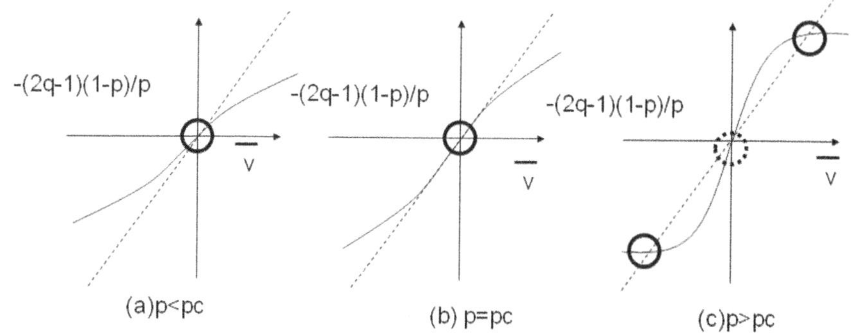

Fig. 4.2 Solutions of the self-consistent equation (Eq. (4.19)) for $q = 1/2$. This is the state equation of the Ising model. (**a**) $p \leq p_c$ (**b**) $p = p_c$ and (**c**) $p > p_c$. Below the critical point p_c, we obtain solution (**a**). We refer to this phase as the one-peak phase. In contrast, above the critical point, we obtain three solutions. Two of them are stable, and one is unstable (**c**). We refer to this phase as the two-peaks phase

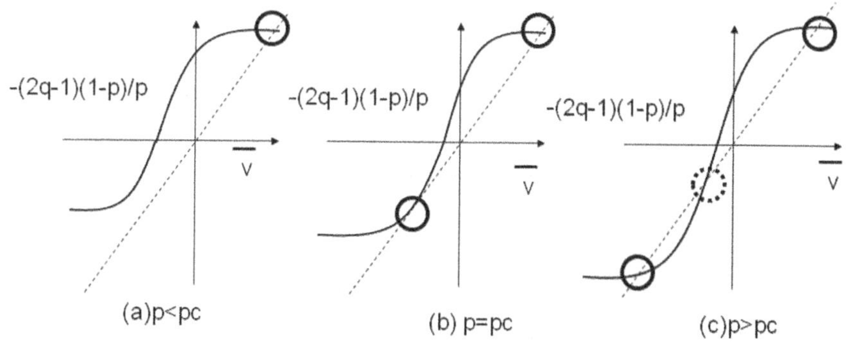

Fig. 4.3 Solutions of the self-consistent equation Eq. (4.19). This is the state equation of the Ising model. (**a**) $p \leq p_c$ (**b**) $p = p_c$, and (**c**) $p > p_c$. Below the critical point p_c, we obtain solution (**a**). We refer to this phase as the one-peak phase. In contrast, above the critical point, we obtain three solutions. Two of them are stable and one is unstable (**c**). We refer to this phase as the two-peaks phase. In the Ising model, the physical solution is the right side one only. In case of (**b**) there is the provability that the process converges at the tangent point. The probability corresponds to the gap, and the phase transition is the discontinuous phase transition

In the Ising model with external fields, the good equilibrium is only a physical solution. This is the difference between our voting model and the Ising model. Ising model is the equilibrium model, and voting model is the non-eqilibrium model. In one-peak phase, the variance of the distribution of Z is 0 in the large τ limit. It is a self-averaging process. However, the variance of the distribution of Z is not 0 in the two-peak phase. It is the non-self-averaging process (Kadanoff 2000). Hence, **the phase transition is between the self-averaging process and non-self-averaging process.**

4.3.3 Critical Exponent Symmetric Case

In this section, we consider the symmetric case, $q = 1/2$. In this case, the external fields are absent. The self-consistent equation (Eq. (4.19)) becomes

$$\bar{v} = \tanh \frac{p\lambda}{2}(\bar{v}). \tag{4.20}$$

The tanh function rises at $\bar{v} = 0$. If $\lambda \leq 2$, there is only one solution $\bar{v} = 0$, in all regions of p. In this case, Z has only one peak, at 0.5, which indicates the one-peak phase. In contrast, if $\lambda > 2$, there are two stable solutions and an unstable solution $\bar{v} = 0$ above p_c. The vote ratio for C_1 achieves a good or bad equilibrium. **This is the so-called spontaneous symmetry breaking.** In one sequence, Z is taken as $\bar{v}p/2 + 1/2$ in the case of a good equilibrium, or as $-\bar{v}p/2 + 1/2$ in the case of a bad equilibrium, where $\bar{v}$ is the solution of Eq. (4.20). This indicates the two-peak phase, and the critical point is $p_c = 2/\lambda$.

Here, we consider the critical exponent near the critical point $p = p_c$. We expand Eq. (4.20),

$$\bar{v} = \tanh \frac{p\lambda}{2}(\bar{v}) = \frac{p}{p_c}\bar{v} - \frac{1}{3}(\frac{p\bar{v}}{p_c})^3 \ldots. \tag{4.21}$$

We obtain

$$\bar{v} = \pm\sqrt{3(\frac{p_c}{p})^3(\frac{p}{p_c} - 1)}. \tag{4.22}$$

Here, we introduce

$$t = \frac{p - p_c}{p_c} << 1. \tag{4.23}$$

We obtain

$$\bar{v} = \pm\sqrt{-3\frac{t}{(1+t)^2}} \sim \pm\sqrt{3\frac{|p - p_c|}{p_c}} \sim \pm|p - p_c|^{\beta}. \tag{4.24}$$

Therefore, the critical exponent $\beta = 1/2$.

4.3.4 Phase Transition in Super and Normal Diffusion Phases

In this section, we consider the phase transition of convergence. This type of transition has been studied when herders are analog herders which we discussed in this chapter. Analog herders exhibit weaker herd behavior than digital herders. Depending on the convergence behavior, there are three phases. We expand Y_τ around the solution $\bar{v}$.

$$Y_\tau = \bar{v}\tau + W_\tau. \tag{4.25}$$

Here, we set $Y_\tau \gg W_\tau$. This indicates $\tau \gg 1$. We rewrite Eq. (4.15) using Eq. (4.25) as follows:

$$\begin{aligned}
dY_\tau &= \bar{v}d\tau + dW_\tau \\
&= \tanh[\frac{p\lambda}{2\tau}(\bar{v}\tau + \frac{(2q-1)(1-p)}{p}\tau) + \frac{p\lambda}{2\tau}W_\tau]d\tau + \sqrt{\epsilon}dB_\tau, \\
&\sim \bar{v}d\tau + \frac{p\lambda}{2\tau\cosh^2\frac{p\lambda}{2}(\bar{v} + \frac{(2q-1)(1-p)}{p})}W_\tau d\tau + \sqrt{\epsilon}dB_\tau, \\
&= \bar{v}d\tau + \frac{p\lambda(1-\bar{v}^2)}{2\tau}W_\tau d\tau + \sqrt{\epsilon}dB_\tau.
\end{aligned} \tag{4.26}$$

We use the relation Eq. (4.19) and consider the first term of the expansion. Hence, we obtain

$$dW_\tau = \frac{p\lambda(1-\bar{v}^2)}{2\tau}W_\tau d\tau + \sqrt{\epsilon}dB_\tau. \tag{4.27}$$

Using Sect. 4.7 Appendix B, we obtain the phase transition of convergence. The critical point p_{vc} is the solution of

$$p_{vc} = \frac{1}{\lambda(1-\bar{v}^2)}, \tag{4.28}$$

and Eq. (4.19).

At the critical point, the gradient of the tangent line at $\bar{v}$ is $1/2$. If the gradient of the tangent line at $\bar{v}$ is under $1/2$, the distribution converges as in a binomial distribution. We define γ as $\mathrm{V}(Z) = \tau^{-\gamma}$, where $\mathrm{V}(Z)$ is the variance of Z. The voting rate for C_1 converges as $\mathrm{V}(Z) = \tau^{-1}$: $\gamma = 1$. We refer to this phase as the normal diffusion phase. If the gradient of the tangent line at $\bar{v}$ is $1/2$ or above $1/2$, the voting rate converges at a slower speed than in a binomial distribution. (See Sect. 4.7 Appendix B in detail). We refer to these phases as the super diffusion phases. In one phase, the voting rate for C_1 converges to $\log(\tau)/\tau$, and the voting rate converges to $\gamma = p/p_{vc} - 2$.

In summary, the convergence of $Z(t)$ is

$$V(Z(t)) \propto \begin{cases} t^{-1} & p < p_{vc}, \\ t^{2-p/p_{vc}} & p > p_{vc}, \\ \frac{\log(t)}{t} & p = p_{vc}, \end{cases} \tag{4.29}$$

where $V(Z(t))$ is the variance of $Z(t)$.

4.3.5 *Phase Diagam*

Symmetric Case

We consider the case $\lambda > 2$. In this case, we observe a phase transition. If $\lambda \leq 2$, we do not observe the phase transition and only observe a part of the phases, as described below.

In the one-peak phase $p \leq p_c = 2/\lambda$, the only solution is $\bar{v} = 0$. p_c is the critical point of the phase transition. The first critical point of convergence is $p_{vc1} = 1/\lambda$. When $p \leq p_c$, p_{vc1} is the solution of Eqs. (4.20) and (4.28). If $0 < p < p_{vc1}$, the voting rate for C_1 becomes $1/2$, and the distribution converges as in a binomial distribution. If $p_c > p \geq p_{vc1}$, candidate C_1 also gathers $1/2$ of all votes in the scaled distributions. However, the voting rate converges slower than in a binomial distribution. We refer to these phases as the super diffusion phases. There are two phases, $p = p_{vc1}$ and $p_c > p > p_{vc1}$; these phases differ in terms of their convergence speed.

Above p_c, in the two-peak phase, we obtain two stable solutions that are not 0. At p_c, $\bar{v}$ moves from 0 to either of these two stable solutions. In one sequence, the votes converge to one of these stable solutions. If $p_c < p \leq p_{vc2}$, the voting rate for C_1 becomes $\bar{v}p/2 + 1/2$ or $-\bar{v}p/2 + 1/2$, and convergence occurs at a rate slower than that in a binomial distribution. Here, $\bar{v}$ is the solution of Eq. (4.20). We refer to this phase as the super diffusion phase. p_{vc2} is the second critical point of convergence from the super to the normal diffusion phase, and it is the solution of the simultaneous equations Eqs. (4.20) and (4.28) when $p > p_c$. We can estimate $p_{vc2} \sim \frac{2.5}{\lambda}$ by approximation.[3] In fact, with higher terms, we can estimate $p_{vc2} \sim \frac{2.7}{\lambda}$ (see Fig. 4.4(a)). In contrast, if $p > p_{vc2}$, the voting rate for C_1 becomes $\bar{v}p/2 + 1/2$ or $-\bar{v}p/2 + 1/2$. However, the distribution converges as in a binomial distribution. This is the normal diffusion phase.

We show the phase diagram for the symmetric case, $q = 0.5$ in Fig. 4.4(a). γ is the speed of convergence. $\gamma = 0$ **is the boundary between the one-peak and the two-peak phases.** The region below $\gamma = 0$ is the one peak phase, whereas the region above $\gamma = 0$ is the two-peak phase. $\gamma = 1$ is between the normal diffusion phase and the super diffusion phase.

In the one-peak phase, the region below $\gamma = 1$ is the normal diffusion phase and that above $\gamma = 1$ is the super diffusion phase. In the two-peak phase, two solutions can be observed. We do not observe a difference between the two solutions, and the region above $\gamma = 1$ is the normal diffusion phase.

In Fig. 4.4(a), the curves $\gamma = 1$, $Z > 1/2$ in the one-peak phase and the two-peak phase are separated by the curve $\gamma = 0$.

[3] Using $\tanh x \simeq x - 1/3x^3$, Eq. (4.20), and the condition of the critical point, we can obtain $p_{vc2} \simeq 2.5/\lambda$.

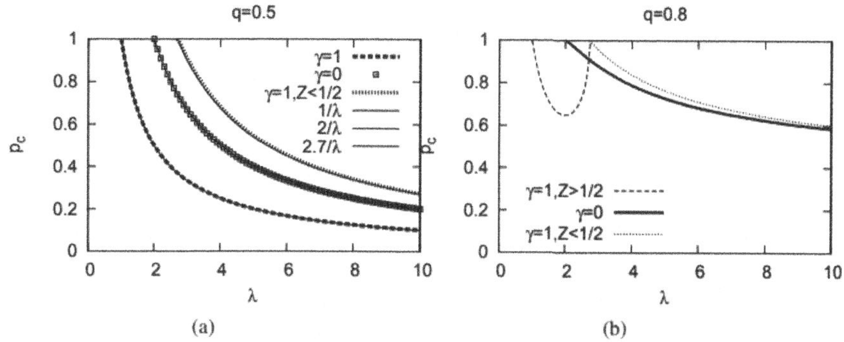

Fig. 4.4 Phase diagram of space λ and p when (**a**) $q = 0.5$ and (**b**) $q = 0.8$. γ denotes the speed of convergence. $\gamma = 0$ is the boundary between the one-peak phase and the two-peak phase. The region below $\gamma = 0$ is the one-peak phase. The region above $\gamma = 0$ is the two-peak phase. $\gamma = 1$ is the between the normal diffusion phase and the super diffusion phase. In the one-peak phase, the region below $\gamma = 1$ is the normal diffusion phase. In the two-peak phase, two solutions are obtained. Each solution generally exhibits a different speed of convergence. When $q = 0.5$, there is no difference between the two solutions and the region below $\gamma = 1$ is the super diffusion phase. When $q = 0.8$, there is a difference between the good and bad equilibria. $Z > 1/2$ is the good equilibrium and $Z < 1/2$ is the bad equilibrium. The left side of $\gamma = 1$, $Z > 1/2$ in the two-peak phase is the phase in which a good equilibrium is a super diffusion phase. The left side of $\gamma = 1$, $Z < 1/2$ is the phase in which a bad equilibrium is a super diffusion phase

Asymmetric Case

As discussed above, when $q = 0.5$, there is no difference between the good and bad equilibria. However, when $q = 0.8$, we observe a difference. $Z > 1/2$ indicates the good equilibrium and $Z < 1/2$ indicates the bad equilibrium. Figure 4.4(b) shows the phase diagram in space λ and p when $q = 0.8$. γ is the speed of convergence. $\gamma = 0$ **is the boundary between the one-peak phase and the two-peak phase.**

The region below $\gamma = 0$ is the one-peak phase, and that above $\gamma = 0$ is the two-peak phase. $\gamma = 1$ is between the normal diffusion and the super diffusion phases. In the one-peak phase, the region below $\gamma = 1$ is the normal diffusion phase and that above $\gamma = 1$ is the super diffusion phase. In the two-peak phase, two solutions can be observed.

When $q = 0.8$, we observe a difference in the speed of convergence between good and bad equilibria. $Z > 1/2$ is the good equilibrium and $Z < 1/2$ is the bad equilibrium. The left side of $\gamma = 1$, $Z > 1/2$ in the two-peak phase is the phase in which a good equilibrium is the super diffusion phase. The left side of $\gamma = 1$, $Z < 1/2$ is the phase in which a bad equilibrium is the super diffusion phase.

When $q \neq 0.5$, the curves $\gamma = 1, Z > 1/2$ in the one-peak phase and the two-peak phase are deformed and meet on the curve $\gamma = 0$ as Fig. 4.4(b). Hence, the region surrounded by $\gamma = 1, Z > 1/2$ is the super diffusion phase of the good equilibrium.

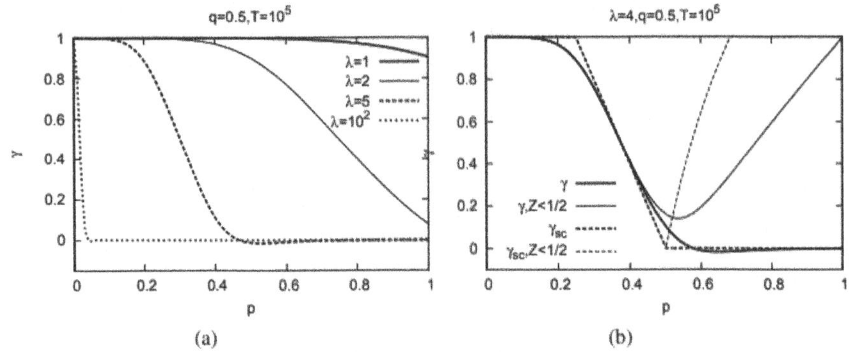

Fig. 4.5 Convergence of distribution in the one-peak phase and the two-peak phase for the $q = 0.5$ case. The horizontal axis represents the ratio of herders p, and the vertical axis represents the speed of convergence. $\gamma = 1$ is the normal phase, and $0 < \gamma < 1$ is the super diffusion phase. (**a**) is for the one-peak phase. The critical point of the convergence of the transition from normal diffusion to super diffusion is $p_{vc1} \sim 1/\lambda$, and the critical point of the phase transition is $p_c \sim 2/\lambda$. (**b**) is for the two-peak phase. The solid lines are obtained from the numerical simulations. γ, $Z < 0$ is the speed of convergence of the region $Z < 0$, which indicates a bad equilibrium. The dotted line γ_{sc} represents the theoretically obtained curve and γ_{sc}, $Z < 0$ represents the theoretically obtained curve for a bad equilibrium. The critical point of phase transition is $p_c \sim 0.5$ and the critical point of the convergence of transition is $p_{vc1} \sim 0.25$ and $p_{vc2} \sim 0.675$

4.3.6 Linear Limit

Finally, we comment on the analog herder case. As discussed in Sect. 4.2, if we set $\lambda = 2$ and use only the linear terms in Eq. (4.3), we can obtain the analog herder case. In this limit, we can solve Eq. (4.19) and obtain $\bar{v} = 2q - 1$. Equation (4.19) has only one solution in the entire region p. Hence, we do not observe a phase transition, which is consistent with the conclusions of Chap. 3. On the other hand, in the super and normal diffusion phase transitions, $p_{vc1} = 1/2$ does not depend on q because the gradient of the linear term is constant.

4.4 Numerical Simulations of the Symmetric Case

To confirm the analytical results, we performed numerical integration of the master equation (4.5). We perform simulations for the symmetric independent voter case, i.e., $q = 1/2$.

4.4.1 One-Peak Phase: $q = 1/2$

Figure 4.5(a) shows the convergence of the distribution in the one-peak phase. We integrated the master equation up to $t = 10^5$. As discussed in the previous sections, the critical point of convergence of this transition is $p_{vc1} \sim 1/\lambda$ and the critical point of phase transition is $p_c \sim 2/\lambda$. At the critical point of the phase transition, the distribution splits into two and the exponent γ becomes 0.[4]

4.4.2 Two-Peak Phase: $q = 1/2, \lambda = 4$

Figure 4.5(b) shows the convergence of the distribution in the two-peak phase. Here, we consider the case where $q = 0.5$ and $\lambda = 4$. In this phase, the voting sequence converges to one of the peaks. We cannot determine which peak is chosen in finite time. We divide the distribution into two parts, $Z > 0$ or $Z < 0$, which correspond to the two regions around the peaks.

Above the critical point of phase transition $p_c \sim 2/\lambda = 0.5$, the speed of convergence in the region $Z < 0$ is slower than that in a binomial distribution.[5] This indicates the super diffusion phase. In the super diffusion phase, the distance between the two peaks is short and the influence of the other peak reduces the convergence speed. After the transition of convergence to $p_{vc2} = 0.675$, the speed of convergence is as in a binomial distribution $\gamma = 1$. This indicates a normal diffusion phase.

The critical point of convergence, p_{vc2}, in the two-peak phase is larger than that predicted by theory. We calculated the variance in the region $Z < 0$ using the data from $t = 9.99 \times 10^4$ to $t = 10^5$. The region may be too wide for use in our assumptions, as described in Eq. (4.25). To fill the gap between the numerical calculations and the theory, we need to perform simulations for a large t.

4.5 Summary

We investigated the nonlinear Pólya model and its rich phases. In the continuous limit, we could obtain the stochastic differential equations. The model exhibits two types of phase transitions. One is the phase transition, which is similar to the phase transition of the Ising model. In fact, we showed that the stationary condition of our model is the same as the equation of state for the Ising model. The difference is that our model is the non-equilibrium and the Ising model is the equilibrium model. As the ratio of herders increased, the model featured a phase transition beyond which

---

[4] Here, we estimate γ from the slope of $V(Z(t))$ as $\gamma = \log\{V(Z(t-\Delta t))/V(Z(t))\}/\log\{t/(t-\Delta t)\}$.

[5] In the case $q = 0.5$, the convergence in the region $Z > 0$ is identical to that in $Z < 0$ because of the symmetry. We consider only $Z < 0$ here.

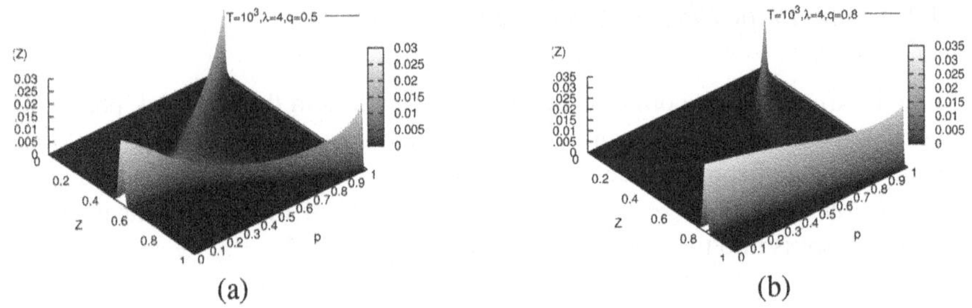

Fig. 4.6 Distributions for (**a**) $q = 0.5$ and (**b**) $q = 0.8$ when the number of votes is 10^3. The three axes are Z, p, and the frequency $P(Z)$. As p increases, the distribution splits into two in both cases. In the region near p_c, both the one-peak and the two-peak phases exhibit super diffusion. In this phase, the speed of convergence is slower than that in the normal diffusion phase. We can confirm that the width of the distribution is greater than that in the region far from p_c

a state where most voters make the correct choice coexists with one where most of them are incorrect. In this transition, the distribution of votes changed from the one-peak phase to the two-peak phase. These two peaks were the two stable solutions out of the three solutions. In the Ising model with an external field, there is only one physical solution out of the three solutions.

The other transition was the transition of convergence between super and normal diffusion. It is the same transition discussed in Chap. 3. In the one-peak phase, if herders increased, the variance converged slower than in a binomial distribution. This represents the transition from normal diffusion to super diffusion. In the two-peak phase, sequential voting converged to one of the two peaks. In the two-peak phase, if the herders increased, the variance converged as in a binomial distribution. This is the transition from super diffusion to normal diffusion, which is opposite to the transition in the one-peak phase.

Figure 4.6 shows the distributions when (a) $q = 0.5$ and (b) $q = 0.8$. As p increases, the distribution splits into two in both cases. We can think of this as the splitting of a particle into two parts by the interactions. In the region near p_c, both the one peak and the two-peak phases exhibit super diffusion. In this phase, the speed of convergence is slower than that in the normal diffusion phase. We believe that the particles are deformed by these interactions. When $q = 0.5$, a particle is continuously divided and it is the continuous phase transition. When $q = 0.8$, a second particle appears far from the original particle and is divided discontinuously and the discontinuous phase transition.

4.6 Appendix A Derivation of Eq. (4.3) Using Bayes' Theorem

We estimate the probability that candidate C_1 is correct using the previous votes. We assume that the $(t+1)$th voter is a herder. The voter estimates the posterior distribution that C_1 is the correct candidate using Bayes' theorem. The voter can see t votes.

The herders estimate the probability that C_i, where $i = 0, 1$, is the correct candidate based on the previous votes. $Pr(C_i)$ is the probability that the voter estimates that C_i is the correct candidate. We set the prior distribution as $Pr(C_0) = Pr(C_1) = 1/2$. Here, we assume that the herders believe that all voters are independent and estimate the percentage of correct answers as $\hat{q}$. The posterior distribution is given by

$$\begin{aligned} Pr(C_1|c_1 = k) &= \frac{1}{2}\frac{t!}{k!(t-k)!}\hat{q}^k(1-\hat{q})^{t-k}, \\ Pr(C_0|c_1 = k) &= \frac{1}{2}\frac{t!}{k!(t-k)!}(1-\hat{q})^k\hat{q}^{t-k}, \end{aligned} \tag{4.30}$$

where $c_1 = k$ indicates that the number of votes for C_1 is k before the voter votes. We obtain

$$\frac{Pr(C_1|c_1 = k)}{Pr(C_0|c_1 = k)} = (\frac{\hat{q}}{1-\hat{q}})^{2k-t} = e^{2\lambda(k-\frac{t}{2})/t}, \tag{4.31}$$

where $\lambda = t \log \frac{\hat{q}}{1-\hat{q}}$. Here, λ increases as t increases. Hence, the herder can calculate the probability that the candidate C_1 is correct when the number of votes for C_1 is k:

$$Pr(C_1|c_1 = k) = \frac{1}{2}[\tanh \lambda(\frac{k}{t} - \frac{1}{2}) + 1]. \tag{4.32}$$

In the region $t >> 1, \lambda \to \infty$, a voter believes the public perception, and the behavior of herders is similar to that of digital herders without $\hat{q} = 1/2$. When $\hat{q} = 1/2$, the herders estimate that the previous votes are not useful to estimate the correct answer and $\lambda = 0$.

4.7 Appendix B Proof of Eq. (4.28)

We consider the stochastic differential equation

$$\mathrm{d}x_\tau = (\frac{l x_\tau}{\tau})\mathrm{d}\tau + \sqrt{\epsilon}\mathrm{d}B_\tau, \tag{4.33}$$

where $\tau \geq 1$. If we set $l = p\lambda(1-\bar{v}^2)/2$, Eq. (4.33) is identical to Eq. (4.27). Let μ_1 and σ_1^2 be the mean and the variance of x_1. If x_1 is Gaussian ($x_1 \sim \mathrm{N}(\mu_1, \sigma_1^2)$) or deterministic ($x_1 \sim \delta_{\mu_1}$), the law of x_τ ensures that the Gaussian is in accordance with the density

$$p_\tau(x) \sim \frac{1}{\sqrt{2\pi}\sigma_\tau} e^{-(x-\mu_\tau)^2/2\sigma_\tau^2}, \tag{4.34}$$

where $\mu_\tau = \mathrm{E}(x_\tau)$ is the expected value of x_τ and $\sigma_\tau^2 \equiv \nu_\tau$ is its variance. If $\Phi_\tau(\xi) = \log(\mathrm{e}^{\mathrm{i}\xi x_\tau})$ is the logarithm of the characteristic function of the law of x_τ, we have

$$\partial_\tau \Phi_\tau(\xi) = \frac{l}{\tau}\xi \partial_\xi \Phi_\tau(\xi) - \frac{\epsilon}{2}\xi^2, \tag{4.35}$$

and

$$\Phi_\tau(\xi) = \mathrm{i}\xi \mu_\tau - \frac{\xi^2}{2}\nu_\tau. \tag{4.36}$$

By identifying the real and imaginary parts of $\Phi_\tau(\xi)$, we obtain the dynamics of μ_τ as

$$\dot{\mu}_\tau = \frac{l}{\tau}\mu_\tau. \tag{4.37}$$

The solution for μ_τ is

$$\mu_\tau = x_1 \tau^l. \tag{4.38}$$

The dynamics of ν_τ are given by the Riccati equation

$$\dot{\nu}_\tau = \frac{2l}{\tau}\nu_\tau + \epsilon. \tag{4.39}$$

If $l \neq 1/2$, we obtain

$$\nu_\tau = \nu_1 \tau^{2l} + \frac{\epsilon}{1-2l}(\tau - \tau^{2l}). \tag{4.40}$$

If $l = 1/2$, we obtain

$$\nu_\tau = \nu_1 \tau + \epsilon \tau \log \tau. \tag{4.41}$$

We can summarize the temporal behavior of the variance as

$$\nu_\tau \sim \frac{\epsilon}{1-2l}\tau \qquad \text{if} \quad l < \frac{1}{2}, \tag{4.42}$$

$$\nu_\tau \sim (\nu_1 + \frac{\epsilon}{2l-1})\tau^{2l} \qquad \text{if} \quad l > \frac{1}{2}, \tag{4.43}$$

$$\nu_\tau \sim \epsilon \tau \log(\tau) \qquad \text{if} \quad l = \frac{1}{2}. \tag{4.44}$$

Here, we introduce the rescaled variables

$$\tilde{\nu}_\tau \equiv \frac{\nu_\tau}{\tau^2}.$$

The solution for $\tilde{\nu}_\tau$ is

$$\tilde{\nu}_\tau \sim \frac{\epsilon}{1-2l}\tau^{-1} \qquad \text{if} \quad l < \frac{1}{2}, \tag{4.45}$$

$$\tilde{\nu}_\tau \sim (\nu_1 + \frac{\epsilon}{2l-1})\tau^{2l-2} \qquad \text{if} \quad l > \frac{1}{2}, \tag{4.46}$$

$$\tilde{\nu}_\tau \sim \epsilon\frac{\log(\tau)}{\tau} \qquad \text{if} \quad l = \frac{1}{2}. \tag{4.47}$$

This model has three phases. If $l > 1/2$ or $l = 1/2$, x_τ/τ converges slower than in a binomial distribution. These phases are the super diffusion phases. If $0 < p < 1/2$, x_τ/τ converges as in a binomial distribution. This is the normal phase.

References

Hisakado M and Mori S, Phase transition and information cascade in a voting model *J. Phys. A***43(31)** 3152 7 (2010)

Hisakado M and Mori S, Digital herders and phase transition in a voting model *J. Phys.* A**44(27)** 275204 (2011)

Hisakado M, and Mori S, Two kinds of phase transitions in a voting model. Journal of Physics A: Mathematical and Theoretical, **45(34)**, 345002 (2012)

Hod S and Keshet U, Phase transition in random walks with long-range correlations, *Phys. Rev. E.* **70** 015104 (2004)

Kadanoff L, Statistical Physics, World Scientific Publishing (2000)

Mori S, Hisakado M, and Takahashi T, Phase transition to two-peaks phase in an information cascade voting experiment, PHYSICAL REVIEW E **86(2)** 026109-026118 (2012)

Mori S, and Hisakado M, Information Cascade Experiment: General Knowledge Quiz, Applications of Data-Centric Science to Social Design: Qualitative and Quantitative Understanding of Collective Human Behavior, 167–179 (2019)

Mori S, and Hisakado M, Information Cascade Experiment: Urn Quiz. Applications of Data-Centric Science to Social Design: Qualitative and Quantitative Understanding of Collective Human Behavior, 181–191 (2019)

Chapter 5
Extreme Nonlinear Pólya Urn Model and Phase Transitions

5.1 Introduction

We continue the study of the nonlinear Pólya urn model which was discussed in Chap. 2. In this chapter, we investigate the extreme case and especially the phase transition. We can analyze the phase transition using the exact solutions for this model, especially the critical exponent. In the voting model, the herder is the extreme herder. There are two types of voters—herders and independents—and two candidates. Analog herders vote for each candidate with the probabilities proportional to the candidates' votes. It is the same as Pólya urn. Herders vote for each candidate with the probabilities, $f(z)$, where z is the number of votes for candidate 1. We can consider extreme herders who always choose the candidate with majority of the previous votes, which is visible to them. We refer to these herders as digital herders and $f(z)$ is the step function. In this chapter, we study them and clarify the phase transition using several methods.

The remainder of this paper is organized as follows. In Sect. 5.2, we introduce our voting model and mathematically define the two types of voters—independents and herders. Voters can see their previous r votes of voters. In this chapter, we study $r = 1$ and $r = \infty$ cases. In Sect. 5.3, we calculate the exact distribution functions of the votes for the case in which voters can see the votes of all voters. We discuss the phase transition using exact solutions. In Sect. 5.4, we discuss the special case, $r = 1$. In this case, we calculate the exact distribution function. In this case, there is no phase transition. In Sect. 5.5, we analyze the model using the mean field approximation for $r = \infty$. We can show that phase transition occurs in this system. In Sect. 5.6, we analyze the two models using block spin transformation. Finally, conclusions are presented in Sect. 5.7. This chapter is based on Hisakado et al. (2011).

M. Hisakado, *Urn Models and Their Applications in Finance*,
https://doi.org/10.1007/978-981-96-3825-3_5

5.2 Voting Model

We model the voting of two candidates, C_0 and C_1; at time t, they have $c_0(t)$ and $c_1(t)$ votes, respectively. At each time step, one voter votes for one candidate; voting is sequential.

We consider the Bernoulli random variables $X(t)$, which corresponds to the vote at time t. $X(t) = 1(0)$ is the vote to candidate 1(0) at t. Then, we obtain

$$c_1(t) = \sum_{i=1}^{t} X(i), \tag{5.1}$$

and

$$c_0(t) = \sum_{i=1}^{t} (1 - X(i)). \tag{5.2}$$

Voters are allowed to see r previous votes for each candidate when they vote so that they are aware of the public perception. If $r > t$, voters can see t previous votes for each candidate. At time t, the number of votes for C_0 and C_1 are $c_0^r(t)$ and $c_1^r(t)$, respectively. Hence, we obtain

$$c_1^r(t) = \sum_{i=t-r}^{t} X(i), \tag{5.3}$$

and

$$c_0^r(t) = \sum_{i=t-r}^{t} (1 - X(i)). \tag{5.4}$$

In the limit $r \to \infty$, voters can see all previous votes. Therefore, $c_0^{\infty}(t) = c_0(t)$ and $c_1^{\infty}(t) = c_1(t)$.

There are two types of voters—independents and herders; we assume an infinite number of voters. Independent voters vote for C_0 and C_1 with probabilities $1 - q$ and q, respectively. Their votes are independent from those of others, i.e., their votes are based on their fundamental values. Here, we set C_0 as the incorrect candidate and C_1 as the correct candidate to validate the performance of the herders. We can set $q \geq 0.5$ because we believe that independent voters vote for the correct candidate C_1 rather than the incorrect candidate C_0. In other words, we assume that the intelligence of the independent voters is virtually correct.

On the other hand, herders vote for a majority candidate; if $c_0^r(t) > c_1^r(t)$, herders vote for candidate C_0. If $c_0^r(t) < c_1^r(t)$, the herders vote for candidate C_1. If $c_0^r(t) = c_1^r(t)$, herders vote for C_0 and C_1 with the same probability, i.e.,$1/2$. In Chap. 3, the herders voted for each candidate with probabilities that were proportional to the candidates' votes; they were known as analog herders. The herders in this chapter are known as digital herders (Fig. 5.1).

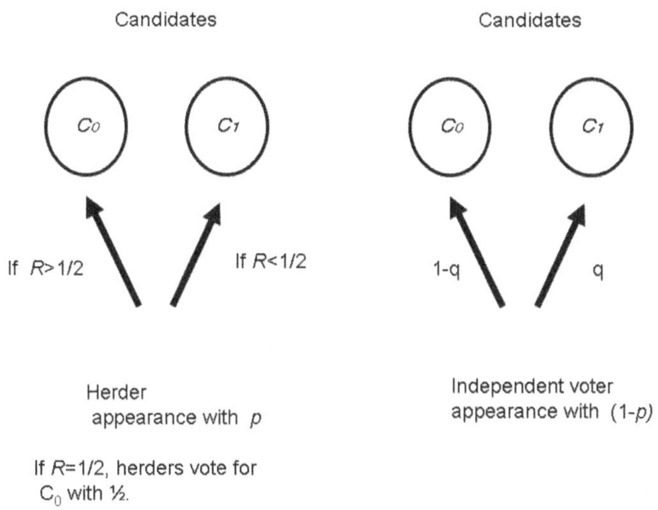

Fig. 5.1 Demonstration of model. $R = c_0^r/\{c_0^r + c_1^r\}$

Independent voters and herders appear randomly and vote. We set the ratio of independent voters to herders as $(1 - p)/p$. In this chapter, we mainly focus on the large t limit. This means the voting of infinite voters.

5.3 Exact Solutions for $r = \infty$ and Phase Transition

5.3.1 Exact Solution

In this section, we study the exact solution for the case $r = \infty$ by using combinatorics. Here, we map the model to the correlated Brownian motion along the edges of a grid with square cells, and count the number of paths. Let m and n be the horizontal and vertical axes, respectively. The coordinates of the lower left corner are (0, 0); this is the starting point. m is the number of voters who vote for C_1, and n is the number of voters who vote for C_0. A path shows the history of the votes. If a voter votes for C_1, the path moves rightward. If a voter votes for C_0, the path moves upward.

We define $P_i(m, n)$ as the probability that the $(n + m + 1)$th voter votes for the candidate C_i, where $i = 0, 1$. The probability of moving upward is as follows.

$$P_0(m,n) = \left\{ \begin{array}{ll} p + (1-p)(1-q) \equiv A & m < n; \\ \frac{1}{2}p + (1-p)(1-q) \equiv B & m = n; \\ (1-p)(1-q) \equiv C & m > n. \end{array} \right\} \tag{5.5}$$

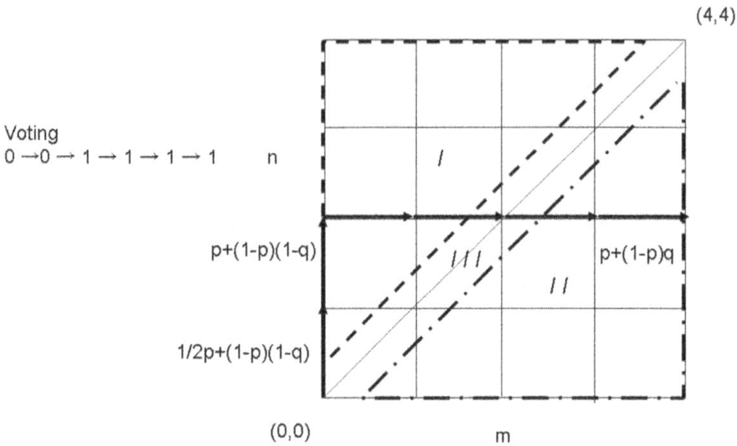

Fig. 5.2 Voting and path. A path shows the history of the votes. If a voter votes for C_1, the path moves rightward. If a voter votes for C_0, the path moves upward. The sample arrowed line shows the voting of 6 voters, 0, 0, 1, 1, 1, 1. We refer to the regions $m < n$, $m > n$, and $m = n$ as I, II, and III, respectively

The probability of moving rightward is $P_1(m, n) = 1 - P_0(m, n)$ for each case. Here, we introduce $X(m, n)$ as the probability that the path passes through the point (m, n). The master equation is

$$X(m, n) = P_1(m - 1, n)X(m - 1, n) + P_0(m, n - 1)X(m, n - 1), \tag{5.6}$$

for $m \geq 0$ and $n \geq 0$, with the initial condition $X(0, 0) = 1$. This defines $X(m, n)$ uniquely. Hereafter, we refer to the region $m < n$ as I, $m > n$ as II, and $m = n$ as III (Fig. 5.2).

First, we consider the case $q = 1$. At this limit, independent voters always vote for only one candidate, C_1 (if we set $q = 0$, independent voters vote only for C_0). The probability is reduced from (5.5) to

$$P_0(m, n) = \left\{ \begin{array}{ll} p & m < n; \\ \frac{1}{2}p & m = n; \\ 0 & m > n. \end{array} \right\} \tag{5.7}$$

In this case, if the path enters II ($m > n$), it can only move rightward. Hence, $n = m - 1$ becomes the absorption wall, where $m, n \geq 0$. There is a difference between the probability (5.7) in I ($m < n$) and that in III ($m = n$). Then, we have to count the number of times the path touches the diagonal.

Using Eq. (5.71), we can calculate the distribution for $m \leq n$. (See Sect. 5.8 Appendix A)

$$X(m, n) = \left\{ \begin{array}{ll} \sum_{k=0}^{m} A_{m,n,k} \frac{p^n(1-p)^m}{2^{k+1}} & m < n; \\ \sum_{k=0}^{m} A_{m,m,k} \frac{p^m(1-p)^m}{2^k} & m = n. \end{array} \right\}, \tag{5.8}$$

where $A_{m,n,k}$ is given by Eq. (5.71) and k is the number of times the path touches the diagonal.

The distribution for $m > n$ can be easily calculated for the absorption wall $n = m - 1$, where $m, n \geq 0$. The distribution for $m > n$ is given by

$$X(m,n) = \sum_{k=0}^{n} A_{n,n,k} \frac{p^n(1-p)^n(1-\frac{1}{2}p)}{2^{k+1}} \qquad m > n. \tag{5.9}$$

We investigate the limit $t \to \infty$. Here, we consider m as a variable; it is the distribution function of the vote for C_1. For large t, we can assume that only the first terms of the summation of Eqs. (5.8) and (5.9) are non-negligible. The first term is the difference of the binomial distributions using Eq. (5.71). For $m/t < 1/2$, the peak of the binomial distribution is $1 - p$ and for $m/t > 1/2$, it is 1. Then, we obtain the distribution in the scaling limit $t = m + n \to \infty$,

$$\frac{m}{t} \Longrightarrow Z. \tag{5.10}$$

The probability measure of Z is

$$\mu = \alpha_- \delta_{1-p} + \alpha_+ \delta_1, \tag{5.11}$$

where δ_x is the delta function. Z is the ratio of voters who vote to C_1 from Eq. (5.10). The distribution has two peaks, one at $Z = 1$ and the other at $Z = 1 - p$. Now, we calculate α_- and α_+, where $\alpha_- + \alpha_+ = 1$. The probability that the path touches the absorption wall $n = m - 1$ is given by

$$\begin{aligned}
\alpha_+ &= \sum_{m=0}^{\infty} X(m,m)(1-\frac{p}{2}) = \sum_{m=0}^{\infty}\sum_{k=0}^{m} A_{m,m,k} \frac{p^m(1-p)^m}{2^k}(1-\frac{p}{2}) \\
&= (1-\frac{p}{2})[1 + \frac{x}{2}C_0(x) + (\frac{x}{2})^2 C_1(x) + (\frac{x}{2})^3 C_2(x) + \cdots] \\
&= (1-\frac{p}{2})[1 + \frac{x}{2}C_0(x) + (\frac{x}{2})^2 \{C_0(x)\}^2 + (\frac{x}{2})^3 \{C_0(x)\}^3 + \cdots] \\
&= (1-\frac{p}{2})[\sum_{k=0}^{\infty} \frac{\{xC_0(x)\}^k}{2^k}] = (1-\frac{p}{2})\frac{1}{1-\frac{xC_0(x)}{2}} \\
&= \frac{4-2p}{3+\sqrt{1-4p(1-p)}} = \frac{4-2p}{3+|1-2p|}.
\end{aligned} \tag{5.12}$$

Then, we can obtain for $p > 1/2$

$$\alpha_- = \frac{2p-1}{p+1}, \tag{5.13}$$

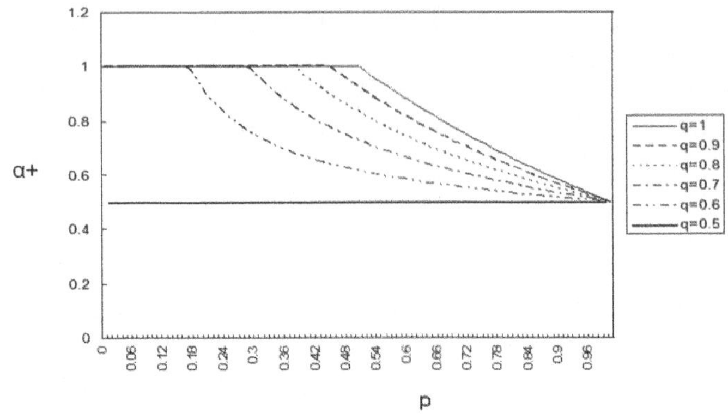

Fig. 5.3 α_+ or $\bar{s}$, i.e., the average votes for candidate C_1 by herders

and $\alpha_- = 0$ for $p < 1/2$. **It is the phase transition which is a absorption phase transition at the critical point** $p_c = 1/2$. $C_k(x)$ is the generating function of the Catalan triangle Eq. (5.76), and $x = p(1-p)$. Here, we use the relations Eqs. (5.75) and (5.78).

5.3.2 *Critical Exponent and Phase Transition*

In Fig. 5.3, we plot α_+ for the some cases including $q = 1$. $\alpha_- = 1 - \alpha_+$ is the order parameter. It is one of non-equilibrium phase transition. Hence, we obtain

$$\alpha_- \propto |p - p_c|^{\beta}, \tag{5.14}$$

where $\beta = 1$ as $p \to p_c$. **The critical exponent** β **is calculated**

$$\beta = \lim_{p \downarrow p_c} \frac{\log \alpha_-}{\log(p - p_c)} = 1, \tag{5.15}$$

which is the same as the universality class of the compact directed percolation (CDP) or the voter model which has absorption phase transition (Liggett 1985; Domany et al. 1984; Essam 1989). The same universality model appears in the contact process model.

Here, we consider the time m dependency of α_+. Using Eq. (5.12),

$$\begin{aligned}
\alpha_+(s) &= \sum_{m=0}^{s}\sum_{k=0}^{m} A_{m,m,k}\frac{p^m(1-p)^m}{2^k}(1-\frac{p}{2}) \\
&\sim \sum_{m=0}^{s} A_{m,m,1}\frac{p^m(1-p)^m}{2^k}(1-\frac{p}{2}) \\
&\sim \sum_{m=0}^{s} \frac{e}{8\sqrt{\pi}} m^{-3/2}(4p(1-p))^m(1-\frac{p}{2}) \\
&= \sum_{m=0}^{s} \frac{e}{8\sqrt{\pi}} m^{-3/2}e^{-m/\xi}(1-\frac{p}{2}),
\end{aligned} \tag{5.16}$$

where

$$\xi = -\frac{1}{\log 4p(1-p)}. \tag{5.17}$$

It is the correlation length and at $p_c = 1/2$, $\xi = \infty$. Note that here we use that $k = 1$ is dominant, $A_{m,m,1} = (2m-2)!/(m!(m-1)!$, and Stirling's formula, $m! \sim \sqrt{2\pi m}(m/e)^m$.

Hence, we can obtain the relation between the order parameter $\alpha_- = 1 - \alpha_+$ and ξ,

$$\alpha_-(s) \sim \int_s^{\infty} ds \frac{e}{8\sqrt{\pi}} s^{-3/2}e^{-s/\xi}(1-\frac{p}{2}) \propto s^{-\alpha}, \tag{5.18}$$

where $\alpha = 1/2$.

About the critical exponent for the divergence of the correlation length ξ is $\nu_{//} = 2$ which is the critical exponent for the correlation length,

$$\xi \propto |p - p_c|^{-\nu_{//}} \tag{5.19}$$

as $p \to p_c$, where we use the expansion of Eq. (5.17),

$$\xi = -\frac{1}{\log(4p(1-p))} = -\frac{1}{\log(1-4\hat{p}^2)} \sim \frac{1}{4\hat{p}^2} \tag{5.20}$$

Here, we use $p = 1/2 - \hat{p}$ and $p \sim p_c = 1/2$.

As in $p < p_c, \alpha_- \sim |p - p_c|^{\beta}$ and $\xi \sim |p - p_c|^{-\nu_{//}}$, we can obtain $\alpha_- \sim \xi^{-\beta/\nu_{//}}$. Then in the limit $s \to \infty$, α_- is

$$\alpha_-(s) \sim s^{-\beta/\nu_{//}} = s^{-\alpha}. \tag{5.21}$$

This is consistent with Eq. (5.18) and **we can obtain the scaling relation,**

$$\beta = \nu_{//}\dot{\alpha}. \tag{5.22}$$

5.3.3 Impact Analysis

For the impact,

$$\begin{aligned} \delta(t_0) = \delta(t - t_0) &= 1 \qquad && t = t_0 \\ &= 0 \qquad && t \neq t_0 \end{aligned} \tag{5.23}$$

the response function $\phi(t - t_0)$ is defined by the difference $\delta c_0(t)$ between $c_0(t)$ and $c_0(t)^+$,

$$\delta c_0(t) = c_0(t)^+ - c_0(t) = \sum_{t_0=1}^{t} \phi(t, t - t_0)\delta(t_0), \tag{5.24}$$

where $c_(t)^+$ is an impact process and $c_(t)$ is a non-impact process and $c_(t) = \sum_i^t X(i)$ and $c_0^+(t) = \sum_i^{t_0}(1 - X(i)) + 1 + \sum_{i=t_0+1}^{t}(1 - X^+(i))$. The second term is the impact at $t = t_0$ and X^+ is the process after the impact. The response function generally depends on t and t_0.

To study the impact, here we consider α_-^+ in Eq. (5.11) which is the distribution of the impact process. We consider the case, $c_0(t_0) = k$ and $p > 1/2$. When $p < 1/2$, the reduced response function is 0, because $\alpha_-^+ = 0$, it is the absorption phase.

We can obtain the recursive relation using the lattice,

$$p\alpha_+(k) + (1 - p)\alpha_+(k - 2) = \alpha_+(k - 1) \tag{5.25}$$

and the initial condition is

$$\alpha_+(0) = \frac{2 - p}{1 + p}. \tag{5.26}$$

The solution of Eq. (5.25) is

$$\alpha_+(k) = \frac{2(1 - p)}{1 + p} + (\frac{1 - p}{p})^k \frac{p}{1 + p}. \tag{5.27}$$

Hence, we obtain

$$\alpha_-(k) = \frac{3p - 1}{1 + p} - (\frac{1 - p}{p})^k \frac{p}{1 + p}. \tag{5.28}$$

The reduced response function is

$$\begin{aligned} \frac{\mathrm{E}(\phi(t, t - t_0))}{t} = \frac{\mathrm{E}(c_0(t)^+) - \mathrm{E}(c_0(t))}{t} &= \sum_k P(k|t_0)\alpha_-^+(k) - \sum_k P(k|t_0)\alpha_-(k) \\ &= \sum_k P(k|t_0)(\frac{1 - p}{p})^k \frac{2p - 1}{1 + p}, \end{aligned} \tag{5.29}$$

where $P(k|t_0)$ is the distribution of the process $k \geq t_0/2$ at $t = t_0$ and $\alpha_-^+(k) = \alpha_-(k+1)$ as the impact. Because $\sum_{k \geq t_0/2} P(k|t_0) > 0$ when $p > 1/2$, the reduced response function does not converge to 0 in the large t limit. Hence, the effect of impact continues as Pólya urn process discussed in Chap. 2.

As a simple example, we consider the impact process with the initial condition,

$$X(1) = 0 \tag{5.30}$$

and the non-impact process with this condition,

$$\begin{aligned} X(1) &= 1 \quad \text{the probability} \quad 1/2, \\ X(1) &= 0 \quad \text{the probability} \quad 1/2. \end{aligned} \tag{5.31}$$

Using Eq. (5.28), we obtain the scaled impact

$$\frac{\mathrm{E}(\phi(t, t-1))}{t} = \frac{\mathrm{E}(c_0(t)^+) - \mathrm{E}(c_0(t))}{t} = \frac{4p-2}{1+p} - \frac{2p-1}{1+p} = \frac{2p-1}{1+p}. \tag{5.32}$$

The impact does not converge to 0. There are two equilibria where these processes converge. We can confirm that **the equilibria are the same, but the distribution changes by the impact.**

5.3.4 *Phase Transition Between the Self-Averaging Process and the Non-Self-Averaging Process*

Here, we consider the coefficients of variables (CV) for the distribution of X_τ when $q = 1$. The expected value is

$$E(X_\tau) = (\alpha_-(1-p) + \alpha_+ 1)\tau = \frac{-2p^2 + 2p + 1}{1+p}\tau, \tag{5.33}$$

for $p > 1/2$ and $E(X_\tau) = \tau$ for $p < 1/2$. The variance is

$$V(X_\tau) = \frac{p^2(2p-1)(2-p)}{(p+1)^2}\tau^2, \tag{5.34}$$

for $p > 1/2$ and $V(X_\tau) \sim \tau^{-1}$ for $p < 1/2$. Then CV can be calculated

$$\lim_{\tau \to \infty} CV(X_\tau)^2 = \lim_{\tau \to \infty} \frac{V(X_\tau)}{E(X_\tau)^2} = \frac{p^2(2p-1)(2-p)}{(-2p^2 + 2p + 1)^2} \neq 0, \tag{5.35}$$

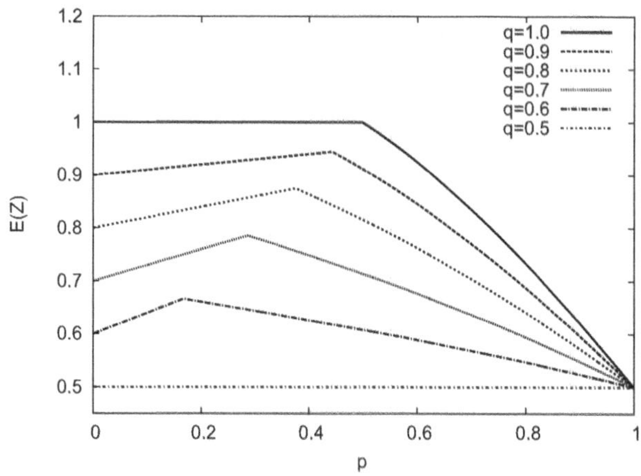

Fig. 5.4 Average vote ratio for the correct candidate C_1 in the case $r = \infty$. The vertical axis is $E(Z) = \bar{Z}$. At the critical point p_c, $\bar{Z} = E(Z)$ is maximum

for $p > 1/2$ and $CV(X_\tau)^2 = 0$ for $p < 1/2$. **We can confirm that this is the phase transition between the self-averaging process and the non-self-averaging process.** In the non-self-averaging process, there are two equilibria. Therefore, we cannot decide which equilibrium is the convergence point.

5.3.5 *General Case*

We are interested in the average votes for C_1 by the herders to validate their performance. We define s as the average vote ratio for the correct candidate C_1 by the herders,

$$s = \frac{Z - (1 - p)}{p}. \tag{5.36}$$

Here, we take the expected values of Eq. (5.36) about several sequences of voting,

$$\bar{Z} = p\bar{s} + (1 - p) = (1 - p)\alpha_- + \alpha_+, \tag{5.37}$$

where $\bar{x}$ means the expected value of x. The second equality can be obtained from Eq. (5.11). Using the relation $\alpha_- + \alpha_+ = 1$, we can obtain $\bar{s} = \alpha_+$. In Fig. 5.4, we plot $\bar{Z}$ for some cases including $q = 1$.

Next, we consider the general q case. In this case, the path crosses the diagonal several times. $n = m - 1$ is no longer the absorption wall, where $m, n \geq 0$. Hence, it is difficult to calculate the exact solution for the general q. However, as in discussion of Eq. (5.11), we obtain the limit shape of the distribution of votes for C_1,

$$\frac{m}{t} \Longrightarrow Z. \tag{5.38}$$

The probability measure of Z is

$$\mu = \alpha_- \delta_{(1-p)q} + \alpha_+ \delta_{p+(1-p)q}, \tag{5.39}$$

where $\alpha_- + \alpha_+ = 1$. When $q = 1$, Eq. (5.39) becomes Eq. (5.11). We can calculate α_+ as

$$\alpha_+ = \tilde{R}_1(1 - R_2)(1 + R_1 R_2 + R_1^2 R_2^2 + \cdots) = \frac{\tilde{R}_1(1 - R_2)}{1 - R_1 R_2}. \tag{5.40}$$

$\tilde{R}_1$ is the probability that the path starts from (0, 0), crosses the diagonal only once, and reaches the wall $n = m - 1$ in II $(m > n)$. R_1 is the probability that the path starts from the wall $n = m + 1$ in I $(m < n)$, crosses the diagonal only once, and reaches the wall $n = m - 1$ in II $(m > n)$. R_2 is the probability that the path starts from the wall $n = m - 1$ in II $(m > n)$, goes across the diagonal only once, and reaches the wall $n = m + 1$ in I $(m < n)$.

For example, the first term of Eq. (5.40) is the path that starts from (0, 0) and passes through I $(m < n)$ or directly enters II $(m > n)$. The path goes across diagonal III $(m = n)$ only once; the first step is rightward. The second term is the path that starts from (0, 0), crosses the diagonal III $(m = n)$ three times, and enters II $(m > n)$. We can calculate $\tilde{R}_1$, R_1, and R_2 similarly to Eq. (5.12) (see Sect. 5.9 Appendix B),

$$\begin{aligned}
\tilde{R}_1 &= \frac{2(1 - B)}{2 - \gamma_1(1 - \sqrt{1 - 4A(1 - A)})}, \\
R_1 &= \frac{(1 - B)\gamma_1(1 - \sqrt{1 - 4A(1 - A)})}{B\{2 - \gamma_1(1 - \sqrt{1 - 4A(1 - A)})\}}, \\
R_2 &= \frac{B\gamma_2(1 - \sqrt{1 - 4C(1 - C)})}{(1 - B)\{2 - \gamma_2(1 - \sqrt{1 - 4C(1 - C)})\}},
\end{aligned} \tag{5.41}$$

where A, B, and C are given by (5.5), $\gamma_1 = B/A$, and $\gamma_2 = (1 - B)/(1 - C)$.

In general, we can calculate the exact value of p_c:

$$p_c = 1 - \frac{1}{2q}. \tag{5.42}$$

As q increases, p_c increases. At p_c, the model features a phase transition beyond which a state where most agents make the correct forecasts coexists with one where most of them are incorrect. Thus, the effectiveness of herding decreases as q decreases. In the limit $q = 0.5$, the phase transition disappears. The distribution becomes symmetric in this case.

5.4 Mean Field Approximation

To investigate long-ranged correlations, we analyze in the limit $t \to \infty$. We can rewrite Eq. (5.7) as

$$c_1(t) = k \to k+1 : P_{k,t} = p \cdot \theta(\frac{k}{t-1} - \frac{1}{2}) + (1-p)q, \tag{5.43}$$

where $\theta(x)$ is the step function,

$$\theta(x) = \begin{Bmatrix} 0 & x < 0; \\ \frac{1}{2} & x = 0; \\ 1 & x > 0. \end{Bmatrix} \tag{5.44}$$

In the scaling limit $t = c_0(t) + c_1(t) \to \infty$, we define

$$\frac{c_1(t)}{t} \Longrightarrow Z. \tag{5.45}$$

Z is the ratio of voters who vote for C_1.

We define a new variable Δ_t such that

$$\Delta_t = 2c_1(t) - t = c_1(t) - c_0(t). \tag{5.46}$$

We change the notation from k to Δ_t for convenience. Then, we have $|\Delta_t| = |2k - t| < t$. Thus, Δ_t holds within $\{-t, t\}$. Given $\Delta_t = u$, we obtain the following random walk model:

$$\Delta_t = u \to u+1 : P_{\frac{u+t}{2},t} = p \cdot [\theta\{\frac{u}{2(t-1)}\}] + (1-p)q,$$
$$\Delta_t = u \to u-1 : Q_{\frac{u+t}{2},t} = 1 - P_{\frac{u+t}{2},t}.$$

We now consider the continuous limit $\epsilon \to 0$,

$$X_\tau = \epsilon \Delta_{[t/\epsilon]},$$
$$P(x,\tau) = \epsilon P(\Delta_t/\epsilon, t/\epsilon), \tag{5.47}$$

where $\tau = t/\epsilon$ and $x = \Delta_t/\epsilon$. We obtain X_τ such that it obeys a diffusion equation with small additive noise:

$$\mathrm{d}X_\tau = [(1-p)(2q-1) + p(2\theta(\frac{X_\tau}{2\tau}) - 1)]\mathrm{d}\tau + \sqrt{\epsilon}\mathrm{d}B_\tau. \tag{5.48}$$

Here, we change the variable X_τ to Y_τ by using the expression,

$$pY_\tau = X_\tau - (1-p)(2q-1)\tau. \tag{5.49}$$

We rewrite Eq. (5.48) by using Y_τ:

$$\mathrm{d}Y_\tau = [2\theta(\frac{p}{2\tau}Y_\tau + \frac{(2q-1)(1-p)}{2}) - 1]\mathrm{d}\tau + \sqrt{\epsilon}\mathrm{d}B_\tau. \tag{5.50}$$

Using Eqs. (5.45), (5.46), and (5.49), we obtain the relations of the variables

$$2Z - 1 = \frac{X_\infty}{\tau} = \frac{pY_\infty}{\tau} + (1-p)(2q-1). \tag{5.51}$$

Here we set $Y_\tau = \bar{v}\tau$, we can obtain

$$\bar{v} = 2\theta(\frac{p\bar{v}}{2} + \frac{(2q-1)(1-p)}{2}) - 1. \tag{5.52}$$

It is the self-consistent equation. When we rewrite Eq. (5.52) using Eq. (5.51), we can obtain,

$$\bar{Z} = (1-p)q + p\theta(\bar{Z} - \frac{1}{2}). \tag{5.53}$$

In Eq. (5.51), the limit of tanh function rises vertically at $\bar{v} = -(2q-1)(1-p)/p$. $\bar{v}$ is 1 or -1. At the critical point $p = p_c$, $\bar{v} = -1$ becomes the solution, too. Therefore, the critical condition is

$$\frac{(2q-1)(1-p_c)}{p_c} = 1. \tag{5.54}$$

Hence, a phase transition occurs at

$$p_c = 1 - \frac{1}{2q}. \tag{5.55}$$

Note that at the critical point, the solution $\bar{v} = -1$ is not stable. Hence, the phase transition is continuous.

When $p \le p_c$, we can obtain only one solution $\bar{v} = 1$. Using (5.53), we obtain the ratio of voters for C_1 as $Z = p + (1-p)q$. In this case, Z has only one peak.

When $p > p_c$, we can obtain two stable solutions: $\bar{v} = 1$ and $\bar{v} = -1$. The solution $\bar{v} = -\frac{(2q-1)(1-p)}{p}$ is unstable. The ratio of voters for C_1 is obtained as $Z = p + (1-p)q$ and $Z = q(1-p)$. The votes ratio for C_1, Z has two peaks. $Z = p + (1-p)q$ shows a good equilibrium, whereas $Z = q(1-p)$ shows a bad equilibrium.

We show the images of the solutions in Fig. 5.5. The left figure (a) is the solutions at $p < p_c$. There is one solution.

The right figure (b) shows the solutions at $p > p_c$. There are three solutions. at $p > p_c$. The solutions of both sides are stable solutions.

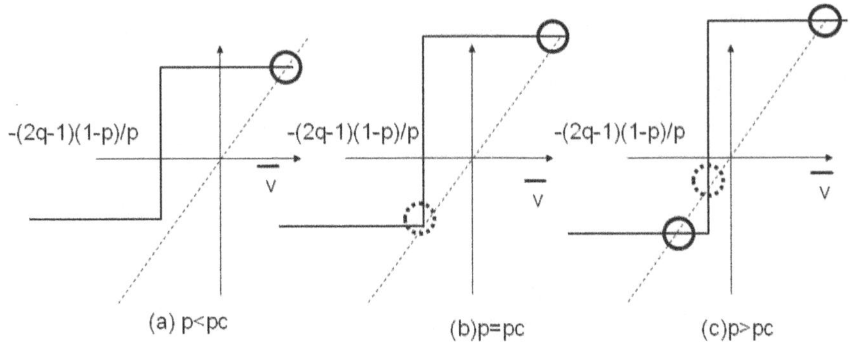

Fig. 5.5 Solution of Eq. (5.53). The figure (**a**) shows the solutions at $p < p_c$. There is only one solution. The figure (**b**) shows the solutions at $p > p_c$. There are three solutions. The solutions on both sides are stable. The figure (**b**) is the critical case. The new solution is not stable and the phase transition is continuous

When the critical case we show in Fig. 5.5, there are two solutions, but the new solution is not the tangent point and is not stable. The convergence point is the right solution. **Hence, it is a continuous phase transition.**

5.5 Exact Solutions for $r = 1$

Here, we discuss the cases $r = 1$ in addition to $p \neq 1$. When $p = 1$, all voters are herders, and the distribution becomes the limit shape of the beta distribution, as discussed in Mori et al. (2010). Herders can see only the vote of the previous voter. We define $P_i(t)$ as the probability that the $(t + 1)$th voter votes for C_i, where $i = 0, 1$. Here, t denotes time.

$$P_0(t) = \begin{cases} p + (1-p)(1-q) \equiv F \;;\; Y_0(t-1) = 1; \\ (1-p)(1-q) \equiv G \quad\;:\; Y_0(t-1) = 0. \end{cases} \tag{5.56}$$

$Y_i(t) = 1$ indicates that at t, the voter votes for C_i. $Y_0(t-1) = 1$ indicates that the previous voter votes for C_0. On the other hand, $Y_i(t) = 0$ indicates that at t, the voter does not vote for C_i. $Y_0(t-1) = 0$ indicates that the previous voter votes for C_1. Thus, $\sum_{l=1}^{t} Y_i(l)$ is the total number of votes for C_i until t. Here, the relation $P_1(t) = 1 - P_0(t)$ holds. The initial distribution is

$$P_0(0) = \frac{1}{2}p + (1-p)(1-q). \tag{5.57}$$

The model was studied as a one-dimensional correlated random walk (Böhm 2000; Konno 2002). Here, we introduce $X(m, n)$ as the probability distribution. m

is the number of voters who vote for C_1 and n is the number of voters who vote for C_0. The master equation is

$$X(m,n) = P_1(t-1)X(m-1,n) + P_0(t-1)X(m,n-1), \tag{5.58}$$

for $m \geq 0$ and $n \geq 0$, with the initial condition $X(0,0) = 1$.

In the limit $t \to \infty$,

$$X(m, t-m) \Longrightarrow N(qt, \sqrt{\frac{F(1-G)}{(1-F)G}t}), \tag{5.59}$$

where $N(\mu, \sigma^2)$ is the normal distribution with mean μ and variance σ^2. (see Theorem 3.1. in Böhm 2000) F and G are given in (5.56). Hence, we can obtain the limit shape of the distribution as

$$\frac{m}{t} \Longrightarrow Z. \tag{5.60}$$

The probability measure of Z is

$$\mu = \delta_q. \tag{5.61}$$

Then, $r = 1$ involves no phase transition, and the majority of voters do not select the incorrect candidate C_0.

The limit distribution is independent of the initial condition $P_0(0)$ because the distribution Z has only one peak.

5.6 Block Spin Transformation

In this section, we consider the block spin transformation. This is the mapping from the r voting model to the correlated random walk which is $r = 1$ voting model.[1] This is the transformation of the parameters from (r, q, p). to (q_r, p_r).

The probabilistic rule of the r voting model is

$$P\left(X(t+1) = 1 | c_1^r(t) = x\right) = (1-p) \cdot q + p \cdot \theta(x - r/2). \tag{5.62}$$

We can map the model to the correlated random walk by applying the real-space renormalization transformation. We group r variables $X(r(t-1)+1), \cdots, X(rt)$ into X^r by the majority rule.

$$X^r(t) = \theta\left(X(r(t-1)+1) + \cdots + X(rt) - r/2\right).$$

[1] The r voting model is discussed in Chap. 6.

rT variables, $X(1), \cdots, X(rT)$ are transformed to T variables $X^r(1), \cdots, X^r(T)$. Here $\theta(x)$ is the step function in Eq. (5.44).

We can modify Eq. (5.62) to

$$P(X(rt+s)=1|X^r(t)=x)=(1-p)q+px, \tag{5.63}$$

where $s = 1, \cdots, r$ and $x \in \{0, 1\}$.

$X^r(t)$ depends only on the previous $X^r(t-1)$. We write the probabilistic rule for $X^r(t)$ as

$$\Pr(X^r(t)=1|X^r(t-1)=x)=(1-p_r)\cdot q_r + p_r \cdot x.$$

This is a real-space renormalization transformation. If $X^r(t-1)=0$, the probability that $X^r(t)$ takes 1 is $(1-p_r)q$. In the modified dynamics of $X(s)$, the probability is given as

$$(1-p_r)q_r = \pi_r((1-p)q) = a. \tag{5.64}$$

Here, $\pi_r(x)$ is defined as $\pi_r(x) = \sum_{n=(r+1)/2}^{r} C_n \cdot x^n(1-x)^{r-n}$ where ${}_rC_n$ is the binomial coefficient. $\pi_r(x)$ gives the majority probability. In the limit $r \to \infty$, $\pi_r(x)$ is $\theta(x-1/2)$.

Likewise, we can obtain

$$(1-p_r)(1-q_r) = \pi_r((1-p)(1-q)) = b. \tag{5.65}$$

We solve Eqs. (5.64) and (5.65) and obtain the following explicit expressions for q_r, p_r:

$$q_r = \frac{\pi_r((1-p)q)}{\pi_r((1-p)q)+\pi_r((1-p)(1-q))}, \tag{5.66}$$

$$p_r = 1-(\pi_r((1-p)q)+\pi_r((1-p)(1-q))). \tag{5.67}$$

We study the transformation $(q, p) \to (q_r, p_r)$, which has five fixed points. We summarize the results in Fig. 5.6. Under the transformation, $(1, 1)$, $(1, 0)$ and $(1/2, 1)$ are stable, and $(1/2, 0)$ is unstable.

When $q > 1/2$, (q, p) on the line $p(q) = 1 - 1/2q$ moves along it under the transformation because (q_r, p_r) also satisfies $p_r = 1 - 1/2q_r$. We call this line the critical line and denote,

$$p_c(q) = 1 - \frac{1}{2q}.$$

This is consistent with Eq. (5.42).

When r is finite, $p_r < 1$ for $p < 1$. Therefore, there is no phase transition as discussed.

In the limit $r \to \infty$ we have $\lim_{r\to\infty}(q_r, p_r) = (1, 1/2)$. If $p < p_c(q)$, we have $\lim_{r\to\infty}(q_r, p_r) = (1, 0)$ which is the one peak phase. If $p > p_c(q)$, we have

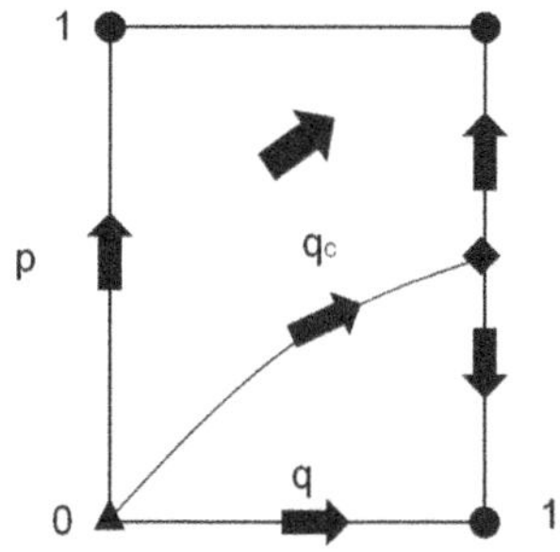

Fig. 5.6 Plot of transformation $(q, p) \to (q_r, p_r)$. Arrows indicate the direction of movement from (q, p) to (q_r, p_r). There are three stable fixed points at $(1, 1)$, $(1, 0)$, $(1/2, 1)$ (filled circles) and two unstable fixed points at $(1/2, 0)$, $(1, 1/2)$ (filled triangle and diamond, respectively)

$\lim_{r\to\infty}(q_r, p_r) = (1, 1)$. which is the two-peaks phase. At the marginal fixed point at $(1, 1/2)$, the direction to $(1, 0)$ is stable and the direction $(1, 1)$ is unstable. This corresponds to the continuous phase transition in the limit $r \to \infty$.

5.7 Summary

We investigated a voting model. We calculated the exact solutions for the special cases $r = 1$ and $r = \infty$. Furthermore, we analyzed the general case using the mean field approximation and numerical simulations. In the case r is infinite, we could find the phase transition and showed the transition point and critical exponent $\beta = 1$. We applied the block spin transformation. In the transformation, we can confirm that there is no phase transition for finite r and there is a phase transition for infinite r.

5.8 Appendix A Proof of Eq. (5.8)

Here, we consider the number of monotonic paths along the edges of a grid with square cells, which do not pass lower the diagonal. Let m and n be the horizontal and the vertical axes, respectively. The coordinates of the lower left corner are (0, 0). A monotonic path is one which starts in the lower left corner, finishes in the upper triangle (m, n), where $0 \leq m \leq n$, and consists entirely of edges pointing rightward or upward.

The number of paths from (0, 0) to (m, n) can be calculated as

$$C_{m,n} = \frac{(n-m+1)(n+m)!}{m!(n+1)!} = \binom{n+m}{n} - \binom{n+m}{n+1}. \tag{5.68}$$

These numbers are known as Catalan triangle.

If the finish point is (m, m), the number of paths becomes the Catalan number

$$C_{m,m} = c_m = \frac{2m!}{m!(m+1)!} = \binom{2m}{m} - \binom{2m}{m+1}. \tag{5.69}$$

Next, we compute the distribution of the number of paths that start in the lower left corner, finish in the upper triangle (m, n), and touch the diagonal k times (Francesco et al. 1997). Let $A_{m,n,k}$ denote the number of paths that touch the diagonal k times. We get a simple recursive relation about $A_{m,n,k}$,

$$A_{m,n,k} = \sum_{j=0}^{m-1} c_j A_{m-j-1,n-j-1,k-1}, \tag{5.70}$$

for $k \geq 0$, $n, m \geq 0$, and $m \geq k$, with the initial condition $A_{0,0,0} = 1$. This defines the numbers $A_{m,n,k}$ uniquely, and it is easy to prove that

$$\begin{aligned} A_{m,n,k} &= \frac{(n-m+k)(n+m-k-1)!}{n!(m-k)!} \\ &= \binom{n+m-k-1}{n-1} - \binom{n+m-k-1}{n}. \end{aligned} \tag{5.71}$$

From Eqs. (5.68) and (5.71), we obtain the relation:

$$A_{m,m,k} = C_{m-k,m-1}. \tag{5.72}$$

The well-known generating function $C_0(x)$ of Catalan numbers is given by

$$\begin{aligned} C_0(x) &= \sum_{n=0}^{\infty} C_{m,m} x^n \\ &= 1 + x + 2x^2 + 5x^3 + 14x^4 + 42x^5 + 132x^6 + \cdots, \end{aligned} \tag{5.73}$$

subject to the algebraic relation

$$x C_0(x)^2 = C_0(x) - 1, \tag{5.74}$$

and we can obtain

$$C_0(x) = \frac{1 - \sqrt{1-4x}}{2x}. \tag{5.75}$$

Here, we obtain the generating function $A_{m,k}(x)$ of $A_{m,m,k}$.

$$A_{m,k}(x) = \sum_{m-k=0}^{\infty} A_{m,m,k}x^{m-k} = \sum_{m-k=0}^{\infty} C_{m-k,m-1}x^{m-k}$$
$$= \sum_{l=0}^{\infty} C_{l,l+k-1}x^{l} = C_{k-1}(x). \tag{5.76}$$

We use Eq. (5.72) for the second equality. $C_j(x) = \sum_{l=1}^{\infty} C_{l,l+j}x^l$ is the generating function of the generalized Catalan number Eq. (5.68). The generating function of the Catalan triangle is given by

$$C_1(x) = \sum_{n=0}^{\infty} C_{m,m+1}x^n$$
$$= 1 + 2x + 5x^2 + 14x^3 + 42x^4 + 132x^5 + 429x^6 + \cdots,$$
$$C_2(x) = \sum_{n=0}^{\infty} C_{m,m+2}x^n$$
$$= 1 + 3x + 9x^2 + 28x^3 + 90x^4 + 297x^5 + 1001x^6 + \cdots,$$
$$C_3(x) = \sum_{n=0}^{\infty} C_{m,m+3}x^n$$
$$= 1 + 4x + 14x^2 + 48x^3 + 165x^4 + 572x^5 \cdots.$$

From Eq. (5.70), we can obtain

$$C_j(x) = C_{j-1}(x)C_0(x). \tag{5.77}$$

Thus, the simple relation between the generating functions is given by

$$C_j(x) = \{C_0(x)\}^{j+1}. \tag{5.78}$$

5.9 Appendix B Derivation of Eq. (5.41)

$\tilde{R}_1$ is the probability that path starts from (0, 0), crosses the diagonal only once, and reaches the wall $n = m - 1$ in II $(m > n)$ (Fig. 5.7).

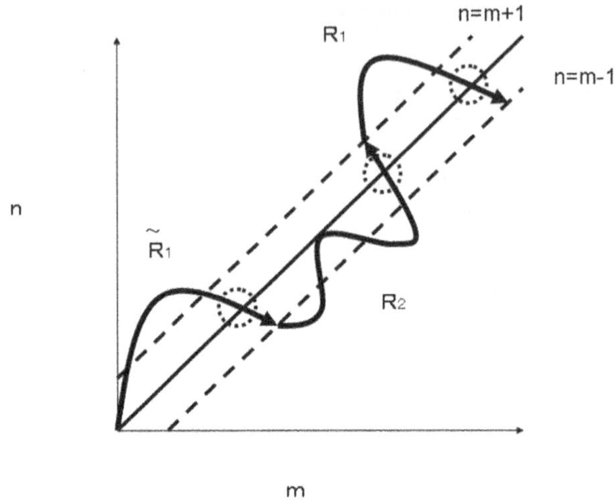

Fig. 5.7 $\tilde{R}_1$, R_1, and R_2. $\tilde{R}_1$ is the probability that path starts from (0, 0), crosses the diagonal only once, and reaches the wall $n = m - 1$ in II $(m > n)$. R_1 is the probability that the path starts from the wall $n = m + 1$ in I $(m < n)$, crosses the diagonal only once, and reaches the wall $n = m - 1$ in II $(m > n)$. R_2 is the probability that the path starts from the wall $n = m - 1$ in II $(m > n)$, crosses the diagonal only once, and reaches the wall $n = m + 1$ in I $(m < n)$

$$
\begin{aligned}
\tilde{R}_1 &= (1-B)[1 + y\gamma_1 C_0(y) + (y\gamma_1)^2 C_1(y) + (y\gamma)^3 C_2(y) + \cdots] \\
&= (1-B)[1 + y\gamma_1 C_0(y) + (y\gamma_1)^2 \{C_0(y)\}^2 + (y\gamma_1)^3 \{C_0(y)\}^3 + \cdots] \\
&= (1-B)[\sum_{k=0}^{\infty} \{\gamma_1 y C_0(y)\}^k] = \frac{1-B}{1-\gamma_1 A C_0(A)} \\
&= \frac{2(1-B)}{2-\gamma_1(1-\sqrt{1-4A(1-A)})},
\end{aligned} \tag{5.79}
$$

where A and B are given by (5.5), $\gamma_1 = B/A$, and $y = A(1-A)$. $C_k(y)$ is the generation function of the Catalan triangle (5.76). Here, we use the relations (5.78) and (5.75). When $q = 1$, (5.79) reduces to (5.12).

R_1 is the probability that the path starts from the wall $n = m + 1$ in I $(m < n)$, crosses the diagonal only once, and reaches the wall $n = m - 1$ in II $(m > n)$.

$$
\begin{aligned}
R_1 &= \frac{1-B}{B}[y\gamma_1 C_0(y) + (y\gamma_1)^2 C_1(y) + (y\gamma_1)^3 C_2(y) + \cdots] \\
&= \frac{1-B}{B}[y\gamma_1 C_0(y) + (y\gamma_1)^2 \{C_0(y)\}^2 + (y\gamma_1)^3 \{C_0(y)\}^3 + \cdots] \\
&= \frac{1-B}{B}[\frac{\tilde{R}_1}{1-B} - 1] = \frac{(1-B)\gamma_1(1-\sqrt{1-4A(1-A)})}{B\{2-\gamma_1(1-\sqrt{1-4A(1-A)})\}}.
\end{aligned} \tag{5.80}
$$

R_2 is the probability that path starts from the wall $n = m - 1$ in II $(m > n)$, acrosses the diagonal only once, and reaches the wall $n = m + 1$ in I $(m < n)$.

$$\begin{aligned} R_2 &= \frac{B}{1-B}[z\gamma_2 C_0(z) + (z\gamma_2)^2 C_1(z) + (z\gamma_2)^3 C_2(z) + \cdots] \\ &= \frac{B}{1-B}[z\gamma_2 C_0(z) + (z\gamma_2)^2\{C_0(z)\}^2 + (z\gamma_2)^3\{C_0(z)\}^3 + \cdots] \\ &= \frac{B\gamma_2(1-\sqrt{1-4C(1-C)})}{(1-B)\{2-\gamma_2(1-\sqrt{1-4C(1-C)})\}}, \end{aligned} \tag{5.81}$$

where C is given by (5.5), $\gamma_2 = (1-B)/(1-C)$, and $z = C(1-C)$.

References

Böhm W, The correlated random walk with boundaries: a combinatorial solution, *J. Appl. Prob.* **37** 470–479 (2000).

Di Francesco P, Golinelli O, Guitter E, Meande, folding, and arch statistics, *Math. Comput. Modelling* **26N8** 97–147, (1997).

Domany E and Kinzel W, Equivalence of cellular automata to Ising models and directed percolation, *Phys. Rev. Lett.* **53** 311, (1984).

Essam J W, Directed compact percolation: cluster size and hyperscaling, *J. Phys. A* **22**, 4927, 198, (1989).

Hisakado M and Mori S, Phase transition and information cascade in a voting model, *J. Phys. A.* **43** 315207 (2010).

Hisakado M and Mori S, Digital herders and phase transition in a voting model *J. Phys. A* **22** 275204, (2011).

Konno N, "Limit theorems and absorption problems for quantum random walks in one dimension" *Quant. Inf. Comp.* **2** 578–595 (2002).

Lang W, "On polynominals related to power of the generating function of Catalan's numbers" *Fib. Quart* **38** 408 (2000).

Liggett T M, "Interacting particle systems" (Springer, New York) (1985).

Mori S and Hisakado M, Exact scale invariance in mixing of binary candidates n voting *J. Phys. Soc. Jpn* **79** 034001–034008 (2010).

Chapter 6
Equilibrium Process and Detailed Balance Condition

6.1 Introduction

We continue to discuss the voting model with two candidates. In this chapter, we consider the case where a voter can see the latest r previous votes. **This process is an equilibrium process because of the detailed balance condition.** The urn model corresponds to the number of balls in the urn being constant. There is no phase transition when r is finite and the solution oscillates between two equilibria. Furthermore, we show relations among this voting model, Kirman's ant colony model, and the kinetic Ising model.

Here, we introduce the infinite range model. It is one of the most popular models in statistical physics which explains the phase transition, and plays the important role in this chapter. In the model, spins interact with all other spins. Hamiltonian is

$$H(\sigma) = -\frac{J}{N}\sum_{i>j}\sigma_i\sigma_j - h\sum_{i=1}^{N}\sigma_i, \tag{6.1}$$

where N is the number of spins, σ is the spin with the value ± 1, J is the parameter of interaction, and h is the outer field. Here, we define the average of spins, as an order parameter, $m = 1/N\sum\sigma_i$.

In large $N \to \infty$ limit the self-consistent equation becomes

$$m = \tanh\beta(Jm + h), \tag{6.2}$$

where $\beta = 1/k_BT$. k_B is the Bolzman constant and T is the temperature. For infinite range model, we obtain the strict solution from the self-consistent equation. In the symmetric case, under the transition temperature T_C, there are two solutions. One of the solutions is selected and is called spontaneous symmetry braking. In contrast, above T_c, there is only one solution. This simple model represents the phase transition.

M. Hisakado, *Urn Models and Their Applications in Finance*,
https://doi.org/10.1007/978-981-96-3825-3_6

When there is an outer field h, the model becomes an asymmetric model and there is no phase transition.

In this chapter, we discussed the equilibrium process. Nonlinear Pólya model which we discussed in Chap. 4 is the non-equilibrium model. The large r limit corresponds to the large N limit of the infinite range model. We can observe the phase transition in both models. In the case of non-equilibrium, we can obtain the same self-consistent equation. In this case, the solution that is against the outer field may be selected. This is the characteristic point of the non-equilibrium model.

The remainder of this chapter is organized as follows. In Sect. 6.2, we introduce our voting model and discuss the relation to the urn model. In Sect. 6.3, we discuss the case where there are digital herders and independents. In Sect. 6.4, we verify the transitions between voting choices through numerical simulations. In Sect. 6.5, we discuss the case where there are analog herders and independents and show the relation between our voting model and Kirman's ant colony model. In Sect. 6.6, we discuss the relation between our voting model and the kinetic Ising model. In the final section, we present the summaries. This chapter is based on Hisakado et al. (2015).

6.2 Voting Model

We model the voting of two candidates, C_0 and C_1. The voting is sequential, and at time t, C_0 and C_1 have $c_0(t)$ and $c_1(t)$ votes, respectively. In each time step, one voter votes for one candidate. Hence, at time t, the tth voter votes, after which the total number of votes is t.

We consider the Bernoulli random variables $X(t)$, which corresponds to the vote at time t. $X(t) = 1(0)$ is the vote to the candidate $1(0)$ at t. Then, we obtain

$$c_1(t) = \sum_{i=1}^{t} X(i), \tag{6.3}$$

and

$$c_0(t) = \sum_{i=1}^{t} (1 - X(i)). \tag{6.4}$$

Voters are permitted to see only r previous votes for each candidate; thus, they are aware of public perception. r is a constant number. [1]

Hence, we obtain

$$c_1^r(t) = \sum_{i=t-r}^{t} X(i), \tag{6.5}$$

[1] In the case $r \to \infty$, the tth voter can see $c_0(t-1)$ and $c_1(t-1)$ votes at time t. This model is discussed in Chap. 5.

and

$$c_0^r(t) = \sum_{i=t-r}^{t} (1 - X(i)). \tag{6.6}$$

In the limit $r \to \infty$, voters can see all previous votes. Therefore, $c_0^\infty(t) = c_0(t)$ and $c_1^\infty(t) = c_1(t)$.

We assume an infinite number of two types of voters—independents and herders. The independents vote for C_0 and C_1 with probabilities $1 - q$ and q, respectively. Their votes are independent of those of others, that is, their votes are based on their own fundamental values.

Here, we set C_0 as the incorrect candidate and C_1 as the correct candidate to validate the performance of the herders. We can set $q \geq 0.5$ because we believe that independents vote for C_1 rather than for C_0. In other words, we assume that the intelligence of the independents is virtually accurate.

In contrast, the herders' votes are based on the number of previous r votes. At time t, the information of r previous votes is the number of votes for C_0 and C_1:$c_0^r(t)$ and $c_1^r(t)$, respectively. Hence, $c_0^r(t) + c_1^r(t) = r$ holds. If $r > t$, voters can see t previous votes for each candidate. For the limit $r \to \infty$, voters can see all previous votes. We define the number of all previous votes for C_0 and C_1 as $c_0^\infty(t) \equiv c_0(t)$ and $c_1^\infty(t) \equiv c_1(t)$.

Now we define the majority's correct decision. If the ratio to candidate C_1 who is correct is $c_1/t > 1/2(< 1/2)$, we define the majority as correct (incorrect). This ratio is important for evaluating the performance of herders. In this chapter, we discuss three types of herders, digital, analog, and tanh-type herders. We define z_r as $z_r = c_1^r/r$. The probability that a herder who refers z_r votes to the candidate C_1 is defined as $f(z_r)$. Digital herders always choose the candidate with the majority of the previous r votes, which is visible to them. In this case, $f(z_r) = \theta(z_r - 1/2)$, where θ is a heaviside step function. Analog herders vote for each candidate with probabilities proportional to the candidates' votes. Thus, $f(z_r) = z_r$. The other herder is the tanh-type herder, which is an intermediate between analog and digital herders. In this case, $f(z_r) = 1/2(\tanh(z_r - 1/2) + 1)$.

Independents and herders appear randomly and vote. We set the ratio of independents to herders as $(1 - p)/p$. In this study, we mainly focus on the upper limit of t. This refers to the voting of infinite voters.

The tanh-type herder model includes the two types of noise (Young 2011). One is independent. They appear with probability $1 - p$ and vote on correct candidates with probability q. If q is high, the error is reduced. They choose action uniformly at random. The model is called "uniform error process". The other is included in the herders' response function. The reason for the name is that the log probability of choosing candidate C_1 is linear. In the limit of no noise case, the herder becomes digital. In the limit of the noisy case, he/she becomes the analog.[2]

[2] Noisy means weak herding and the response function is not steep.

6.3 Digital Herder

In this section, the herder is a digital herder and hence votes for the majority candidate; if $c_0^r(t) > c_1^r(t)$, herders vote for candidate C_0. If $c_0^r(t) < c_1^r(t)$, herders vote for the candidate C_1. If $c_0^r(t) = c_1^r(t)$, the herders vote for C_0 and C_1 with the same probability, that is, 1/2.

Chapter 5, we discussed the exact solutions. In this chapter, we try to discuss this using stochastic partial differential equations. First, we consider an approximation to introduce the partial differential equations. The voter selects r votes randomly from the latest previous r voters with overlap. The process can be written as follows,

$$
\begin{aligned}
c_1(t) = k \rightarrow k+1 &: P_{k,t:l,t-r} = p\pi((k-l)/r) + (1-p)q, \\
c_1(t) = k \rightarrow k &: Q_{k,t:lt-r} = 1 - P_{k,t},
\end{aligned} \tag{6.7}
$$

where $c_1(t-r) = l$. $P_{k,t}$ and $Q_{k,t}$ are the probabilities of the process. The sum of $P_{k,t}$ and $Q_{k,t}$ is 1. This means that at time $(t-r)$, the number of votes for C_1 is $c_1(t-r) = l$. Here, we define $\pi(Z)$ as the majority probability of the binomial distribution of Z. $\pi(Z)$ can be calculated as follows:

$$
\pi(Z) = \frac{(2n+1)!}{(n!)^2}\int_0^Z x^n(1-x)^n dx = \frac{1}{B(n+1, n+1)}\int_0^Z x^n(1-x)^n dx, \tag{6.8}
$$

where $r = 2n+1$.

For convenience, we define a new variable Δ_t such that

$$
\Delta_t = 2c_1(t) - t = c_1(t) - c_0(t). \tag{6.9}
$$

We change the notation from k and l to Δ_t and Δ_{t-r} for convenience. Given $\Delta_t = u$ and $\Delta_{t-r} = s$, we obtain the following random walk model:

$$
\begin{aligned}
\Delta_t = u \rightarrow u+1 &: P_{u,t:s,t-r} = \pi(\frac{1}{2} + \frac{u-s}{2r})p + (1-p)q, \\
\Delta_t = u \rightarrow u-1 &: Q_{u,t:s,t-r} = 1 - P_{u,t}.
\end{aligned}
$$

We now consider the continuous limit $\epsilon \rightarrow 0$,

$$
\begin{aligned}
X_{\hat{\tau}} &= \epsilon\Delta_{[t/\epsilon]}, \\
P(x, \hat{\tau}) &= \epsilon P(\Delta_t/\epsilon, t/\epsilon),
\end{aligned} \tag{6.10}
$$

where $\hat{\tau} = t/\epsilon$,$\hat{r} = r/\epsilon$ and $x = \Delta_t/\epsilon$. By approaching the continuous limit, we obtain the stochastic partial differential equation:

$$dX_{\hat{\tau}} = \left[(1-p)(2q-1) - p + 2p\frac{(2n+1)!}{(n!)^2}\int_0^{\frac{1}{2}+\frac{X_{\hat{\tau}}-X_{\hat{\tau}-\hat{r}}}{2r}} x^n(1-x)^n dx\right]d\hat{\tau}$$
$$+\sqrt{\epsilon}dB, \tag{6.11}$$

where we used Eq. (6.8). Equation (6.11) depends on $X_{\hat{\tau}-\hat{r}}$ and is the feedback system.

We are interested in the behavior of $X_{\hat{\tau}}$ in the limit $\hat{\tau} \to \infty$. The relation between X_∞ and the voting ratio to C_1,Z is $2Z - 1 = X_\infty/\hat{\tau}$.

We can assume the stationary solution to be

$$X_\infty = \bar{\upsilon}\hat{\tau} + (1-p)(2q-1)\hat{\tau}, \tag{6.12}$$

where $\bar{\upsilon}$ is constant. Substituting Eq. (6.12) into Eq. (6.11), we obtain

$$\bar{\upsilon} = -p + \frac{2p\cdot(2n+1)!}{(n!)^2}\int_0^{\frac{1}{2}+\frac{(1-p)(2q-1)}{2}+\frac{\bar{\upsilon}}{2}} x^n(1-x)^n dx. \tag{6.13}$$

Equation (6.13) is self-consistent with the permission for overlapping. If we set $n = 2r+1 \to \infty$, which is the second term of the RHS, the herders behave as digital herders and Eq. (6.13) becomes the strict form.

When $r = 1, 2$, there is only one solution of Eq. (6.13) in the range p. However, when $r \geq 3$, there is only one solution in $p < p_c$ and three solutions in $p > p_c$, where p_c is the critical p. The middle solution is unstable, while the other two solutions are stable. Because there exist both good and bad equilibria, phase transition may occur.

In the case where $r \to \infty$, the description is correct. In fact, there is a phase transition when r is infinite. Voting rates converge to one of two stable points. However, when r is finite, the solution oscillates between good and bad equilibria. Hence, there is no phase transition. We elaborate on this below.

Here, we consider a random walk between the two states, $c_1^r(t)/r > 1/2$ (good equilibrium) and $c_0^r(t)/r > 1/2$ (bad equilibrium). It is a coarse-grained votes and corresponds to the block spin transformation (Kadanoff et al. 1996).[3] We define the hopping probability from the state $c_0^r(t)/r > 1/2$ to $c_1^r(t)/r > 1/2$ as a and that from the state $c_1^r(t)/r > 1/2$ to $c_0^r(t)/r > 1/2$ as b. Note that a and b are not functions of t. When $t > r$, the transition matrix A of the random walk is

$$A = \begin{pmatrix} 1-a & a \\ b & 1-b \end{pmatrix}. \tag{6.14}$$

The random walk of the two states, $\tilde{X}_n$, is defined as the transition matrix A when $t > r$ and the initial condition $\tilde{X}_0 = 0$. If $r > t$, voters can see t previous votes for each candidate.

[3] See Sect. 5.6.

The model was studied as a one-dimensional correlated random walk (Böhm 2000; Konno 2002). For the limit $t \to \infty$,

$$\tilde{X}_\infty \Longrightarrow N(\frac{a-b}{a+b}t, \frac{4ab(2-(a+b))}{(a+b)^3}t), \tag{6.15}$$

where $N(\mu, \sigma^2)$ is the normal distribution with mean μ and variance σ^2. (See Theorem 3.1, in Böhm 2000). a and b are given in Eq. (6.14).

If consecutive independent voters choose candidate $C_1(C_0)$ when $c_0^r(t)/r > 1/2$ $(c_1^r(t)/r > 1/2)$, the state changes from $c_0^r(t)/r > 1/2\,(c_1^r(t)/r > 1/2)$ to $c_1^r(t)/r > 1/2(c_0^r(t)/r > 1/2)$. Thus, independent voters behave as a switch for hopping. When the independents who vote $C_1(C_0)$ are the majority, the state hops from $c_0^r(t)/r > 1/2(c_1^r(t)/r > 1/2)$ to $c_1^r(t)/r > 1/2(c_0^r(t)/r > 1/2)$. Hence, the hopping rates a and b are estimated to be

$$\begin{aligned} a &= \pi[(1-p)q] = \frac{(2n+1)!}{(n!)^2}\int_0^{(1-p)q} x^n(1-x)^n dx \sim (1-p)^{\frac{r+1}{2}} q^{\frac{r+1}{2}}, \\ b &= \pi[(1-p)(1-q)] \sim (1-p)^{\frac{r+1}{2}}(1-q)^{\frac{r+1}{2}}, \end{aligned} \tag{6.16}$$

where the approximations are in $p \sim 1$. In the case where $r = 1$, $a = (1-p)q$ and $b = (1-p)(1-q)$. We obtained an identical solution in Chap. 5.

In the finite r case, the hopping rates a and b do not decrease as t increases, and the state oscillates between good and bad equilibria. Hence, the distribution of $X_{\hat{t}}$ becomes normal, and there is no phase transition. Voting rates converge to $(1-p)q + pa/(a+b) \sim (1-p)q + pq^{\frac{r+1}{2}}/(q^{\frac{r+1}{2}} + (1-q)^{\frac{r+1}{2}})$. The first term is the number of votes by independent voters and the second term is the number of votes by digital herders. erders' votes oscillate between good and bad equilibria in Eq. (6.13). As r increases, the stay in the good equilibrium becomes longer. The ratio of stay in the good equilibrium to that in the bad equilibrium is $a/b \sim (\frac{q}{1-q})^{r+1/2}$.

When $r = \infty$, we consider two cases. One is the case where r increases with t. In addition, in this case, r increases rapidly, that is, the annealing schedule is not sufficiently slow (Geman et al. 1984). Voters can refer to all historical votes. The hopping rates a and b decrease exponentially as t increases, and $\bar{v}$ converges to the solutions of the self-consistent equation, Eq. (6.13). Hence, above p_c, which is the critical p, the voting rate could converge to a bad equilibrium. Thus, this is the phase transition.

The other case is where r increases slowly, that is, the annealing schedule is sufficiently slow. First, we set finite r. After sufficient votes, the state frequently oscillates between good and bad equilibria. We increase r after several oscillations of the states. We continue this process until r reaches ∞. Thus, we deem the annealing schedule to be sufficiently slow. (See Sect. 6.4, where the schedule $r \sim \log t$.) The ratio of stay in the good equilibrium to that in the bad equilibrium is $a/b \sim (\frac{q}{1-q})^{r+1/2} \to \infty$. This means that the state is always in good equilibrium, and the voting rate converges only to good equilibrium $(1-p)q + pa/(a+b) \to (1-p)q + p$ when r is large.

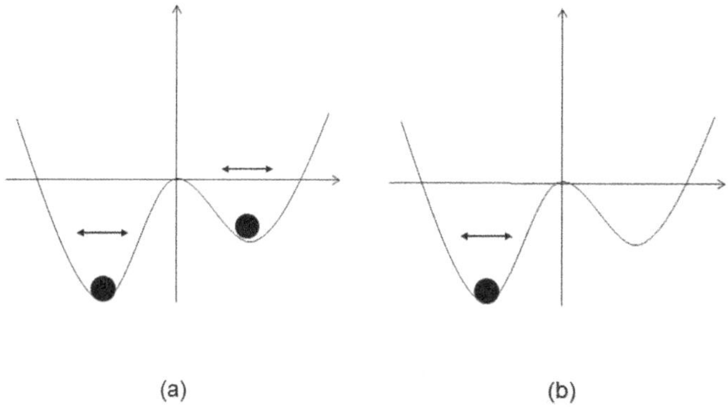

Fig. 6.1 Illustration of voting rate convergence using the analogy of the physical potential. The ball stops at the bottom of the potential, which corresponds to the convergence of the voting rate. The deeper (shallower) potential corresponds to a good (bad) equilibrium. **a** is the case where $r = \infty$. As the annealing schedule is not sufficiently slow, the voting rate may converge to a bad equilibrium. **b** is the case where the annealing schedule is sufficiently slow. If we increase r slowly, the voting rate converges only to a good equilibrium

In Fig. 6.1, we depict the convergence as r increases. (a) is the case where r increase rapidly, or the annealing schedule is not sufficiently slow. $\bar{v}$ converges to one of the solutions of the self-consistent equation, Eq. (6.13). Therefore, the average correct ratio $E(c_1(t)/t)$ decreases above p_c as p increases. The average correct ratio is the ratio of the number of votes for candidate C_1 to all votes. Further, (b) is the case where r and p increase slowly from finite r. In this case, the voting rate converges only to good equilibrium.

6.4 Numerical Simulations and Simulated Annealing

To confirm the analytical results of Sect. 6.3, we performed numerical and Monte Carlo (MC) integration of the master equation for the digital herder case. For the Monte Carlo study, we solved the master equation 10^6 times and calculated the average value of the ratios.

In Fig. 6.2, we show the average vote ratio for the correct candidate vs. p for $q = 0.6$ and $r = 5$, 101 and $r = \infty$. For the time horizon t, we set $t = 10^2$, 10^4, and 10^6 to observe the limit behavior $t \to \infty$. We also plot the theoretical results for the limit value as

$$\lim_{t\to\infty} \mathrm{E}(c_1/t) = (1-p)q + p\frac{a}{a+b}, \tag{6.17}$$

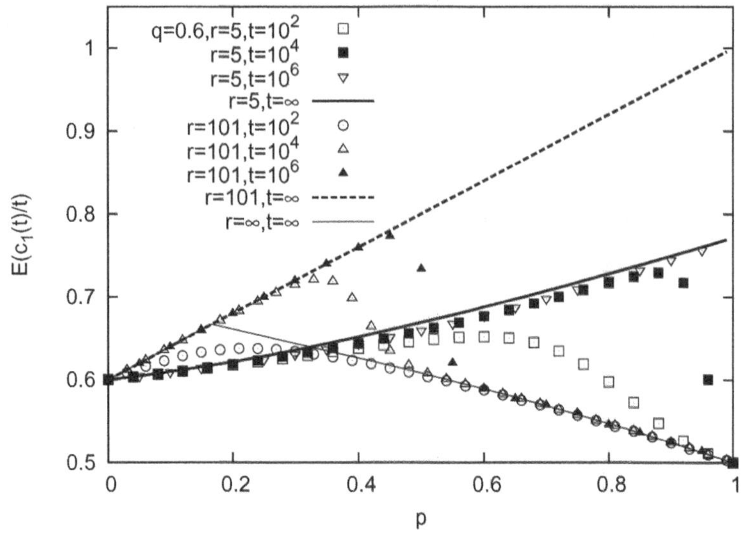

Fig. 6.2 $E(c_1(t)/t)$ versus p for $r = 5, 101, \infty$ and $t = 10^2, 10^4, 10^6, \infty$. The symbols indicate the results from the numerical studies. The lines represent the theoretical results. Theoretical results for finite r, Eq. (6.17) are plotted as solid ($r = 5$) and broken lines ($r = 101$). For $r = \infty$, the result is given in Chap. 5 and is plotted as a thin solid line. We set $q = 0.6$

where a, b are given in Eq. (6.16). The thin solid line plots the exact result of the limit value. For $p < p_c = 1/6$, the herders make the correct choice and it behaves as $(1 - p)q + p \cdot 1$. Above p_c, the probability that all herders choose the incorrect candidate becomes finite and becomes smaller than $(1 - p)q + p \cdot 1$.

When $r = 5$ and $t = 10^2$, $E(c_1(t)/t)$ slightly increases with p for small p. For large p, the probability that the system is in bad equilibrium becomes large and $E(c_1(t)/t)$ becomes small. As p increases further, the probability that the system escapes from the bad equilibrium becomes negligibly small and it approaches the limit value for $r = \infty$. As t increases from 10^2 to $10^4, 10^6$, the probability that the system is in a bad equilibrium decreases and $E(c_1(t)/t)$ becomes larger. As p approaches 1, as in the $t = 10^2$ case, it approaches the limit value for $r = \infty$. We also see that $E(c_1(t)/t)$ approaches theoretically estimated limit value Eq. (6.17) as t becomes large. For the limit $t \to \infty$, the probability that the herders make the correct choice can be estimated to be $a/(a + b)$. For $r = 101$, we also observe the same feature for $r = 5$. Compared with $r = 5$, for $r = 101$, the peak position of $E(c_1(t)/t)$ on the p-axis becomes smaller and it rapidly approaches the limit value for $r = \infty$. As r increases, the mean oscillation time between the good and bad equilibria increases and the probability that the system can escape from the bad equilibrium decreases. Despite this, the system finally escapes from the bad equilibrium for the limit $t \to \infty$ and $E(c_1(t)/t)$ approaches the theoretically estimated result. The

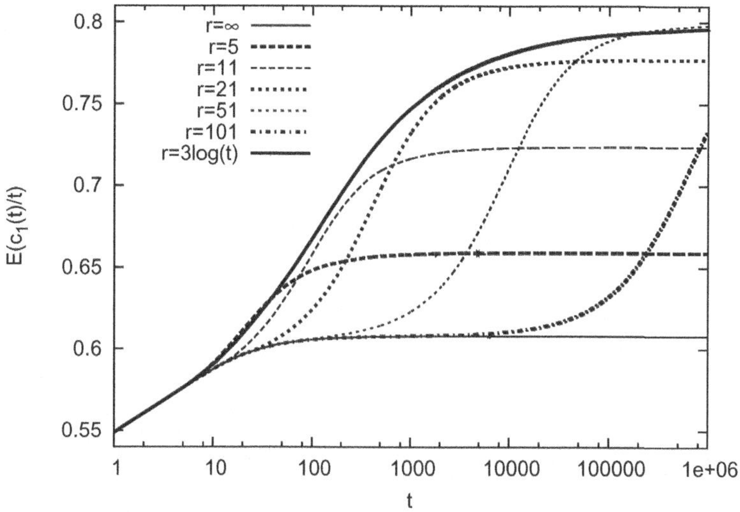

Fig. 6.3 Plots of $E(c_1(t)/t)$ versus t for $r = 5, 11, 21, 51, 101, \infty$ and $r = 3\log t$. We set $q = 0.6$ and $p = 0.5 > p_c = 1/6$

probability that herders make the correct choice $a/(a+b)$ is an increasing function of r, and the slope of $E(c_1(t)/t)$ versus p becomes larger for large r. At $r = 100$ and $q = 0.6$, $a/(a+b)$ is almost one and $E(c_1(t)/t)$ becomes one for the limit $p \to 1$. For a better correct ratio $E(c_1(t)/t)$, r should be large. However, to obtain a high value of $E(c_1(t)/t)$, t should also be large. Otherwise, one cannot necessarily obtain a high correct ratio. Therefore, there needs to be a trade-off between time and the correct ratio.

To obtain the high value for $E(c_1(t)/t)$, r should be large. For large r, for the probability in the bad equilibrium to be small, t should be large, and the system should oscillate several times between the two equilibria. Because the mean cycle of the oscillation behaves as $1/[(1-p)^{(r+1)/2}(1-q)^{(r+1)/2}] = e^{Cr}$ with $C = -\frac{1}{2}\log(1-p)(1-q)$, a good annealing schedule $r(t)$ is estimated as

$$r(t) = \log t/C. \tag{6.18}$$

The annealing schedule $r(t)$ logarithmically depends on t and is very slow.

In Fig. 6.3, we plot $E(c_1(t)/t)$ vs t for $r = 5, 11, 21, 51, 101, \infty$ and $r = r(t)$ in Eq. (6.18). We set $q = 0.6$ and $p = 0.5 > p_c = 1/6$. For $r = \infty$, the bad equilibrium is stable and E$(c_1(t)/t)$ saturates at low value ~ 0.6 for $t \geq 10^2$. For $r = 5$, the herder's probability of correct choice is low and $c_1(t)/t$ soon reaches the maximum value ~ 0.65 at approximately $t = 2 \times 10^2$. For $r = 11$, E$(c_1(t)/t)$ reaches the maximum value for about $t = 10^3$. As r increases, the time t needed for the high value of E$(c_1(t)/t)$ increases. For $r = 51$, $c_1(t)/t$ reaches approximately 0.8 at $t = 10^5$. If the probability that herders choose the correct candidate C_1 is one, $c_1(t)/t$

takes $0.8 = (1-p)q + 1 \cdot p$ for $q = 0.6$ and $p = 0.5$. $r = 51$ is sufficiently large to maximize the limit value $\lim_{t\to\infty} c_1(t)/t$. If we adopt $r(t) = 3\log t$ as the annealing schedule for $r(t)$, $\mathrm{E}(c_1(t)/t)$ increases smoothly to the maximum value 0.8 without saturation at some value between $(1-p)q + p \cdot 0.5 = 0.55$ and the maximum value. If we adopt $r = 51$, at $t = 10^6$, $c_1(t)/t$ reaches its maximum value. However, for $t < 10^5$, $\mathrm{E}(c_1(t)/t)$ with $r = r(t)$ is higher than that for $r = 51$. If we set $r = 51$, the probability of escape from the bad equilibrium is small , and it is necessary to wait for a long time t for $c_1(t)$ to reach the maximum value. In contrast, for $r = r(t)$, r increases slowly and the probability of staying in the bad equilibrium is minimized. Furthermore, unnecessary stay in good equilibrium with medium $\mathrm{E}(c_1(t)/t)$ is also avoided by increasing r and smoothly increasing $c_1(t)$.

6.5 Analog Herder and Kirman's Ant Colony Model

In this section, we consider the case of the analog herder. As mentioned earlier, the herders vote for the candidate with a probability that is proportional to the previous vote ratio that can be referred. Voter can see the latest previous r voters. The transition is

$$\begin{aligned}
c_1(t) = k \to k+1 : P_{k,t:l,t-r} &= (1-p)q + p\frac{k-l}{r} = \frac{q(1-\rho) + \rho(k-l)}{(1-\rho) + \rho r}, \\
c_1(t) = k \to k : Q_{k,t:lt-r} &= 1 - P_{k,t:l,t-r} = \frac{(1-q)(1-\rho) + \rho(r-(k-l))}{(1-\rho) + \rho r}, \qquad (6.19)
\end{aligned}$$

where $c_1(t-r) = l$. $P_{k,t:l,t-r}$ and $Q_{k,t:l,t-r}$ are the probabilities of the process. The voting ratio of C_1 is $c_1(t-r) = l$. We changed the parameters from p to ρ, which is the correlation of rth beta binomial model. The relation between p and ρ is

$$p = \frac{\rho r}{(1-\rho) + \rho r}. \qquad (6.20)$$

Hence, we can map independent voters and herders to the beta binomial distribution. As r increases, $1-p$ decreases as $1/r$ and the ratio of independent voters decreases.

Here, we consider the hopping rate among $r+1$ states $\hat{k} = k - l = 0, 1, \ldots, r$. The dynamic evolution of the process is given by

$$\begin{aligned}
\hat{k} \to \hat{k}+1 : P_{\hat{k},\hat{k}+1,t} &= \frac{r-\hat{k}}{r}\frac{q(1-\rho) + \rho\hat{k}}{(1-\rho) + \rho r}, \\
\hat{k} \to \hat{k}-1 : P_{\hat{k},\hat{k}-1,t} &= \frac{\hat{k}}{r}\frac{(1-q)(1-\rho) + \rho(r-\hat{k})}{(1-\rho) + \rho r}, \\
\hat{k} \to \hat{k} : P_{\hat{k},\hat{k},t} &= 1 - P_{\hat{k},\hat{k}-1,t} - P_{\hat{k},\hat{k}+1,t}. \qquad (6.21)
\end{aligned}$$

This process means that a new vote is added using the r references, where the oldest vote of the r references exists. If we set $\epsilon = \frac{1}{2}(1-\rho)/[(1-\rho)+\rho r]$, $1-\delta = (r-1)\rho/[(1-\rho)+\rho r]$, and $q = 1/2$, we obtain an equivalent of Kirman's ant colony model (Kirman 1993; Alfarano et al. 2005). The ant colony model has been reasonably successful in replicating the statistical features of financial returns, like volatility clustering and power low tails of the return distribution. These are famous characteristics of financial returns in econophysics. Kirman's colony model corresponds to $r \geq 2$ case. In the voting model, there exists a constraint between the parameters,

$$2\epsilon + \frac{r}{r-1}(1-\delta) = 1. \tag{6.22}$$

Here, we define $P_r(\hat{k}, t)$ as a distribution function of the state $\hat{k}$ at time t. The number of all states is $r+1$. Using the fact that the process is reversible, we have

$$\frac{P_r(\hat{k}+1,t)}{P_r(\hat{k},t)} = \frac{P_{\hat{k},\hat{k}+1,t}}{P_{\hat{k}+1,\hat{k},t}} = \frac{r-\hat{k}}{\hat{k}+1}\frac{q(1-\rho)+\rho\hat{k}}{(1-q)(1-\rho)+(r-\hat{k}-1)\rho}. \tag{6.23}$$

It is the detailed balance condition. For the limit $t \to \infty$, we obtain the equilibrium distribution,

$$P_r(\hat{k},\infty) = {}_rC_{\hat{k}}\frac{\Pi_{j=0}^{\hat{k}-1}(q(1-\rho)+(j\rho))\Pi_{j'=0}^{r-\hat{k}-1}((1-q)(1-\rho)+(j'\rho))}{\Pi_{i=0}^{r-1}((1-\rho)+r\rho)}. \tag{6.24}$$

It is the $(r-1)$th beta binomial distribution that is constructed on the lattice.[4] This is the same distribution as that in Kirman's colony model.

From the central limit theorem, as the distribution of states converges to the beta binomial distribution, the voting rate for the candidate C_1 converges,

$$\lim_{t\to\infty}\frac{c_1(t)}{t} = \delta_q, \tag{6.25}$$

where reference number r is finite and δ_q is a delta function. Furthermore, when $r \to \infty$, the distribution of states is the same as the voting rates for candidate C_1, that is, the beta distribution.

Now, we discuss the case where r increases slowly, that is, annealing schedule insufficiently slow. First, we set a finite r. After sufficient votes, the distribution of the state converges to the rth beta binomial distribution. We increase r after the state converges. We continue this process until r reaches ∞. We call this the case where the annealing schedule is sufficiently slow. In this case, the voting rate for candidate C_1 converges to the same point, that is, the independent voter's correct ratio q, as in the finite r case.

---

[4] See Chap. 2.

6.6 Tanh-Type Herder and Kinetic Ising Model

In this section, we identify the relation between our voting model and the infinite-range kinetic Ising model.

The state of the Ising model is denoted by the vector $\boldsymbol{\sigma} = (\sigma_1, \ldots, \sigma_{r+1})$ with $\sigma_j = \pm 1$.

The Hamiltonian is defined as

$$H(\boldsymbol{\sigma}) = -J \sum_{i \neq j}^{r+1} \sigma_i \sigma_j, \tag{6.26}$$

where J is the exchange interaction.

We define F_j as a spin flip operator on the j-th site: $F_j\boldsymbol{\sigma}$ is the state in which the j-th spin is flipped from $\boldsymbol{\sigma}$ with the other spins are fixed. The Markov chain is characterized by a transition probability, $w_j(\boldsymbol{\sigma})$per unit time from $\boldsymbol{\sigma}$ to $F_j\boldsymbol{\sigma}$. Let $p(\boldsymbol{\sigma}, t_n)$ be the probability distribution for finding the spin state $\boldsymbol{\sigma}$ at time t_n. The discrete master equation is written as follows:

$$p(\boldsymbol{\sigma}, t_{n+1}) = p(\boldsymbol{\sigma}, t_n) - [\sum_j w_j(\boldsymbol{\sigma})]p(\boldsymbol{\sigma}, t_n)\Delta t + [\sum_j w_j(F_j\boldsymbol{\sigma})]p(F_j\boldsymbol{\sigma}, t_n)\Delta t, \tag{6.27}$$

where the second and third terms in the RHS are the ongoing and incoming probabilities, respectively.

For the master equation, Eq. (6.27), we consider the sufficient condition that $p(\boldsymbol{\sigma}, t_n)$ converges to the equilibrium distribution $\pi(\boldsymbol{\sigma})$ as $t_n \to \infty$ is that the transition probability $w_j(\boldsymbol{\sigma})$ satisfies a detailed balance condition:

$$w_j(\boldsymbol{\sigma})\pi(\boldsymbol{\sigma}) = w_j(F_j\boldsymbol{\sigma})\pi(F_j\boldsymbol{\sigma}). \tag{6.28}$$

This condition is known as reversibility. From this condition, the transition probability is given by

$$\frac{w_j(\boldsymbol{\sigma})}{F_j w_j(\boldsymbol{\sigma})} = \frac{\pi(F_j\boldsymbol{\sigma})}{\pi(\boldsymbol{\sigma})} = \frac{\exp[-\sigma_j \beta h_j]}{\exp[\sigma_j \beta h_j]}, \tag{6.29}$$

where $h_j = J \sum_{i \neq j}^{r+1} \sigma_i$ and an inverse temperature β, in units where the Boltzmann constant is 1. **It is the detailed balance condition.**

Here, we set

$$w_j(\boldsymbol{\sigma}) = \frac{1}{2}(1 - \sigma_j \tanh \beta h_j). \tag{6.30}$$

We define the total number of $\sigma_i = 1$, where $i \neq j$, is $\hat{c}_1$ and the total number of $\sigma_i = -1$, where $i \neq j$, is $\hat{c}_{-1}$. Hence, $\hat{c}_1 + \hat{c}_{-1} = r$. The transition is given by

$$\sigma_j = 1 \to -1 : w_j(\boldsymbol{\sigma}) = \frac{1}{2}(1 - \tanh \beta J(\hat{c}_1 - \hat{c}_{-1})),$$
$$\sigma_j = -1 \to 1 : F_j w_j(\boldsymbol{\sigma}) = \frac{1}{2}(1 + \tanh \beta J(\hat{c}_1 - \hat{c}_{-1})). \tag{6.31}$$

Note that as $w_j(\boldsymbol{\sigma}) + F_j w_j(\boldsymbol{\sigma}) = 1$, the transition does not depend on the previous state of σ_j and depends on the other spins σ_i where $i \neq j$.

In the ordinary case, an updated spin is chosen randomly. Here, we consider the case in which the updated spin is chosen by the following rules. The ordering of the update is from σ_1 to σ_{r+1}. After the update of σ_{r+1}, we update σ_1 and so on. We repeat this process. Hereafter, we define the updated spin σ_j after n as $\sigma_j^{(n)}$. The initial condition is $\sigma_j^{(0)}$, in which $j = 1, 2, \ldots, r$.

Here, we consider a voting model where all herders are tanh-type herders. The voter can see the latest previous r voters. The transition is given by

$$c_1(t) = k \to k+1 : P_{k,t:l,t-r} = \frac{1}{2}[\tanh \lambda(\frac{k-l}{r} - \frac{1}{2}) + 1]$$
$$c_1(t) = k \to k : Q_{k,t:lt-r} = 1 - P_{k,t:l,t-r}, \tag{6.32}$$

where $c_1(t-r) = l$ and λ is a parameter. (Please see in Chap. 4 for details.) The number of votes for candidate C_1 at time $(t-r)$ is $c_1(t-r) = l$.

We define a new variable Δ_t such that

$$\Delta_t = 2c_1(t) - t = c_1(t) - c_0(t). \tag{6.33}$$

For convenience, we change the notation from k to Δ_t as in Sect. 6.3. Given $\Delta_t = u$ and $\Delta_{t-r} = s$, we obtain the following random walk model:

$$\Delta_t = u \to u+1 : P_{u,t:s,t-r,t} = \frac{1}{2}(1 + \tanh \frac{\lambda(u-s)}{2r}),$$
$$\Delta_t = u \to u-1 : Q_{u,t:s,t-r,t} = \frac{1}{2}(1 - \tanh \frac{\lambda(u-s)}{2r}). \tag{6.34}$$

Given that the voter who voted at time $(t-r-1)$ chose candidate $C_0(C_1)$, the probability that another voter at time t votes for candidate $C_1(C_0)$ is $P_{u,t:s,t-r,t}$ $(Q_{u,t:s,t-r,t})$. Hence, Eq. (6.34) means that the hopping rate is $P_{u,t:s,t-r,t}$ and $Q_{u,t:s,t-r,t}$. This is equivalent to Eq. (6.31), where $\beta J = \lambda/2r$, $\hat{c}_1 - \hat{c}_{-1} = u - s$. We use the relations $\hat{c}_1 = k - l$, $\hat{c}_{-1} = r - (k-l)$, $u = 2k - t$, and $s = 2l - (t - r)$. **Therefore, the voting model satisfies the the detailed balance condition, Eq.** (6.29).

If we consider the row of spins $\sigma_1^{(0)} \ldots \sigma_{r+1}^{(0)} \sigma_1^{(1)} \ldots \sigma_{r+1}^{(n)} \sigma_1^{(n+1)} \ldots$ where $n = 0, 1, 2 \ldots$ as the voting row, **the voting model is equivalent to the infinite-range kinetic Ising model.** This is why we can find the same mean field approximation equation as that in the Ising model in limit $r \to \infty$ case. Thus, for the finite r case, we confirm that there is no phase transition and that it is not appropriate to use the mean field approximation.

When $r \to \infty$, the kinetic Ising model is the voting model when the annealing schedule is sufficiently slow and the state is always in good equilibrium. Hence, the Ising model with an external field has no phase transition, as discussed in Sects. 6.3 and 6.4. In fact, we can confirm that as r increases, the maximum point of $E(c_1(t)/t)$ approaches the maximum point of $r = 3\log(t)$, the slow annealing, in Fig. 6.3. On the other hand, when the annealing schedule is not sufficiently slow, the voting model has a phase transition. This means that the voting rate could converge to a bad equilibrium. The phase transition is observed when the size of r is infinite.

In addition, the voting model becomes the Potts model, if there are several candidates above two. We explain the relation to the Potts model in Sect. 6.8 Appendix A. This is a simple extension of this section. While in this section, we have discussed the Ising model without external fields, we discuss how prior distribution and independent voters correspond to the outer fields of the Ising model in Sect. 6.9 Appendix B.

6.7 Summary

In this chapter, we investigated the case that voters can obtain information from previous r voters. In this case, the process is in equilibrium.

First, we investigated the case where there are only digital herders and independent voters. Through numerical simulations and analysis, we have shown that for finite r, no phase transition occurs and that there is only one peak phase. From the viewpoint of annealing, if the annealing schedule is sufficiently slow, that is, $r \sim \log t$, the voting rate converges only to the good equilibrium. The correct ratio $E(c_1(t)/t)$ increases as r increases. On the other hand, if the annealing schedule is not sufficiently slow, that is, $r \sim t$, phase transitions occur. Furthermore, above p_c, there are good and bad equilibria. The voting rates converge to one of two equilibria, but we cannot estimate which of them is selected. The correct ratio decreases when a phase transition exists. Therefore, there exists a trade-off between time and the correct ratio.

Next, we considered analog herders and independent voters. We show that the states of the reference votes in this model are equivalent to Kirman's ant colony model. In this case, the states follow the beta binomial distribution (BBD), and for large r, and this becomes the beta distribution.

In the case of tanh-type herders and independents, we have shown that the model is equivalent to the finite-size kinetic Ising model. When $r \to \infty$, the voting model behaves as the infinite-range kinetic Ising model when the annealing schedule is sufficiently slow. Hence, the Ising model with an external field has no phase transition, because we can obtain the good equilibrium. On the other hand, when the annealing schedule is not sufficiently slow, the voting model with the external fields exhibits phase transition. In this case, we may obtain the bad equilibrium.

6.8 Appendix A Potts Model and Voting Model

In this Appendix, we consider the extended. In this chapter, the voters choose between two candidates, here we consider that the voters select among K candidates C_i where $i = 1.2. \dots, K$. We assume that the correct candidate is candidate C_1. As the urn mode, this is on of the multicolor urn model. In this book, multicolor urn model is discussed in Chap. 7. Voters estimate the probability that C_i, where $i = 1, 2, \dots, K$, is the correct candidate using the original information and the information on $(r-1)$ previous votes. Voters do not distinguish between original information and reference votes. Hence, the voters have r votes as information. $Pr(C_i)$ is the probability that the voter estimates that C_i is the correct candidate. We set the prior distribution as $Pr(C_i) = 1/K$. Here, we assume that voters estimate the percentage of correct answers as $\hat{q}$. The sum of previous voters' votes and the original information for candidate C_i is k_i. The total votes are r, $r = \sum_{i=1}^{K} k_i$. The posterior distribution is

$$Pr(C_i|c_1^r = k_1, \dots, c_K^r = k_K) = \frac{1}{K}\frac{r!}{(k_1)!\cdots(k_K)!}\hat{q}^{k_i}(1-\hat{q})^{r-k_i}, \tag{6.35}$$

where $c_i^r = k_i$ indicates that the number of votes with original information for C_i is k_i before the individual votes. Then, we obtain

$$\frac{Pr(C_{j_1}|c_1^r = k_1, \dots, c_K^r = k_K)}{Pr(C_{j_2}|c_1^r = k_1, \dots, c_K^r = k_K)} = (\frac{\hat{q}}{1-\hat{q}})^{k_{j_1}-k_{j_2}} = \frac{e^{\lambda k_{j_1}/r}}{e^{\lambda k_{j_2}/r}}, \tag{6.36}$$

where $\lambda = r \log \frac{\hat{q}}{1-\hat{q}}$.

The state of the Potts model is denoted by the vector $\boldsymbol{\sigma} = (\sigma_1, \dots, \sigma_{r+1})$ with $\sigma_j = 1, 2, \dots, K$. The Hamiltonian is defined as

$$H(\boldsymbol{\sigma}) = -J\sum_{i\neq j}^{r+1}\delta(\sigma_i, \sigma_j), \tag{6.37}$$

where J is the exchange interaction.

Let $p(\boldsymbol{\sigma}, t_n)$ be the probability distribution for finding the spin state $\boldsymbol{\sigma}(\sigma_j = j_1)$ at time t_n. We define $F_j(j_1, j_2)$ as a spin flip operator on the j-th site to be the state that the j-th spin is flipped from j_1 to j_2 with the other spins fixed, where the j-th spin of $\boldsymbol{\sigma}$ is j_2:$F_j(j_1, j_2)\boldsymbol{\sigma}(\sigma_j = j_1) = \boldsymbol{\sigma}(\sigma_j = j_2)$.

We define the transition probability $w_j(\boldsymbol{\sigma})$per unit time from $\boldsymbol{\sigma}$ to $F_j(j_1, j_2)\boldsymbol{\sigma}$. $w_j(\boldsymbol{\sigma})$ is the probability of that σ_j hops from j_1 to j_2. $w_j(F_j(j_1, j_2)\boldsymbol{\sigma})$ is the probability that σ_j hops from j_2 to j_1.

We consider that $p(\boldsymbol{\sigma}, t_n)$ converges to the equilibrium distribution $\pi(\boldsymbol{\sigma})$ as $t_n \to \infty$ is when transition probability $w_j(\boldsymbol{\sigma})$ satisfies a detailed balance condition:

$$w_j(\boldsymbol{\sigma})\pi(\boldsymbol{\sigma}) = w_j(F_j(j_1, j_2)\boldsymbol{\sigma})\pi(F_j(j_1, j_2)\boldsymbol{\sigma}). \tag{6.38}$$

From this condition, the transition probability is given by

$$\frac{w_j(\boldsymbol{\sigma})}{F_j(j_1, j_2)w_j(\boldsymbol{\sigma})} = \frac{\pi(F_j(j_1, j_2)\boldsymbol{\sigma})}{\pi(\boldsymbol{\sigma})} = \frac{\exp[\beta(h_j(j_2) - h_j(j_1))]}{\exp[\beta(h_j(j_1) - h_j(j_2))]}$$
$$= e^{2\beta(h_j(j_2)-h_j(j_1))}, \tag{6.39}$$

where $h_j(\sigma j) = J\sum_{i\neq j}^{r+1}\delta(\sigma_j, \sigma_i)$ and an inverse temperature β, in the units where the Boltzmann constant is 1. $h_j(\sigma j)$ is the number of $\sigma = \sigma_j$ without σ_j.

From Eqs. (6.36) and (6.39), if we set $\lambda/2r = \beta J$, $k_{j_1} = h_j(j_1)$ and $k_{j_2} = h_j(j_2)$, we can identify that these processes are the same. As r increases, the voting ratio converges to an equilibria.

In Araújo et al. (2010), the real vote data are analyzed using Potts model, and it is shown that the voting behavior adheres to the model from the macro perspective using the real voting data.

6.9 Appendix B Outer Field of the Ising Model

In this Appendix, we discuss the prior distribution and how independent voters play the role of outer fields in the Ising model. We consider the case where two candidates C_1 and C_0. We assume the correct candidate to be candidate C_1.

Voters estimate the probability that C_i, where $i = 0, 1$, is the correct candidate using the $(r-1)$ previous votes and original information. Voters do not distinguish between the original information and reference votes. Hence, voters have r votes as information. $Pr(C_i)$ is the probability that the voter estimates that C_i is the correct candidate. The sum of previous voters' votes and the original information for candidate C_1 is $c_1^r = k$. the total number of votes for the candidate C_0 and C_1 is r. We set the prior distribution as $Pr(C_1) = \hat{p}$. Here, we assume that voters estimate the percentage of correct answers as $\hat{q}$.

We obtain, as in this chapter,

$$\frac{Pr(C_1|c_1^r = k)}{Pr(C_0|c_1^r = k)} = (\frac{\hat{p}}{1-\hat{p}})(\frac{\hat{q}}{1-\hat{q}})^{2k-t} = e^{2\lambda(k-\frac{r}{2})/r+2h}, \tag{6.40}$$

where $\lambda = t\log\frac{\hat{q}}{1-\hat{q}}$ and $h = \frac{1}{2}\log\frac{\hat{p}}{1-\hat{p}}$. We obtain the correspondence,

$$\frac{\hat{h}}{J} = \frac{h}{\beta J} = \frac{2r}{\lambda}h, \tag{6.41}$$

where $\beta J = \lambda/2r$. $\hat{h} = h/\beta$ corresponds to the outer fields in the Ising model.

References

Alfarano S, Lux T, and Wagner F, Estimation of agent-based models: The case of asymmetric herding model *Comp. Eco.* **26(1)** 19–49 (2005). (1996).

Araújo N A, Andrade Jr J S, and Herrmann H J, Tactical voting in plurality elections, PloS one, **5(9)**, e12446 (2010).

Böhm W, The correlated random walk with boundaries: a combinatorial solution, *J. Appl. Prob.* **37** 470–479, (2000).

Geman S and Geman D, Stochastic relaxation, Gibbs, distribution, and the Bayesian restoration of images, *IEEE Trans. PAMI* **6** 721–741, (1984).

Hisakado M and Mori S, Two kinds of phase transitions in a voting model, *J. Phys. A* **45** 345002–345016, (2012).

Hisakado M and Mori S, Information cascade, Kirman's ant colony model and Ising model, *Physica A* **417** 63–75 (2015).

Kadanoff L, Scling laws for Ising model near T_c, *Physics.* **2** 263, (1996).

Kirman A. "Ants, rationality, and recruitment" *Quart. J. Eco.* **108** 137–156, (1993).

Konno N, Limit theorems and absorption problems for quantum random walks in one dimension, *Quant. Inf. Comp.* **2** 578–595, (2002).

Young H P, The dynamics of social innovation, *Proceedings of the National Academy of Sciences* **108 No.4** 21285–21291, (2001).

References

Chapter 7
Multicolor Urn Process and Equilibrium

7.1 Introduction

In this chapter, we extend the Pólya urn model to the multicolor urn. The Ewens sampling formula is a one-parameter probability distribution on the set of all partitions of an integer (Ewens 1990). The Pitman sampling formula is a two-parameter extension of the Ewens sampling formula (Pitman 2006). The Pitman-Yor process (Pitman and Yor 1997) and a generalized Pólya urn Yamato and Shibuya (2001) are the non-equilibrium stochastic processes that derive the Pitman sampling formula (Pitman 2006). These processes permit new entries of individuals and an increasing number of them. In a similar non-equilibrium process in which the number of species or vertices increases, a power-law distribution can be obtained (Yule 1925; Simon 1955; Barabási and Albert 1999).

We introduced the sequential voting model in previous chapters. At each time step t, a voter chooses one of the two candidates. In addition, the t-th voter can see the previous r votes, and, thus, is given access to public perception. When the voters vote for a candidate with a probability that is proportional to the previous referable votes and there are two candidates, the model can be considered as Kirman's ant colony model (Kirman 1993). In Chap. 6, the beta-binomial distribution was derived as the equilibrium distribution of the r referable votes in the stationary state of the process. If we assume that voters can refer to all votes, the process becomes non-equilibrium. The equilibrium distribution and the probability distribution in the non-equilibrium process are the same.

We extend Kirman's ant colony model when the number of candidates is greater than two and not a fixed case. The model is a finite reference version of the Pitman-Yor process and the generalized Pólya urn model. We derive the Pitman sampling formula as the equilibrium distribution.

The remainder of this paper is organized as follows. In Sect. 7.2, we introduce a voting model and derive the Pitman sampling formula as the equilibrium distribution

M. Hisakado, *Urn Models and Their Applications in Finance*,
https://doi.org/10.1007/978-981-96-3825-3_7

of votes. In Sect. 7.3, we present the fixed number of the candidates case. In Sect. 7.4, we present the characteristics of several parameters. We conclude with the last section. This chapter is based on Hisakado (2018).

7.2 Voting Model

We examine choice behavior using a voting model with candidates $C_1, C_2, \ldots$. At time t, candidate C_j has $c_j(t)$ votes. At each time step, the voter votes for one candidate, and the voting is sequential. Thus, at time t, the tth voter votes, after which the total number of votes is t. Voters are allowed to see the r previous votes for each candidate, where r is a constant and, thus, voters are aware of public perception. Candidates are allowed to enter and exit. The voter votes for new candidate C_i with probability $(\theta + K_r\alpha)/(\theta + r)$, where r is the number of referred votes and K_r is the number of candidates who have more than one vote in the last r votes. α and θ are parameters. If a candidate does not have more than one vote in the last r votes, he/she exits. i in C_i is the number of candidates who have appeared in the past plus one. The number of candidates at $t = 1$ is one, and there is only one candidate C_1.

In terms of the Chinese restaurant process or Hoppe's urn process, we describe the voting process as follows Hoppe (2025). First, there is an urn with θ black balls in it. In each step, one ball is taken out from the urn, and two balls are placed back into the urn. In the first turn, a black ball is drawn, and a ball of color 1 and the black ball are placed back into the urn. In subsequent turns, if the drawn ball is black, a ball of another color that has not appeared in the past and a black ball are returned to the urn, and if the ball taken out is not black, the ball is duplicated, and the two balls are placed back into the urn. The difference between this voting model and the Chinese restaurant process is that the voter refers to only the recently added r balls and black balls. We show the process in Fig. 7.1. In this chapter, we consider the case where the number of balls in the urn is r. In the voting model, it corresponds to voters being allowed to see the r previous votes.

We illustrate the parameter space in Fig. 7.2. θ is the parameter that controls the overall probability of adding a new candidate, and α is the parameter that adjusts the entry probability of new candidates according to the number of candidates $K_r = K$. When $\alpha < 0$, the constraint $\theta = -\alpha K$ exists.

We consider the case in which voters are analog herders. If $c_j(t) \geq 1$, the transition is

$$c_j(t) = k \to k + 1 : P_{j,k,t:l,t-r} = \frac{-\alpha + (k - l)}{\theta + r}, \tag{7.1}$$

where $P_{j,k,t:l,t-r}$s are the probabilities of the process. The number of votes for C_j at $(t - r)$ is $c_j(t - r) = l$. Hence, if $(k - l) = 0$, candidate C_j exits the system.

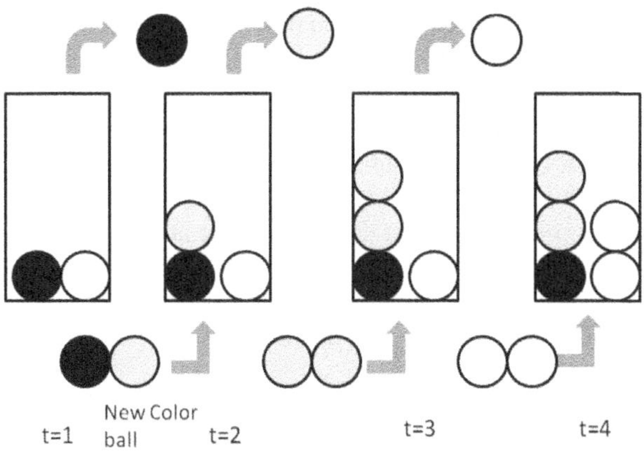

Fig. 7.1 Hoppe' urn process. In the process the number of ball increases as the time goes by

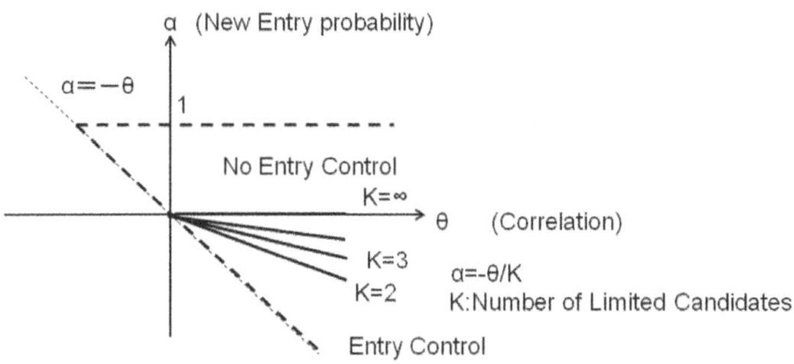

Fig. 7.2 Parameter space α and θ. α is the parameter that adjusts the entry probability of new candidates to the number of candidates, K_r. θ is the parameter that controls the overall probability of adding a new candidate. The dotted line is excluded from the parameter space. When $\alpha < 0$, the constraint $\theta = \alpha K$ exists, and K_r cannot exceed K

The process by which a new candidate C_j entering is

$$c_j(t) = 0 \to 1 : P_{j,k,t:l,t-r} = \frac{K_r \alpha + \theta}{\theta + r}, \tag{7.2}$$

where the number of candidates with more than 0 votes is K_r.

When $\alpha \geq 0$, from Eqs. (7.1) and (7.2), the constraints $\theta + \alpha > 0$ and $1 > \alpha \geq 0$ exist (see Fig. 7.2.) There is no upper limit on the number of candidates. When $\alpha > 0$, the probability of a new entry increases with K_r. When $\alpha = 0$, the probability of a new entry is constant. When $\alpha < 0$, from Eq. (7.2), the constraint $\alpha K + \theta = 0$ exists. The probability of a new entry decreases with an increase in K_r, and K is the upper limit of K_r. The number of candidates with more than one vote does not exceed K. This model is similar to the model that does not allow candidate entry, which is further discussed in the next section.

The distribution of $c_j(t)$ as the partition of integer t follows the Pitman sampling formula in the generalized Pólya urn model (Pitman 2006; Yamato and Shibuya 2001). This is a non-equilibrium process, and the number of votes increases. We focus not on snapshot $c_j(t)$ but on the time series of state $c_j(t) - c_j(t-r)$. This is an equilibrium process, and the number of total votes is constant.

We consider a hopping rate among $(r+1)$ states $\hat{k}_j = k - l$, $\hat{k}_j = 0, 1, \ldots, r$, and, here, we focus on the state. At each t, the vote at time $(t-r)$ is deleted, and a new vote is obtained. $\hat{k}_j$ is the number of votes that candidate C_j received in the previous r votes.

First, we consider the case $\hat{k}_j > 1$. The transition is

$$\begin{aligned}
\hat{k}_j \to \hat{k}_j + 1 : P_{\hat{k}_j,\hat{k}_j+1,t} &= \frac{r-\hat{k}_j}{r}\frac{-\alpha+\hat{k}_j}{\theta+r-1}, \\
\hat{k}_j \to \hat{k}_j - 1 : P_{\hat{k}_j,\hat{k}_j-1,t} &= \frac{\hat{k}_j}{r}\frac{(\theta+\alpha)+(r-\hat{k}_j-1)}{\theta+r-1}, \\
\hat{k}_j \to \hat{k}_j : P_{\hat{k}_j,\hat{k}_j,t} &= 1 - P_{\hat{k},\hat{k}-1,t} - P_{\hat{k},\hat{k}+1,t}.
\end{aligned} \tag{7.3}$$

$P_{\hat{k}_j,\hat{k}_j\pm1,t}$ and $P_{\hat{k}_j,\hat{k}_j,t}$ are the probabilities of the process. $P_{\hat{k}_j,\hat{k}_j\pm1,t}$ is the product of the probabilities of exit and entry.

We consider hopping from candidate C_i to C_j.

$$\begin{aligned}
\hat{k}_i \to \hat{k}_i - 1, \hat{k}_j \to \hat{k}_j + 1 : P_{\hat{k}_i\to\hat{k}_i-1,\hat{k}_j\to\hat{k}_j+1,t} &= \frac{\hat{k}_i}{r}\frac{-\alpha+\hat{k}_j}{\theta+r-1}, \\
\hat{k}_i - 1 \to \hat{k}_i, \hat{k}_j + 1 \to \hat{k}_j : P_{\hat{k}_i-1\to\hat{k}_i,\hat{k}_j+1\to\hat{k}_j,t} &= \frac{\hat{k}_j+1}{r}\frac{-\alpha+\hat{k}_i-1}{\theta+r-1}.
\end{aligned} \tag{7.4}$$

Here, we define $P_r(\hat{k}, t)$ as the distribution function of state $\hat{k}$ at time t. The number of all states is $(r+1)$. Considering that the process is reversible, in equilibrium, we have

$$\frac{P_r(\hat{k}_i, \hat{k}_j, t)}{P_r(\hat{k}_i - 1, \hat{k}_j + 1, t)} = \frac{\hat{k}_j+1}{\hat{k}_i}\frac{-\alpha+\hat{k}_i-1}{-\alpha+\hat{k}_j}. \tag{7.5}$$

It is the detailed balance condition. We separate the indexes i and j and obtain

$$\begin{aligned}
\frac{P_r^i(\hat{k}_i, t)}{P_r^i(\hat{k}_i - 1, t)} &= \frac{-\alpha+\hat{k}_i-1}{\hat{k}_i}c, \\
\frac{P_r^j(\hat{k}_j + 1, t)}{P_r^j(\hat{k}_j, t)} &= \frac{-\alpha+\hat{k}_j}{\hat{k}_j+1}c,
\end{aligned} \tag{7.6}$$

where c is constant.

In equilibrium, the number of candidates with $\hat{k}_j > 0$ is K_r. We ignore candidates with $\hat{k}_j = 0$ and change the number of candidates and votes from $C_j, \hat{k}_j$ to $\tilde{C}_m, \tilde{k}_m$, where $m = 1, \ldots, K_r$, where $\tilde{k}_m > 0$.

We can write the distribution as

$$P_r(\tilde{\boldsymbol{a}}, \infty) = \binom{\theta + r - 1}{r}^{-1} \prod_{m=1}^{K_r} \frac{(1-\alpha)^{[\tilde{k}_m]}}{\tilde{k}_m!} P_r(\tilde{\boldsymbol{a}} = \underbrace{(1, 1, \ldots, 1)}_{K_r}, \infty), \qquad (7.7)$$

where $\tilde{\boldsymbol{a}} = (\tilde{k}_1, \ldots, \tilde{k}_{K_r})$ and $x^{[n]} = x(x+1)\cdots(x+n-1)$, which is the Pochhammer symbol.

Given Eq. (7.6), we obtain the equilibrium condition between $\tilde{\boldsymbol{a}} = \underbrace{(1, 1, \ldots, 1)}_{K_r} = \boldsymbol{1}$ and $\tilde{\boldsymbol{a}} = \underbrace{(0, 0, \ldots, 0)}_{K_r} = \boldsymbol{0}$:

$$\frac{P_r^i(\tilde{\boldsymbol{a}} = (\overbrace{1, \ldots, 1}^{n}, 0, \ldots, 0), \infty)}{P_r^i(\tilde{\boldsymbol{a}} = (\underbrace{1, \ldots, 1}_{n-1}, 0, \ldots, 0), \infty)} = \frac{\theta + (n-1)\alpha}{n} c, \qquad (7.8)$$

where $n = 1, \ldots, K_r$. Hence, we obtain

$$P_r(\tilde{\boldsymbol{a}} = \boldsymbol{1}, \infty) = \prod_{m=1}^{K_r} \frac{\theta + (m-1)\alpha}{m} P_r(\tilde{\boldsymbol{a}} = \boldsymbol{0}, \infty) = \frac{(\theta)^{[K_r:\alpha]}}{K_r!}, \qquad (7.9)$$

where $x^{[n:\alpha]} = x(x+\alpha)\cdots(x+(n-1)\alpha)$. Therefore, we can write Eq. (7.7) as

$$P_r(\tilde{\boldsymbol{a}}, \infty) = \binom{\theta + r - 1}{r}^{-1} \frac{\theta^{[K_r:\alpha]}}{K_r!} \prod_{j=1}^{r} \left(\frac{(1-\alpha)^{[j-1]}}{j!} \right)^{a_j}, \qquad (7.10)$$

where a_j is the number of candidates with j votes. Therefore, the number of candidates $\sum_{j=1}^{r} a_j = K_r$ and the number of votes $\sum_{j=1}^{r} j a_j = r$ are related. Hereafter, we use a partition vector $\hat{\boldsymbol{a}} = (a_1, \ldots, a_r)$.

We consider the partitions of the integer K_r. To normalize the result, we add the following combination term, $K_r!/a_1! \cdots a_r!$:

$$P_r(\hat{\boldsymbol{a}}, \infty) = \frac{r! \theta^{[K_r:\alpha]}}{\theta^{[r]}} \prod_{j=1}^{r} (\frac{(1-\alpha)^{[j-1]}}{j!})^{a_j} \frac{1}{a_j!}. \qquad (7.11)$$

Equation (7.11) is simply a Pitman sampling formula (Pitman 2006). In the limit $\alpha = 0$, we obtain the Ewens sampling formula (Ewens 1990):

$$P_r(\hat{\boldsymbol{a}}, \infty) = \frac{r!}{\theta^{[r]}} \prod_{j=1}^{r} (\frac{\theta}{j})^{a_j} \frac{1}{a_j!}. \tag{7.12}$$

The Ewens sampling formula for the equilibrium process is presented in Aoki (2002).

7.3 Fixed Number of Candidates Case

We model the voting of K candidates, $C_1 \dots C_K$. At time t, candidate C_j has $c_j(t)$ votes. We consider the case in which the number of candidates K is fixed, that is, no new entries are allowed. In each time step, a voter votes for one candidate; the voting is sequential. Hence, at time t, the tth voter votes, after which the total number of votes is t. Voters are permitted to see r previous votes for each candidate and are aware of public perception. r is a constant number. We consider the case in which all voters vote for the candidate with a probability proportional to the previous vote ratio, which is visible to voters.

The transition is

$$c_j(t) = k \to k+1 : P_{j,k,t:l,t-r} = \frac{\frac{q_j(1-\rho)}{\rho} + (k-l)}{\frac{1-\rho}{\rho} + r} = \frac{\beta_j + (k-l)}{\theta + r}, \tag{7.13}$$

where $c_j(t-r) = l$, ρ is the correlation coefficient, and q_j is the initial constant of the jth candidate. ρ is the correlation of the beta binomial model. The constraint $\sum_{j=1}^{K} q_j = 1$ exists. We define $\theta = (1-\rho)/\rho$ and $\beta_j = q_j(1-\rho)/\rho$. $P_{j,k,t:l,t-r}$ denotes the probabilities of the process. The voting ratio of C_j at $t-r$ is $c_j(t-r) = l$. We consider the case $\beta_j \geq 0$ from $P_{j,k,t:l,t-r} > 0$ and the constraint $\sum_j \beta_j = \theta$. When $\beta_j = \beta$, the constraint becomes $\beta K = \theta$. In this reduction the model corresponds to the case, $-\alpha = \beta$. In next section, we use negative α for the classification.

We consider the hopping rate among $(r+1)$ states $\hat{k}_j = k - l, \hat{k}_j = 0, 1, \dots, r$. In each step of t, the vote at time $(t-r)$ is deleted, and a new vote is obtained. $\hat{k}$ is the number of votes for candidate C_j obtained in the latest r votes. In the case $K = 2$, the model becomes Kirman's ant colony model (Kirman 1993).[1] The dynamic evolution of the process is given by

[1] Kirman's ant colony model is discussed in Chap. 6.

$$\hat{k}_j \to \hat{k}_j + 1 : P_{\hat{k}_j,\hat{k}_j+1,t} = \frac{r-\hat{k}_j}{r}\frac{\beta_j+\hat{k}_j}{\theta+r-1},$$
$$\hat{k}_j \to \hat{k}_j - 1 : P_{\hat{k}_j,\hat{k}_j-1,t} = \frac{\hat{k}_j}{r}\frac{(\theta-\beta_j)+(r-1-\hat{k}_j)}{\theta+r-1},$$
$$\hat{k}_j \to \hat{k}_j : P_{\hat{k}_j,\hat{k}_j,t} = 1 - P_{\hat{k},\hat{k}-1,t} - P_{\hat{k},\hat{k}+1,t}.$$

$P_{\hat{k}_j,\hat{k}_j\pm 1,t}$ are the probabilities of the process and the products of probabilities of exit votes and new entry votes.

We consider hopping from candidate C_i to C_j.

$$\hat{k}_i \to \hat{k}_i - 1, \hat{k}_j \to \hat{k}_j + 1 : P_{\hat{k}_i\to\hat{k}_i-1,\hat{k}_j\to\hat{k}_j+1,t} = \frac{\hat{k}_i}{r}\frac{\beta_j+\hat{k}_j}{\theta+r-1},$$
$$\hat{k}_i - 1 \to \hat{k}_i, \hat{k}_j + 1 \to \hat{k}_j : P_{\hat{k}_i-1\to\hat{k}_i,\hat{k}_j+1\to\hat{k}_j,t} = \frac{\hat{k}_j+1}{r}\frac{\beta_i+\hat{k}_i-1}{\theta+r-1}.$$

Here, we define $P_r(\hat{k}, t)$ as a distribution function of state $\hat{k}$ at time t. The number of all states is $(r+1)$. Given that the process is reversible, we have

$$\frac{P_r(\hat{k}_i,\hat{k}_j,t)}{P_r(\hat{k}_i-1,\hat{k}_j+1,t)} = \frac{\hat{k}_j+1}{\hat{k}_i}\frac{\beta_i+\hat{k}_i-1}{\beta_j+\hat{k}_j}. \tag{7.14}$$

It is the detail balance condition. We can separate the indexes i and j and obtain

$$\frac{P_r^i(\hat{k}_i,t)}{P_r^i(\hat{k}_i-1,t)} = \frac{\beta_i+\hat{k}_i-1}{\hat{k}_i}c$$
$$\frac{P_r^j(\hat{k}_j+1,t)}{P_r^j(\hat{k}_j,t)} = \frac{\beta_j+\hat{k}_j}{\hat{k}_j+1}c, \tag{7.15}$$

where c is a constant. Using Eq. (7.15) sequentially, in the limit $t \to \infty$, we obtain the equilibrium distribution, which can be written as

$$P_r(\boldsymbol{a},\infty) = \binom{\theta+r-1}{r}^{-1}\prod_{j=1}^{K}\binom{\beta_j+\hat{k}_j-1}{\hat{k}_j}, \tag{7.16}$$

where $\boldsymbol{a} = (\hat{k}_1, \hat{k}_2, \ldots, \hat{k}_K)$. The distribution is written as

$$P_r(\boldsymbol{a},\infty) = \frac{r!}{\theta^{[r]}}\prod_{i=1}^{K}\frac{\beta_i^{[\hat{k}_i]}}{\hat{k}_i!}, \tag{7.17}$$

where $x^{[n]} = x(x+1)\cdots(x+n-1)$. **This is the Dirichlet multinomial distribution.**

Here, we set $\beta_j = \beta$. The relation $\beta = -\alpha$ exists, where α is the parameter used in the main text. We write Eq. (7.16) as

$$P_r(\hat{\boldsymbol{a}}, \infty) = \binom{\theta + r - 1}{r}^{-1} \prod_{j=1}^{r} \binom{\beta + j - 1}{j}^{a_j}, \tag{7.18}$$

where a_j is the number of candidates for whom j voters voted, and $\hat{\boldsymbol{a}} = (a_1, \ldots, a_r)$. Hence, the relations $\sum_{i=1}^{r} a_i = K_r < K$ and $\sum_{i=1}^{r} i a_i = r$ exist. Here, we define K_r as the number of candidates who have more than one vote. $(K - K_r)$ candidates have no vote.

We consider the partitions of the integer K_r. To normalize the result, we add the term of the combination: $K!/a_1! \cdots a_r!(K - K_r)!$. We obtain

$$\begin{aligned} P_r(\hat{\boldsymbol{a}}, \infty) &= \frac{K!}{a_1! \cdots a_r!(K - K_r)!} \binom{\theta + r - 1}{r}^{-1} \prod_{j=1}^{r} \binom{\beta + j - 1}{j}^{a_j} \\ &= \frac{r! \theta^{[K_r:-\beta]}}{\theta^{[r]}} \prod_{j=1}^{r} (\frac{(1 + \beta)^{[j-1]}}{j!})^{a_j} \frac{1}{a_j!}, \end{aligned} \tag{7.19}$$

where $x^{[n:-\beta]} = x(x - \beta) \cdots (x - (n - 1)\beta)$. We use the relation $\theta = K\beta$. Equation (7.19) is simply the Pitman sampling formula.

In the limit $\beta \to 0$ and $K \to \infty$, subject to a fixed $\theta = \beta K$, we can obtain the Ewens sampling formula. In this case, the sum of the probabilities that a candidate with zero votes can obtain one vote is $\theta/(\theta + r)$. This case is the same as $\alpha = 0$ in Sect. 2.

7.4 Four Regions in the Parameter Space

In this section, we characterize four regions in the parameter space. The parameter θ denotes the intensity of the correlations, and α refers to the intensity of the competition. In the upper plane, newcomers increase with a rise in α. In the lower plane, the probability of the minimum growth increases with a decrease in α.

We consider both an increase and a decrease in $\hat{k}_j$. The probability of a decrease in votes for candidate j is $\hat{k}_j/r$, and that of an increase in votes for candidate j is $(-\alpha + \hat{k}_j)/(\theta + r - 1)$, as shown in Eq. (7.1). We consider the condition in which the probability of an increase is greater than that of a decrease:

$$\alpha \leq -\frac{\hat{k}_j}{r}(\theta - 1). \tag{7.20}$$

In this region, the number of votes increases on average.

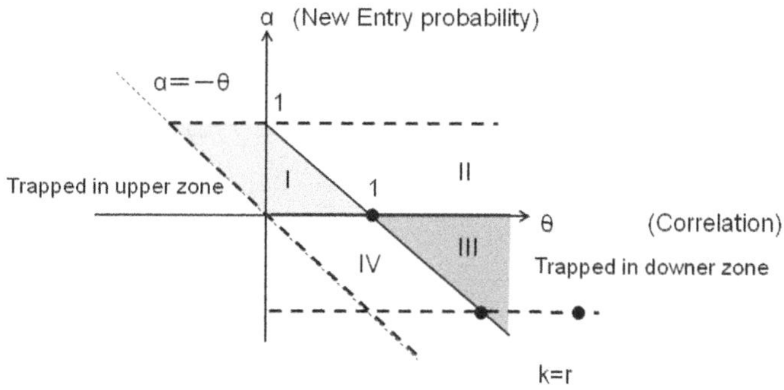

Fig. 7.3 Four regions in parameter space

We can solve Eq. (7.20) for the equal case. The solution is

$$\hat{k} = \bar{k} = \frac{r\alpha}{1-\theta}. \tag{7.21}$$

Under the condition $\bar{k} > 0$, there is a solution in $\alpha > 0, \theta < 1$ which corresponds to zone I, and in $\alpha < 0, \theta > 1$ which corresponds to zone III in Fig. 7.3. In these cases, the wall and the trapped zone in the parameter space.

We divide the parameter space into four regions, $I \sim IV$, as shown in Fig. 7.3. To clarify the regions, we define zone $U_{\hat{k}_j}$ as the region where the probability of an increase in votes is larger than that of a decrease. $U_{\hat{k}_j}$ is defined in Eq. (7.20). U_r is $\alpha \leq -(\theta - 1)$, and U_0 is $\alpha \leq 0$.

We define zone I, where $\alpha > 0$ and $\alpha < (1 - \theta)$, as U_r. In zone I, $U_r \supset U_{r-1} \supset \cdots \supset U_0$. We consider the case in which the parameter set $\boldsymbol{x} = (\theta, \alpha)$ is $\boldsymbol{x} \in U_{\bar{k}}$ and $\boldsymbol{x} \notin U_{\bar{k}-1}$. In this case, $\hat{k}_j \geq l$ is the increasing zone and $\hat{k}_j < \bar{k}$ the decreasing zone. If a candidate has more than $\bar{k}$ votes, he/she can increase the number of votes and maintain his/her position. However, it is difficult to increase the number of votes if the candidate has fewer than $\bar{k}$ votes. We show the average trend in Fig. 7.4. In this region, the leader in the trapped zone has an advantage. In contrast, the competition intensifies for newcomers as α increases.

We define zone II as $\alpha > 0$ and $\alpha \geq (1 - \theta)$. In zone II, $\boldsymbol{x} \notin U_0, \ldots, U_r$. Furthermore, it is difficult to increase the number of votes for every candidate and to be a stable leader. The zone becomes increasingly competitive with an increase in α. In other words, it is possible to adjust the competitive intensity and protect newcomers by adjusting α, which denotes the number of newcomers.

In the plane in the lower half, there is a capacity limit and no newcomers. We define zone III as $\alpha < 0$ and $\alpha > (1 - \theta)$. When $\alpha < 0$, $U_0 \supset U_1 \supset \cdots \supset U_r$. We consider the case in which the parameter set $\boldsymbol{x} = (\theta, \alpha)$ is $\boldsymbol{x} \in U_{\bar{k}}$ and $\boldsymbol{x} \notin U_{\bar{k}+1}$. In this case, $\hat{k}_j > l$ is the decreasing zone and $\hat{k}_j \leq l$ the increasing zone. It is easy to

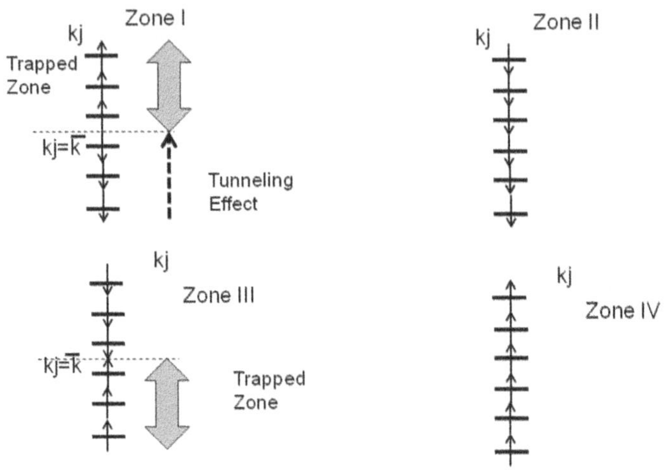

Fig. 7.4 The average increase and decrease in each region

increase the number of votes to $\hat{k}_j = \bar{k}$, but it is difficult to increase the number of votes above $\hat{k}_j = \bar{k} + 1$. In this region, it is also difficult to be a stable leader.

We define zone IV as $\alpha < 0$ and $\alpha \le (1 - \theta)$. In this zone, $\boldsymbol{x} \in U_0, \ldots, U_r$. It is easy to increase the number of votes for each candidate. In addition, this zone is competitive when the number of members is fixed.

Next, we consider the Ewens sampling formula on the θ axis. When $\alpha = 0$ and $\theta = 1$, the probabilities of both increases and decreases become $\hat{k}_j / r$. For any $\hat{k}_j$, the probabilities of an increase and decrease are equal. Thus, the correlation becomes $\rho = 1/2$ (see the previous section), and there is a uniform random permutation. The probability of an increase or decrease is proportional to the number of partitions.

When $\alpha = 0$ and $\theta < 1$ within the boundaries of zone IV, if candidates can enter, the number of votes easily increases. In this zone, the correlation is high. When $\alpha = 0$ and $\theta > 1$ within the boundaries of zone II, it is difficult to increase the number of votes. There is a low correlation. In summary, numerous candidates have few votes.

7.5 Summary

In this chapter, we discuss choice behavior using a voting model comprising voters and candidates. Voters vote for a candidate with a probability that is proportional to the ratio of previous votes, which is visible to the voters. In addition, voters can obtain information from a finite number r of the most recent previous voters.

In the large t limit, the system becomes equilibrium, and the partition of r votes follows the Pitman sampling formula. Kirman's ant colony model is a special case that corresponds to the number of states $K = 2$. The equilibrium probability distribution and non-equilibrium probability distribution for $t = r$ are the same.

Recently, the monopoly of information is a hot topic. A few largest companies gather most of the data. In Fig. 7.2. we divided four regions, I, II, III, and IV, in the parameter space. Here, we assume that the candidates are enterprises and the voters are users. The lower half is the membership region, with high barriers to entry. Recently, barriers to entry have become lower and lower for enterprises in the internet era. In this mean. the upper half region is the trend. When $\theta < 1$, a larger size appears in the trapped zone. For users, enterprises that gather larger amounts of data are useful platforms because of convenience. This phenomenon is called the network effect. We can observe this in zone I. The monopoly will be advanced hereafter.

References

Aoki M, *Modeling Aggregate Behavior and Fluctuations in Economics (Cambridge University Press, London, 2002).*

Barabási A and Albert R, *Science* **286** (1999) "Emergence of scaling in random networks" 509–512.

Ewens W J, *inPopulation genetics theory-The past and future, ed. S. Lessard (Kluwer Academic Publishers, London, 1990).*

Hisakado M, Mori S, *J. Phys. Soc. Jpn.* **87 (2)** (2018) "Pitman-Yor process and an empirical study of choice behavior" 024002-2419.

Hoppe F. *Math. Biology* "Polya-like Urns and the Ewens Sampling Formula" **20** 91–94.

Kirman A, *Quart. J. Eco.***108** (1993) "Ants, rationality, and recruitment" 137–156.

Pitman J and Yor M, *Ann. Prob.* **25(2)** (1997) "The two-parameter Poisson-Dirichlet distribution derived from a stable subordinator" 855–900.

Pitman J, *Combinatorial Stochastic Processes (Springer-Verlag, Berlin, 2006).*

Simon H A,: *Biometrica* **42** (1955) "On a class of skew distribution functions" 425–440.

Yamato H and Shibuya M, *RIMS* **1240** (2001) "Pitman's model in random partition" 64–73.

Yule G, *Philos. Trans. R. Soc. London B* **213** (1925) "A mathematical theory of evolution, based on the calculations of Dr. J. C. Willis, F.R.S." 21.

Chapter 8
Network and Urn Process

8.1 Introduction

Complex networks, especially networks with hubs, have been extensively studied. The degree distribution in the networks is important because it is related to the features that define scale-free networks. Scale-free networks have hubs with many links, and they are crucial in many applications (Barabási 2016; Newman 2018).

As a model for a scale-free network, Barabási and Albert (BA) model is well known (Barabáshi and Albert 1999). It is an evolving network model. New nodes are linked to the network based on the preferential attachment process, where the probability that a node is selected by a new node is proportional to its degree. The preferential attachment process of the BA model is related to Pólya urn process.

Here, we discuss several networks from a unified perspective using the extended feedback system of the urn process. The system includes not only the positive but also the negative feedback using parameter, ω. This shows the strength of the feedback.

We propose a model that connects the random graph, Barabási-Albert (BA) model, and the lattice model with a parameter ω, which represents the weights of the incoming and outgoing links. Our model is an evolving network model, and newly added nodes are more likely to be linked to other nodes with high "popularity". Popularity is defined as the sum of the number of incoming links and the product of the number of outgoing links with ω. When ω is positive, the evolving mechanism is the preferential attachment process in the BA model, which exhibits positive feedback. When $\omega = 1$, the weights of the incoming and outgoing links are equal, and the model takes the form of the BA model. When ω is negative, there is a limit on the number of links from newly added nodes in the evolution of the network. In this case, the model incorporates negative feedback. The negative feedback effects of the networks have been reported in Jennen and Verwijmeren (2009).

To understand ω more deeply, we introduce the fitness model, which is an extension of the BA model (Barabáshi et al. 1999; Bianconi and Barabáshi 2001). Fitness is introduced at each node to adjust popularity, and it can be transformed into "energy"

M. Hisakado, *Urn Models and Their Applications in Finance*,
https://doi.org/10.1007/978-981-96-3825-3_8

using the Boltzmann weight. The fitness decreases with increasing energy, and the degree distribution can thus be interpreted in terms of quantum statistics. When ω is positive, the number of outgoing links is not limited, as in the case of Bose statistics. When ω is negative, the number of outgoing links is limited, as in the case of Fermi statistics, where an energy level can accommodate one particle at most.

We obtain a distribution that connects the Bose and Fermi statistics using the parameter ω. When ω is negative, Fermi degeneracy can be observed in the network at low temperatures. Nodes with high fitness have the maximum number of links, and nodes with low fitness have no links except for the incoming links. When ω is positive, Bose-Einstein condensation (BEC) occurs in the network, and a small number of hubs with low energy (high popularity) are connected to most of the nodes in the network. In other words, the "winner takes all" scenario occurs. Furthermore, ω is related to the threshold of the BEC, and the BEC phase becomes wider as ω increases. When $\omega = 0$, which corresponds to the random network, we obtain the Boltzmann distribution.

The remainder of this paper is organized as follows. In Sect. 8.2, we introduce the networks with parameter ω which represents the feedback. In Sect. 8.3, we discuss the relation between the quantum statistics and networks. Finally, the conclusions are presented in Sect. 8.4. This chapter is based on Hisakado and Mori (2021).

8.2 Networks

We begin this section by discussing the generation of networks. The proposed model is an evolving network model, and we consider the case where a newly added node is selected at most r different nodes based on popularity. The process is sequential as in the BA model. When the i-th node enters the network at time $t = i$, it selects r nodes for the connections, as shown in Fig. 8.1. For node i, the links from the selected r nodes are incoming links, as shown by the incoming arrows in Fig. 8.1. For the r nodes selected by node i, the links to the node i are outgoing links. In Fig. 8.1, we draw the links with outgoing arrows. We denote the number of incoming and outgoing links of node i as k_i^{IN} and k_i^{OUT}, respectively. The degree of node i is given by $k_i = k_i^{IN} + k_i^{OUT}$. The popularity l_i of node i is defined as

$$l_i = \bar{\omega} \cdot k_i^{IN} + \omega \cdot k_i^{OUT}. \tag{8.1}$$

Here, $\bar{\omega}$ and ω are the weights for k_i^{IN} and k_i^{OUT}, respectively. When $\omega = \bar{\omega} = 1$, the model attains the form of the BA model. Hereafter, we set $\bar{\omega} = 1$ for normalization without loss of generality.

The initial state of the model is a complete network with $r + 1$ nodes, and we label as node $i, i = 1, \cdots, r + 1$. We set $k_i^{IN} = i - 1$ and $k_i^{OUT} = r + 1 - i$. We add node $t \geq r + 2$ and link it with r nodes sequentially. The probability that node $i < t$ is selected by node t is

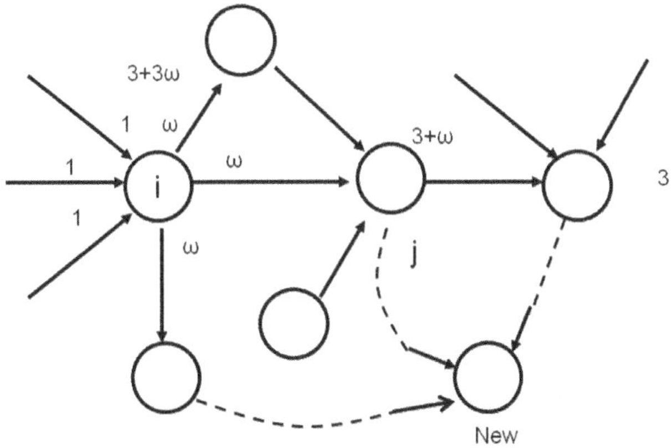

Fig. 8.1 Sample graph with $r = 3$. Node i selects three nodes and is selected by three nodes. The arrows represent the directions from the selected nodes to the selecting nodes. $k_i^{IN} = k_i^{OUT} = 3$, and $l_i = 3\omega + 3$. Node j selects three nodes and is selected by a node. $k_j^{IN} = 3$, $k_j^{OUT} = 1$, and $l_j = 3 + \omega$. l_j changes from $3 + \omega$ to $3 + 2\omega$ when a new node joins the network and selects node j. k_j^{OUT} increases from 1 to 2

$$P(\text{node } i \text{ is selected by node } t) = p_i(t) = \frac{\text{Max}(l_i, 0)}{\sum_{s=1}^{t-1} \text{Max}(l_s, 0)} \tag{8.2}$$

We note that when l_i becomes negative or zero, node i cannot be selected. This occurs for negative values of ω and $k_i^{OUT} \geq \frac{k_i^{IN}}{|\omega|}$. The maximum value of k_i^{OUT} is $\lceil k_i^{IN}/|\omega| \rceil$, where $\lceil x \rceil$ is the ceiling function, and k_i^{OUT} cannot exceed $\lceil \frac{k_i^{IN}}{|\omega|} \rceil$. We denote the number of nodes with positive $p_i(t), i = 1, \cdots, t-1$ by r'. If r' is less than or equal to r, all nodes are linked by node t, and we set $k_t^{IN} = r'$. If $r' > r$, r different nodes are selected from r' nodes with probability $p_i(t)$, and we set $k_t^{IN} = r$. k_i^{OUT} for the selected nodes increases by unity.

$\omega = 1$ and $\omega = 0$ correspond to the BA model and the random network, respectively. This is because when $\omega = 0$, all nodes are selected with equal probability. We consider that the range of negative ω is $-1 \leq \omega < 0$. When $\omega > 0$, the popularity and the probability selected by new node increase. **This is called positive feedback or preferential attachment.** When $\omega < 0$, the popularity and the probability selected by new node decrease. This is the negative feedback,

Samples of the network are presented in Fig. 8.2. In Fig. 8.2(a), we present the case of $r = 1$, which corresponds to tree networks, and in (b), we present the case of $r = 3$.

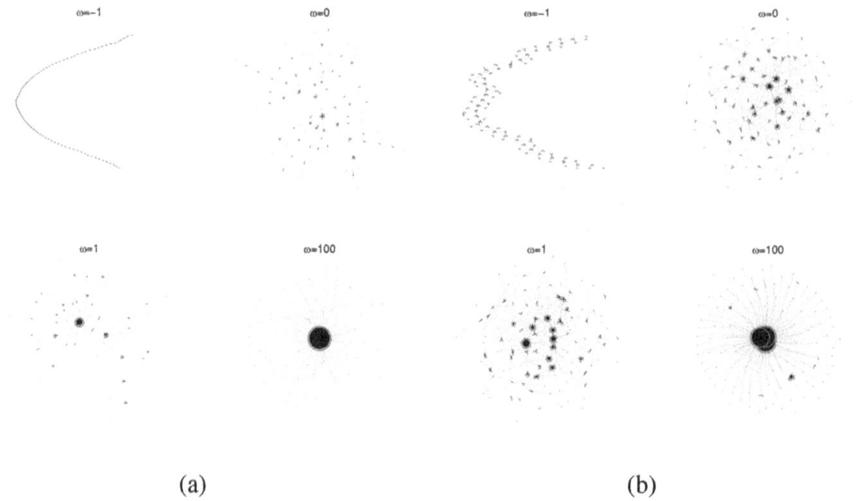

Fig. 8.2 Sample networks with $\omega \in \{-1, 0, 1, 100\}$ in 100 steps. (**a**) $r = 1$. In this case, the network becomes a tree network. (**b**) $r = 3$. $\omega = 1$ corresponds to the BA model. When $\omega > 0$, the network is scale-free. When $\omega = 0$, the network is a random network. When $\omega = -1$, the network corresponds to an extended lattice

8.2.1 Relation to the Urn Model

In this subsection, we consider this network using the urn problem. The urn corresponds to the node, and outgoing and incoming links correspond to the balls. We show the image in Fig. 8.3. One urn joins as time passes. When the i-th urn enters the network at time $t = i$, r balls are taken out from all urns with a probability that is proportional to the popularity. When the urn joins, r balls which corresponds to the outgoing link are in the urn. The weighted number of balls corresponds to popularity.

In the network, a new node (urn) connects to the selected nodes (urns) by incoming and outgoing links. When the node (urn) is selected, the outgoing links (balls) are added to it. This affects ω to popularity of the urn. In the negative ω case, there is a cap on the number of balls in the urn. In contrast, positive ω case, there is no cap.

8.2.2 Degree Distribution

We consider the degree distribution of the networks. The number of links in node i is k_i. When $\omega = 1$ for the BA model, $l_i = k_i$. The sum of the popularity is $\sum_j l_j = (\omega + 1)r[t - 1/2(r + 1)] \sim (\omega + 1)rt$ for large t. Note that when ω is negative, popularity decreases as the number of links to the node increases. However, the total popularity does not decrease when $\omega > -1$, and new nodes can join. We can obtain the differential equation for the expected value of k_i as

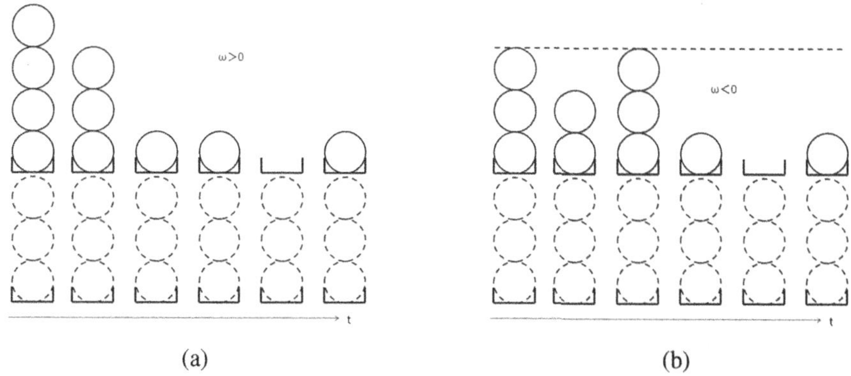

Fig. 8.3 (**a**) $\omega > 0$ (**b**) $\omega < 0$ cases. The horizontal axis represents time t and the new joined node. The dotted (real) balls correspond to the incoming(outgoing) arrows

$$\frac{dk_i}{dt} = \frac{(k_i - r)\omega + r}{(1+\omega)t}. \tag{8.3}$$

Note that when ω is negative, if $k_i^{OUT} > k_i^{IN}/|\omega|$, $l_i = k_i^{IN} - |\omega| k_i^{OUT} < 0$ and $p_i = 0$. We will discuss this effect in more detail later.

Case When $\omega > 0$

Equation (8.3) can be solved with the initial condition $k_i(t_i) = r$, and we obtain,

$$k_i = \frac{r}{\omega}\left(\frac{t}{t_i}\right)^{\omega/(1+\omega)} - r\frac{1-\omega}{\omega}.$$

We estimate the average probability that node i is selected as

$$\frac{l_i}{(1+\omega)t} = \frac{r}{1+\omega}\left(\frac{1}{t_i}\right)^{\frac{\omega}{1+\omega}} t^{-1/(1+\omega)}.$$

The average frequency of selection of node i, called memory, decreases as a power law with t, and the index is $1/(1+\omega)$. We summarize the power index of memory $1/(1+\omega)$ in Fig. 8.4(a). As ω increases, the power index decreases, and memory becomes longer. In the limit $\omega \to \infty$, the memory covers all history equally, and the power index of memory becomes 0. We refer to the cases where the power index is larger than 1 and smaller than 1 as the cases of intermediate memory and long memory, respectively. This classification depends on whether the integral of memory over t is finite (Florescu et al. 2016).

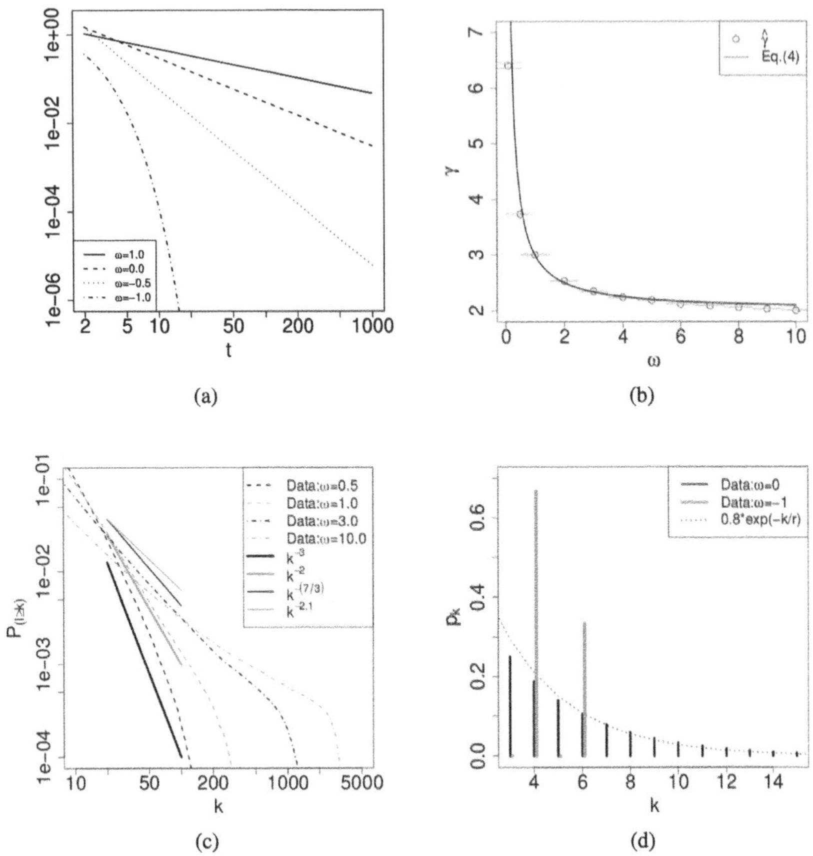

Fig. 8.4 (**a**) Plot of average probability of choice of node i (memory) versus t. The power index of the decay of the memory is $1/(1+\omega)$. (**b**) Plot of γ versus $\omega > 0$, which is the power index of the degree distribution in Eq. (8.4). We show the estimates of γ and its standard errors using symbols ($\circ$) and error bars. We sample 100 networks of 10^4 nodes for each $\omega \in \{0.1, 0.5, 1, 2, 3, 4, 5, 10\}$, and estimate γ using the empirical degree distributions. (**c**) Empirical degree distributions for $\omega = 0.5, 1, 3, 10$ for the data in (**b**). The straight lines are the power-law equations with the theoretical power index $\gamma = 2 + 1/\omega$. (**d**) Empirical distribution for $\omega = 0, -1$

We can obtain the cumulative degree distribution of the network as

$$P[k_i(t) < k] = P\left[(\frac{r}{\omega})^{(1+\omega)/\omega} \frac{t}{(k + r\frac{1+\omega}{\omega})^{(1+\omega)/\omega}} < t_i \right]$$
$$= 1 - (\frac{r}{\omega})^{(1+\omega)/\omega} \frac{1}{(k + r\frac{1-\omega}{\omega})^{(1+\omega)/\omega}}.$$

Note that $k_i \geq r$. Hence, the degree distribution is

$$p(k) = \frac{\partial P[k_i(t) < k]}{\partial k} \propto k^{-\frac{1+2\omega}{\omega}}. \quad (8.4)$$

In this region, the degree distribution exhibits the power law. When $\omega = 1$, the network corresponds to the BA model with distribution k^{-3}. When $\omega \to \infty$ and $\omega = 1/2$, the networks have distributions k^{-2} and k^{-4}, respectively.

We show the power index in Fig. 8.4(b). We also show numerical estimates of γ with 100 sample networks of 10^4 nodes with $r = 3$.[1] We estimate γ and its standard error using the maximum likelihood estimator (Newman et al. 2005). This is similar to the extended BA model (Barabási 2016), which has a discount factor for each node. If a discount factor is introduced, the transition from the lattice to the scale-free network can be observed. However, these networks do not include tree network, whereas the proposed networks include the tree network. The minimum power index is 2 when the network is graphical. This means that a degree sequence can be converted into a simple graph (Barabási 2016). Figure 8.4(c) shows the empirical degree distribution for the data in Fig. 8.4(b) with $\omega = 0.5, 1, 3, 10$. When $\omega = 0.5$ or 10, the discrepancy from the power-law behavior becomes obvious. This may be attributed to the finite size effect. It becomes difficult to observe the power-law behavior for a small $\omega << 1$ and a large $\omega >> 1$.

We obtain the distribution, $P[k_i(t) < k]$ from a sample of the network in the large t limit. **The sample average converges to the ensemble average. This is the self-averaging process of the distribution as a network.** Another self-averaging process of the distribution can be seen in the random matrix theory. The distribution of eigenvelues of a matrix, sample average, converges to the ensemble average. For each node, the process is a non-self-averaging process because each process is the Pólya's urn process when $\omega > 0$.

Case When $\omega = 0$

We solve Eq. (8.3) with the initial condition $k_i(t_i) = r$ when $\omega = 0$ to obtain

$$k_i(t) = r \log(\frac{t}{t_i}) + r.$$

We estimate the probability that node i is selected as

$$\frac{l_i}{t} = \frac{r}{t}.$$

[1] Here, we calculate the ensemble because the number of nodes is small.

The selected frequency decreases according to the power law, and the index is 1. Therefore, the memory is long, but shorter than that in the case when $\omega > 0$. We can obtain the cumulative degree distribution of the network as

$$P[k_i(t) < k] = P[t_i > te^{-\frac{k}{r}}] = 1 - e^{-\frac{k}{r}+1}.$$

The degree distribution is

$$p(k) \propto e^{-\frac{k}{r}}. \tag{8.5}$$

The degree distribution decays exponentially (Fig. 8.4(d)), and the model corresponds to a random network.

Case When $-1 < \omega < 0$

In this case, there is a probability that popularity l_i becomes negative and $p_i = 0$. Equation (8.3) becomes an approximated equation when $1/|\omega|$ is not an integer. When $1/|\omega|$ is an integer, $p_i \geq 0$ for all i.

We solve Eq. (8.3) with the initial condition $k_i(t_i) = r$ for $\omega < 0$ as

$$k_i(t) = r + r\frac{1}{|\omega|}\left(1 - \left(\frac{t_i}{t}\right)^{\frac{|\omega|}{1-|\omega|}}\right).$$

Note that in this case, $r \leq k_i \leq r + \lceil r/|\omega| \rceil$. We estimate the probability that node i is selected as

$$\frac{l_i}{(1+\omega)t} = \frac{r}{1+\omega} t_i^{-\frac{\omega}{1+\omega}} \cdot t^{-\frac{1}{1+\omega}}.$$

The selected frequency of node i decreases according to the power law with t, and the index is $1/(1+\omega)$. When $-1 < \omega < 0$, the power index is greater than 1, and the memory becomes intermediate. We obtain the cumulative degree distribution of the network as

$$\begin{aligned} P[k_i(t) < k] &= P\left[(\frac{|\omega|}{r})^{\frac{1-|\omega|}{|\omega|}} (\frac{1+|\omega|}{|\omega|} r - k)^{\frac{1-|\omega|}{|\omega|}} t > t_i \right] \\ &= \left(1 - \frac{|\omega|}{r}(k-r)\right)^{\frac{1-|\omega|}{|\omega|}}. \end{aligned} \tag{8.6}$$

In this region, the distribution is the power decay in the range r to $r + \lceil r/|\omega| \rceil$. The power index is $2 - 1/|\omega|$. When $\omega = -1/2$, the power index is zero, and the distribution becomes a uniform distribution between r and $3r$. We show the empirical degree distribution in Fig. 8.5 for $r = 3$, and confirm the uniform distribution for $\omega = -1/2$. When $\omega = -0.9$ and $\omega = -0.999$, $k = 6 = 3 + \lceil 3/|\omega| \rceil$. After node i is selected by four new nodes, l_i becomes negative, and $p_i = 0$. We can observe this effect to a larger extent for $\omega = -0.9$ than for $\omega = -0.999$.

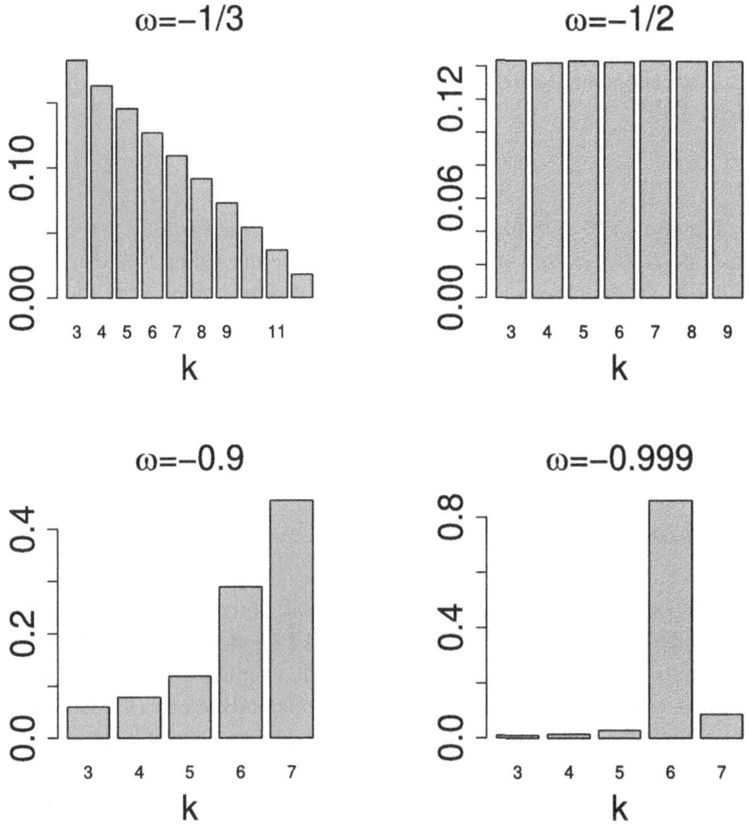

Fig. 8.5 Empirical degree distribution for $r = 3, \omega = -1/3, -1/2, -0.9$, and -0.999. We sample 10^3 networks of 10^4 nodes

Limit $\omega \to -1$

Before we discuss the limit $\omega \to -1$, we discuss the initial configuration of the network. If the initial condition of the network before its evolution is a complete graph, the limit $\omega \to -1$ is trivial. The number of linkable nodes, r' is r or $r-1$ and the network becomes an extended lattice. Here, we consider a more general initial configuration of the network. In the limit $\omega \to -1$, Eq. (8.3) is undefined. The differential equation for the expected value of k_i becomes

$$\frac{dk_i}{dt} = \frac{-(k_i - k_i^{IN}) + k_i^{IN}}{C} = \frac{-(k_i - 2k_i^{IN})}{C}, \tag{8.7}$$

where C is a constant that depends on the initial condition. For example, C is the number of nodes with no links at time 0. We note that node t cannot necessarily be linked with r nodes when it enters the network because the number r of nodes with $p_i > 0$ might be less than r. The solution of Eq. (8.7) with $k(t_i) = k_i^{IN}$ is

$$k_i = 2k_i^{IN} - k_i^{IN} e^{-\frac{1}{C}(t-t_i)}.$$

k_i increases exponentially from k_i^{IN} to $2k_i^{IN}$. We estimate the probability that node i is selected as

$$\frac{l_i}{C} \sim \frac{r}{C} e^{-\frac{1}{C}(t-t_i)}.$$

Here, we assume $k_i^{IN} = r$ for most nodes, which is true for $\omega > -1$. The selected frequency decreases exponentially, and memory becomes short. We can confirm this in Fig. 8.4(a). In the case $C \to 0$, the network becomes an extended lattice, which can be confirmed in Fig. 8.4(d). The degree distribution is the superposition of delta functions at $k = 2r$ and $2(r - 1)$. Most of the nodes are selected by the previous r or $r - 1$ nodes and the numbers of incoming and outgoing arrows are equal. We discussed the lattice case in Chap. 6.

8.2.3 Character of the Network

We study the average distance of the network $L(\omega)$ and the cluster coefficient $C(\omega)$ in Fig. 8.6(a). We sample 10^3 networks with 10^3 nodes and $r = 3$, and estimate the average distance and cluster coefficients. The average distance $L(\omega)$ decreases rapidly as ω increases from $\omega = -1$, which corresponds to an extended lattice. The cluster coefficient $C(\omega)$ decreases more slowly than $L(\omega)$, and it becomes minimum at $\omega \sim 0$. Regions with large cluster coefficients and small average distances are observed for $\omega \sim -1$ and $\omega > 0$. These properties are observed in several real-world networks (Barabási 2016). We compared average distance and cluster coefficient with those in the Watts–Strogatz (WS) model, which uses a parameter p instead of ω (Watts and Strogatz 1998). p is the probability of rewiring the extended lattice. When $p = 0$ and $p = 1$, the network becomes the extended lattice and the random network, respectively. There is a difference between our model and the WS model in the regions $p \sim 0$ and $\omega \sim -1$ near the extended lattice where the cluster coefficient is large, and the average distance is small. The region with this property is wider in the WS model than in the proposed model. We also study the system size dependence of the region in Fig. 8.6(b). As the number of nodes N increases from 10^3 to 10^4, the region moves leftward. This result shows that the region disappears in the limit $N \to \infty$. In the WS model, the curve of the cluster coefficient does not move with N and only the curve for the average distance moves leftward. The region becomes wider with an increase in N. Near the random network $p \sim 1$ and $\omega \sim 0$, we can confirm a small cluster coefficient and a small average distance.

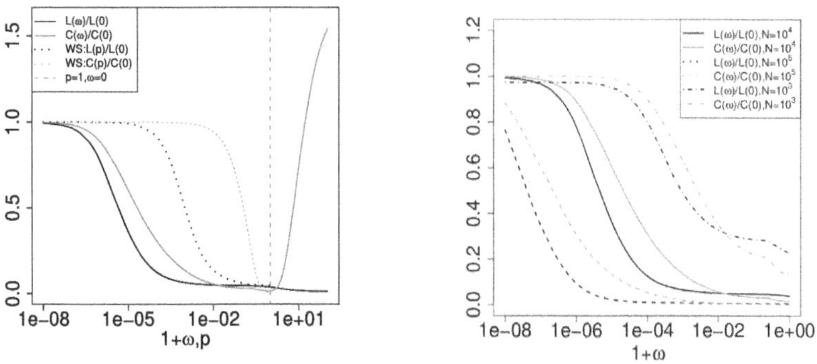

(a) (b)

Fig. 8.6 Average distance $L(\omega)$ and cluster coefficient $C(\omega)$ of the networks. (**a**) $N = 10^3$, $r = 3$ and we collect 10^3 sample networks. WS is the Watts–Strogatz model. (**b**) $N = 10^2, 10^3, 10^4$ and we collect 10^3, 10^3 and 10^2 networks, respectively

8.3 Quantum Statistics and Networks

In this section, we introduce the fitness model for the nodes (Barabáshi et al. 1999; Bianconi and Barabáshi 2001) to establish the relation between the degree distribution and quantum statistics. The fitness is the weight of popularity l_i, and the popularity of node i is transformed as $\hat{l}_i = e^{-\beta\epsilon_i} l_i$. Here, $e^{-\beta\epsilon_i}$ is the fitness, β is the inverse temperature, and ϵ_i is the associated energy level of node i in quantum statistical mechanics. The equation for the expected amount of k_i in Eq. (8.3) is modified as

$$\frac{dl_i}{dt} = \omega r \frac{e^{-\beta\epsilon_i} l_i}{\sum_i e^{-\beta\epsilon_i} l_i}. \tag{8.8}$$

Here, we assume the solution of this equation as

$$l_i = r\left(\frac{t}{t_i}\right)^{f(\epsilon_i)}, \tag{8.9}$$

which satisfies the initial condition, $l_i = r$ at $t = t_i$. We substitute Eq. (8.9) into Eq. (8.8) and obtain

$$f(\epsilon_i) = \omega \frac{rte^{-\beta\epsilon_i}}{\sum_i e^{-\beta\epsilon_i} l_i}, \tag{8.10}$$

$$\sum_i e^{-\beta\epsilon_i} l_i \sim \int e^{-\beta\epsilon} g(\epsilon) d\epsilon \int_1^t l_i dt_i = \int e^{-\beta\epsilon} g(\epsilon) r \frac{t - t^f}{1 - f(\epsilon)} d\epsilon, \tag{8.11}$$

where $g(\epsilon)$ is the distribution of energy level ϵ_i.

In the large-t limit, we calculate Eq. (8.11) using $f(\epsilon) < 1$,

$$\sum_i e^{-\beta\epsilon_i} l_i \sim rt \int g(\epsilon) \frac{e^{-\beta\epsilon}}{1 - f(\epsilon)} d\epsilon = rte^{-\beta\mu}, \tag{8.12}$$

where

$$e^{-\beta\mu} = \int g(\epsilon) \frac{e^{-\beta\epsilon}}{1 - f(\epsilon)} d\epsilon. \tag{8.13}$$

μ becomes the chemical potential in quantum statistics. Thus, we can rewrite Eq. (8.10) using Eq. (8.12) as

$$f(\epsilon) = \omega e^{-\beta(\epsilon-\mu)}. \tag{8.14}$$

Then, we use Eq. (8.13) to obtain

$$1 = \int g(\epsilon) \frac{1}{e^{\beta(\epsilon-\mu)} - \omega} d\epsilon = I(\omega, \beta, \mu). \tag{8.15}$$

Equation (8.15) is the normalization. $1/(e^{\beta(\epsilon-\mu)} - \omega)$ corresponds to the distribution function of quantum statistics. This is an extension of the Bose and Fermi distributions and is similar to the fractional exclusion statistics (Haldane 1991; Bianconi 2002; Cinardi et al. 2002). When $\omega = 1(-1)$, it corresponds to the Bose (Fermi) distribution that connects the Bose distribution with the Fermi distribution.

8.3.1 Boson-Type Network: $\omega > 0$

We rewrite the distribution function in Eq. (8.15) as follows:

$$\frac{1}{e^{\beta(\epsilon-\mu)} - \omega} = \frac{1}{\omega(e^{\beta(\epsilon-\mu+\delta_\mu)} - 1)} = \frac{1}{\omega(e^{\beta(\epsilon-\mu')} - 1)}, \tag{8.16}$$

where $\omega = e^{-\beta\delta_\mu}$ and $\mu' = \mu - \delta_\mu$. When $\omega = 1$, which corresponds to the BA model, $\delta_\mu = 0$, and we obtain the Bose distribution.

We consider the BEC of the network under the condition of large β. We can rewrite Eq. (8.15) using Eq. (8.16) as

$$1 = \frac{1}{\omega} \int g(\epsilon) \frac{1}{e^{\beta(\epsilon-\mu')} - 1} d\epsilon = \frac{I'(\omega, \beta, \mu')}{\omega} = I(\omega, \beta, \mu), \tag{8.17}$$

where $I'(\omega, \beta, \mu') = \int g(\epsilon)/(e^{\beta(\epsilon-\mu')} - 1) d\epsilon$. We can obtain the condition of the BEC in Eq. (8.15) as

$$I'(\omega, \beta, \mu') = \omega. \tag{8.18}$$

$I'(\omega, \beta, \mu')$ increases as μ'. If $I'(\omega, \beta, 0) > \omega$, we obtain the solution $\mu'(< 0)$ for Eq. (8.18). If $I'(\omega, \beta, 0) \leq \omega$, no solution exists. $\omega - I'(\omega, \beta, \mu')$ of nodes are connected to the maximum fitness node that corresponds to the hub in the network. This corresponds to the "the winner takes all" or the BEC of the network.

To confirm this condition, we consider the case where the distribution of energy level ϵ_i is

$$g(\epsilon) = A\epsilon^{\theta}, \tag{8.19}$$

where θ is a parameter for the power distribution, and $A = (\theta + 1)/\epsilon_{max}^{\theta+1}$. ϵ_{max} is the maximum energy of the fitness model.

The condition of the BEC in Eq. (8.18) is

$$I'(\omega, \beta, 0) = \frac{\theta + 1}{(\beta\epsilon_{max})^{\theta+1}} \int_0^{\beta_{max}} dx \frac{x^{\theta}}{e^x - 1} < \omega \tag{8.20}$$

In the large-time limit, the lower bound of the transition temperature can be obtained as

$$T_{BEC} > \epsilon_{max} \left(\frac{\omega}{\zeta(\theta + 1)\Gamma(\theta + 2)} \right)^{1/(\theta+1)}, \tag{8.21}$$

where $\Gamma(x)$ is the gamma function and $\zeta(x)$ is the zeta function. We can observe the BEC when $\theta > 0$.

As ω increases, the lower bound of the transition temperature increases. In the limit $\omega \to \infty$, the phase is the BEC phase. In the limit $\omega \to 0$, which corresponds to the random network limit, the BEC phase disappears. In summary, ω is related to the threshold of the BEC phase by Eq. (8.21). As ω is increased, lower energy levels are occupied. In Fig. 8.7(a), we show a depiction of the Bose statistics.

8.3.2 *Classical Network: $\omega = 0$*

When $\omega = 0$, which corresponds to the random network case, the density function becomes $e^{-\beta(\epsilon-\mu)}$. This corresponds to the Boltzmann distribution.

8.3.3 *Fermion-Type Network: $-1 \leq \omega < 0$*

We revise the distribution function in Eq. (8.15) as

$$\frac{1}{e^{\beta(\epsilon-\mu)} + |\omega|} = \frac{1}{|\omega|(e^{\beta(\epsilon-\mu+\delta_\mu)} + 1)} = \frac{1}{|\omega|(e^{\beta(\epsilon-\mu')} + 1)}, \tag{8.22}$$

where $|\omega| = e^{-\beta\delta_\mu}$ and $\mu' = \mu - \delta_\mu$. When $0 < |\omega| \leq 1$, $\delta_\mu \geq 0$.

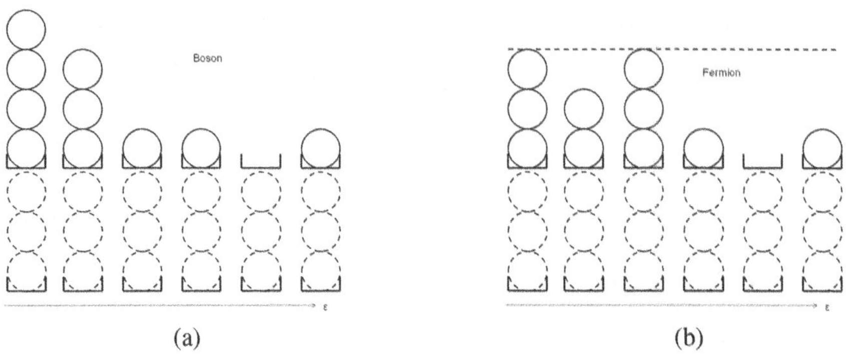

Fig. 8.7 (**a**) Bose statistics. (**b**) Fermi statistics. The horizontal axis corresponds to ϵ. The dotted (real) circles correspond to the incoming (outgoing) arrows. In the Fermion case, the occupied number is limited

We consider the case when $\beta >> 1$, i.e., the low-temperature limit. Equation (8.22) becomes a step function at $\epsilon = \mu'$. When $\epsilon < \mu'(\epsilon > \mu')$, Eq. (8.22) becomes $1/|\omega|(0)$. The maximum occupied number, which corresponds to the number of outgoing arrows, is $-/|\omega|$ per incoming arrow. ω corresponds to the change in the chemical potential and the maximum occupied number in Eq. (8.22). The lower ϵ nodes are connected to the other $r + \lceil r/|\omega| \rceil$ nodes. When a node joins the network, it is connected to r nodes, and then the node is connected to $\lceil r/|\omega| \rceil$ nodes with the maximum number of links. On the other hand, the high ϵ nodes are connected to other r nodes that are connected only when the node joins. This is similar to the Fermi degeneracy of the network. As $\omega < 0$ is increased, the maximum occupied number increases, and the Fermi energy μ' decreases. In the limit $\omega \to 0$, corresponding to the limit of the random network, no maximum occupied number. In the limit $\omega = -1$, $\delta_\mu = 0$, we can obtain the Fermi distribution. The maximum occupied number obtained after node joins is one per incoming arrow. In Fig. 8.7(b), we show the Fermi statistics.

8.4 Summary

In this chapter, we discuss several types of networks, including the random graph, Barabási-Albert(BA) model, and lattice networks, using the parameter ω. The parameter ω is related to the preferential attachment of the nodes in the networks and represents the strength of the feedback. In particular, negative ω corresponds to negative feedback.

The parameter ω is also related to memory length, which indicates how old a node is relative to the new node it connects to. When $\omega > 0$ newly added nodes tend to be connected with older nodes, and the degree distribution exhibits power-law behavior.

$\omega = 0$ corresponds to a random network and long memory with a larger power index than the positive ω case. Negative ω corresponds to intermediate memory. The case of $\omega = -1$ corresponds to an extended lattice and short memory.

In addition, we discuss the correspondence between quantum statistics and the degree distribution of the network model. Positive (negative) ω corresponds to Bose (Fermi)-like statistics, and we obtain a distribution that connects Bose and Fermi statistics. When ω is positive, it corresponds to the threshold of the BEC. As ω decreases, the area of the BEC phase narrows. When ω is negative, $r + \lceil r/|\omega| \rceil$ is the maximum occupied number. At low temperatures, we can observe the Fermi degeneracy of the network. The lower energy nodes have the maximum number of links, and the higher energy nodes have no links except for the initial ones. As ω decreases, the maximum occupied number also decreases. In 8.5 Appendix A, we discuss the relation to the percolation on network.

8.5 Appendix A Molloy and Reed Condition and Percolation on Networks

In this Appendix, we consider the Molly and Reed conditions of the networks. Using the Molly and Reed conditions, we calculate the phase transition point of the percolation model. In this Appendix, we calculate the Molly Reed conditions. First, we calculate the second momentum of the degree distribution.

8.5.1 Appendix A.I: A. When $\omega > 0$

The cumulative degree distribution of the network as

$$P[k_i(t) < k] = P\left[(\frac{r}{\omega})^{(1+\omega)/\omega} \frac{t}{(k + r\frac{1+\omega}{\omega})^{(1+\omega)/\omega}} < t_i\right],$$

$$= \frac{1}{r+t}\left[t - (\frac{r}{\omega})^{(1+\omega)/\omega} \frac{t}{(k + r\frac{1-\omega}{\omega})^{(1+\omega)/\omega}}\right],$$

where k_i denotes the number of links of node i. Notably, $k_i \geq r$ because node i has r initial links. Hence, in the limit $t \to \infty$ the degree distribution is

$$p(k) = \frac{\partial P[k_i(t) < k]}{\partial k} = A_B \frac{1+\omega}{\omega}(k + r\frac{1-\omega}{\omega})^{-2-1/\omega}, \tag{8.23}$$

where $A_B = (r/\omega)^{1+1/\omega}$. We can confirm the normalization

$$\int_r^\infty p(k)dk = A_B(r/\omega)^{-1-1/\omega} = 1. \tag{8.24}$$

The first moment was calculated as follows:

$$< k >= \int_r^\infty kp(k)dk = r + r = 2r. \tag{8.25}$$

The second moment was obtained. Here, when $0 < \omega < 1$,

$$< k^2 >= \int_r^\infty k^2 p(k)dk = 3r^2 + \frac{2r^2}{1-\omega}. \tag{8.26}$$

When $\omega > 1$, we obtain

$$< k^2 >= \int_r^\infty k^2 p(k)dk = 3r^2 + \frac{2r^2}{1-\omega} + 2\frac{\omega^2}{\omega-1}(\frac{r}{\omega})^{1+1/\omega}(k_{Max} + r\frac{1-\omega}{\omega})^{1-1/\omega}, \tag{8.27}$$

where k_{Max} is the maximum number of links and goes to infinity. Hence, the second moment is finite when $1 > \omega > 0$ and infinite, when $\omega > 1$.

Hence, under the condition $1 > \omega > 0$, the Molloy–Reed condition is

$$\frac{< k^2 >}{< k >} = \frac{3r}{2} + \frac{r}{1-\omega}. \tag{8.28}$$

When $\omega > 1$, $< k^2 >\to \infty$.

8.5.2 Appendix A.2: B. When $\omega = 0$

The cumulative degree distribution of the network is expressed as follows

$$P[k_i(t) < k] = P[t_i > te^{-\frac{k}{r}}] = \frac{1}{r+t}(t - te^{-\frac{k}{r}}).$$

The degree distribution in the limit $t \to \infty$ is:

$$p(k) = A_R e^{-\frac{k}{r}}, \tag{8.29}$$

where $A_R = e/r$ to satisfy the condition $\int_r^\infty p(k)dk = 1$.

The first moment was calculated as follows:

$$< k >= \int_r^\infty kp(k)dk = 2r. \tag{8.30}$$

Similarly, we obtain the second moment as

$$< k^2 >= \int_r^{\infty} k^2 p(k)dk = 5r^2. \tag{8.31}$$

Then, the Molloy-Reed condition is

$$\frac{< k^2 >}{< k >} = 5r/2. \tag{8.32}$$

8.5.3 Appendix A.3: C. When $-1 < \omega < 0$

The cumulative degree distribution of the network is

$$\begin{aligned} P[k_i(t) < k] &= P\left[(\frac{|\omega|}{r})^{\frac{1-|\omega|}{|\omega|}} (\frac{1+|\omega|}{|\omega|} r - k)^{\frac{1-|\omega|}{|\omega|}} t > t_i \right], \\ &= \frac{1}{r+t} \left(\frac{|\omega|}{r} \right)^{\frac{1-|\omega|}{|\omega|}} \left(\frac{1+|\omega|}{|\omega|} r - k \right)^{\frac{1-|\omega|}{|\omega|}} t. \end{aligned} \tag{8.33}$$

We obtain the degree distribution in the limit $t \to \infty$:

$$p(k) = A_F \frac{1-|\omega|}{|\omega|} (\frac{1+|\omega|}{|\omega|} r - k)^{1/|\omega|-2}, \tag{8.34}$$

where $A_F = (r/\omega)^{1-1/|\omega|}$ to satisfy the condition $\int_r^{r+r/|\omega|} p(k)dk = 1$.
The first moment was calculated as follows:

$$< k >= \int_r^{\infty} kp(k)dk = r + r = 2r. \tag{8.35}$$

Similarly, we obtain the second moment.

$$< k^2 >= \int_r^{\infty} k^2 p(k)dk = 3r^2 + \frac{2r^2}{1+|\omega|}. \tag{8.36}$$

Then, the Molloy-Reed condition is

$$\frac{< k^2 >}{< k >} = \frac{3r}{2} + \frac{r}{1+|\omega|}. \tag{8.37}$$

8.5.4 Appendix A.4: D. Phase Transition in the Percolation Model

We can combine Eqs. (8.26), (8.31), and (8.36).

$$< k^2 >= 3r^2 + \frac{2r^2}{1-\omega}, \tag{8.38}$$

where $1 > \omega > -1$. We can combine Eqs. (8.28), (8.32), and (8.37)

$$\frac{< k^2 >}{< k >} = \frac{3r}{2} + \frac{r}{1-\omega}, \tag{8.39}$$

where $1 > \omega > -1$. When $\omega \geq 1$, $< k^2 > / < k > \rightarrow \infty$.

The critical probability of the percolation model is

$$q_c = 1/[\frac{3r}{2} - 1 + \frac{r}{1-\omega}], \tag{8.40}$$

where $1 > \omega > -1$. In the case of the limit $\omega \rightarrow 1$, we c obtain $q_c = 0$. Then, $\omega \geq 1$, there is no phase transition. In the limit $\omega = -1$, we obtain $q_c = 1/(2r-1)$. In contrast, $\omega = -1$, the network is a lattice, and $q_c = 1$. When $r = 1$, q_c is continuous, $r \neq 1$, q_c is discontinuous. When $\omega > 1$, $q_c = 0$ at $\omega = 1$. We show q_c in Fig. 8.8. Hence, a phase transition is observed in $1 > \omega > -1$.

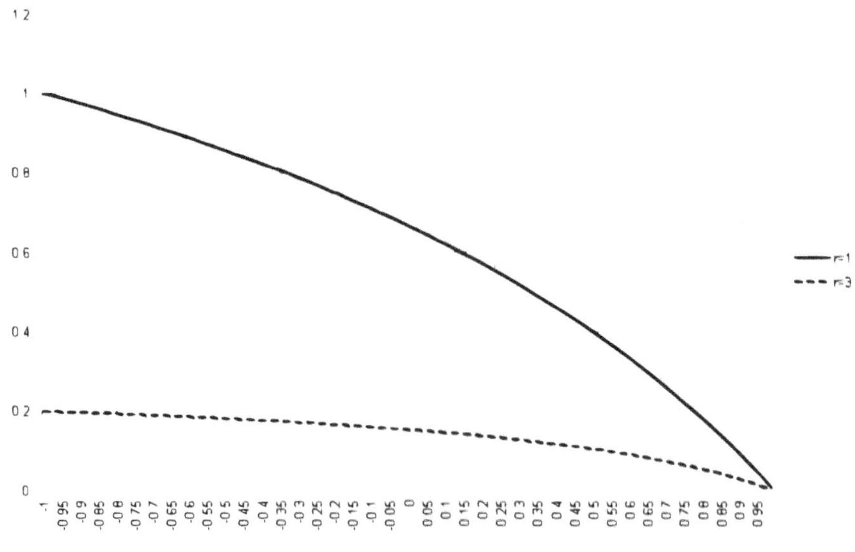

Fig. 8.8 The transition point q_c. The horizontal axis is ω and the vertical axis is q_c There is a discontinuity in $\omega = -1$ without $r = 1$. The phase transition is observed in $> 1\omega > 1$

References

Barabáshi A L, Albert R, Emergence of Scaling in Random Networks, *Science* **286**, 510 (1999).

Barabási A L, Albert R, Jeong H, Mean-field theory for scale-free random networks. *Physica A*, **272(1–2)** 173–187 (1999).

Barabási A L, *Network science* (Cambridge University Press, 2016).

Bianconi G, Barabáshi A L, Competition and multiscaling in evolving networks. *Euro. Phys. Lett*, **54(4)** (2001).

Bianconi G, Barabáshi A L, Bose-Einstein condensation in complex networks. *Phy. Rev. Lett*, **86(24)** 5632 (2001).

Bianconi G, Growing Cayley trees described by Fermi distribution, *Phys. Rev.* **E 66** 036116 (2002).

Cinardi N, Rapisarda A, Bianconi G, Quantum statistics in Network Geometry with Fractional Flavor, *J. Stat. Mech.*, 103403 (2019).

Florescu I, Mariani M C, Stanley H E, Viens F G (Eds.) *Handbook of High-Frequency Trading and Modeling in Finance* (John Wiley & Sons, 2016).

Haldane F D M, "Fractional statistics" in arbitrary dimensions: a generalization of the Pauli principle, *Phys Rev Lett* **67** 937 (1991).

Hisakado M, and Mori S, Quantum statistics and networks by asymmetric preferential attachment of nodes-between bosons and fermions *Journal of the Physical Society of Japan* **90(8)**, 084801 (2021).

Jennen M, Verwijmeren P, Agglomeration Effects and Financial Performance, Urban Studies **47(12)** 2683–2703.

Newman M, Power laws, Pareto distributions and Zipf's law. *Contemporary Phys.* **46**, 323–351 (2005).

Newman M, *Networks* (Oxford University Press, 2018).

Watts D J, Strogatz S H, Collective dynamics of 'small-world' networks. *Nature* **393** 440–442 (1998).

Chapter 9
Between Ant Colony Optimization and Genetic Algorithm

9.1 Introduction

Ant colony optimization (ACO) is a method of optimization that uses herding behavior as a pheromone to optimize a path in a graph, based on the behavior of ants seeking a path between their colony and a source of food (Drigo and Stutzle 2004).

For ants, the probability of moving from state to state depends on the combination of two values, the attractiveness of the move, as computed by some heuristic indicating the a priori desirability of the move and the trail level of the move, indicating how proficient it has been in the past to make that particular move. The trail level represents the posteriori indication of the desirability of that move. Trails are updated usually when all ants have completed their solution, increasing or decreasing the level of trails corresponding to moves that were part of correct or incorrect solutions, respectively. It is a useful algorithm for the traveling salesman problem (TSP), quadratic assignment problem (QAP), and graph coloring problem, which are non-deterministic polynomial time (NP) problems (Dorigo and Gambardella 1997; Maniezzo and Colorni 1999; Dorigo et al. 1999). In Fig. 9.1, we show an example of TSP.

The t-th agent moves from state i to state$j$ with probability

$$p_{ij}(t) = \frac{(\tau_{ij})^a(\xi_{ij})^b}{\sum_j(\tau_{ij})^a(\xi_{ij})^b} \tag{9.1}$$

whereτ_{ij} is the amount of pheromone deposited for the transition from state i to j, $a \geq 1$ is a parameter to control the influence of the pheromone τ_{ij}, ξ_{ij} is some heuristic indicating the priori desirability of that move, typically an inverse of the distance and $b \geq 1$ is a parameter to control the influence of ξ_{ij}.

When all ants have completed a solution, the trails are updated by

$$\tau_{ij}(t+1) \rightarrow (1-\hat{\rho})\tau_{ij}(t) + \hat{\rho}\sum_k \Delta\tau_{ij}(t), \tag{9.2}$$

M. Hisakado, *Urn Models and Their Applications in Finance*,
https://doi.org/10.1007/978-981-96-3825-3_9

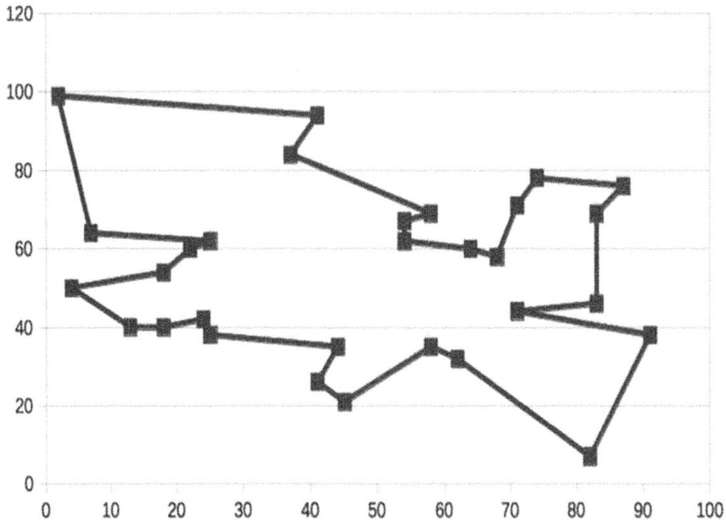

Fig. 9.1 Oliver's 30 cities problem. The problem is what is the shortest possible route that visits each city exactly once and returns to the origin city in the given list of 30 cities and the distances between each pair of cities

where τ_{ij} is the amount of pheromone deposited for a state transition from i to j, $\hat{\rho}$ is the pheromone evaporation coefficient. In this case, the pheromone decreases exponentially and has short memory. $\Delta\tau_{ij}(t)$ is the amount of pheromone deposited by the t-th ant.

$$\begin{aligned} \Delta\tau_{ij,t} &= C/L_t, \quad \text{if ant } t \text{ uses the path } ij \\ &= 0, \qquad \text{otherwise} \end{aligned} \tag{9.3}$$

where L_t is the cost of the t-th ant's tour and C is a constant.

In the previous chapters, we discussed a sequential voting model. At each time step t, one voter chooses either candidate. Using this voting model, we can show the relation between kinetic Ising model and ACO.

The remainder of this chapter is organized as follows. In Sect. 9.2, we discuss the relation between the voting model and ACO. We show the phase transition in ACO. In Sect. 9.3, we discuss the relation between kinetic Ising model and ACO using the voting model. In Sect. 9.4, we discuss the genetic algorithm (GA) and its extension. In Sect. 9.5, we confirm the results using numerical analysis. We compare several models for optimization. Finally, the conclusions are presented in Sect. 9.6. This chapter is based on Hisakado and Hino (2016).

9.2 Ant Colony Optimization and Phase Transition

9.2.1 Stochastic Differential Equation and Phase Transition

In this section, we reconstruct the ant colony optimization as a voting model. We consider N quizzes with two options. Here we consider the case where the herder can see the mark as feedback. The mark is the number of correct answers in N quizzes. The mark is given for each voter. Hereafter, we focus on one of the N quizzes, m-th quiz.

Here, we set C_0 as the incorrect candidate and C_1 as the correct candidate to validate the performance of the votes. We can set $q \geq 0.5$ because we believe that independents vote for the correct candidate C_1 rather than the incorrect candidate C_0.

The correct ratio of t-th voter is η_t. If the t-th voter selects candidate $C_1(C_0)$ for the m-th quiz, the probability that $C_1(C_0)$ is the correct answer of the m-th quiz is η_{1m}. Here, we set the correct candidate is C_1. Herder denotes the average of marks, $\bar{\eta}_{1m}$. As the higher the average of marks is, the higher the correct ratio is. This is the difference from the discussed voting model. We model the voting of two candidates, C_0 and C_1; at time t, C_0 and C_1 have $c_0(t)$ and $c_1(t)$ votes, respectively. In each time step, one voter votes for one candidate; the voting is sequential. Hence, at time t, the tth voter votes, after which the total number of votes is t. Voters are allowed to see all marks for each candidate; thus, they are aware of public perception.

We consider the Bernoulli random variables $X(t)$, which corresponds to the vote at time t. $X(t) = 1(0)$ is the vote to the candidate $1(0)$ at t. Then, we obtain

$$c_1(t) = \sum_{i=1}^{t} X(i), \tag{9.4}$$

and

$$c_0(t) = \sum_{i=1}^{t} (1 - X(i)). \tag{9.5}$$

There are two types of voters—independents and herders; we assume an infinite number of voters. The independents vote for C_0 and C_1 with probabilities $1 - q$ and q, respectively. Their votes are independent of those of others, i.e., their votes are based on fundamental values.

On the other hand, the herders' votes are based on the marks. We use the following functions. If the number of votes is $c_0(t)$ and $c_1(t)$ at time t, herder votes for C_1 at time $t + 1$ with the following probability:

$$f(z) = \frac{(\bar{\eta}_1 z)^a}{(\bar{\eta}_1 z)^a + (\bar{\eta}_0(1 - z))^a}, \tag{9.6}$$

where $z(t) = c_1(t)/(c_1(t) + c_0(t))$. It is the response function similar to the tanh -type herder, which discussed in Chap. 4.

We clarify the reduction in ant colony optimization. In our model, there are only two candidates. Hence, j is 1 or 0, which corresponds to voting for the candidate C_1 or C_0. In Eqs. (9.1) and (9.2), we set $\xi_{ij} = 1$ and $\hat{\rho} = 1$. The former means that we do not use heuristic indicating. The latter indicates that pheromone does not evaporate and has a long memory. In Eq. (9.3) the feedback is C/L_t. Instead, we use η_t which is the mark of the ant t. η_t is the correct ratio in N quizzes. We set the average marks for the candidate $C_1(C_0)$ is $\bar{\eta}_{1m}$ $(\bar{\eta}_{2m})$ in m-th quiz. In the voting model, we omit m.

Voters vote for each candidate with probabilities that are power a of the weighted candidates' votes. The simplest case, $a = 1$, is a type of analog herder. We can map Eq. (9.1) into as follows. The average mark of $\eta_1(\eta_0)$ is $\bar{\eta}_1(\bar{\eta}_0)$ from time 0 to t. There is the constraint $\bar{\eta}_1 + \bar{\eta}_0 = 1$. For simplicity, here $\bar{\eta}_1$ and $\bar{\eta}_0$ are constants, because we consider the case, $t >> 1$. We set the correct candidate as candidate C_1 and $\bar{\eta}_1 > \bar{\eta}_0$. It is appropriate for the quizzes with two options.

We can write the voting process as

$$
\begin{aligned}
c_1(t) = k \rightarrow k+1 : P_{k,t} &= p\frac{(\bar{\eta}_1 k)^a}{(\bar{\eta}_1 k)^a + (\bar{\eta}_0(t-k))^a} + (1-p)q, \\
c_1(t) = k \rightarrow k : Q_{k,t} &= 1 - P_{k,t}.
\end{aligned}
\tag{9.7}
$$

The last term is for dull ants or independents (Shimomura et al. 2010). The appearance ratio is $1-p$. They do not refer to other voters. $P_{k,t}$ corresponds to $p_{ij}(t)$ in Eq. (9.1) and the pheromone of the path $i \rightarrow j$, $\tau_{ij}(t)$ in Eq. (9.2) corresponds to the number of votes c_j, where $j = 1$ or 0.

We define a new variable Δ_t such that

$$
\Delta_t = 2c_1(t) - t = c_1(t) - c_0(t). \tag{9.8}
$$

We change the notation from k to Δ_t for convenience. Then, we have $|\Delta_t| = |2k - t| < t$. Thus, Δ_t holds within $\{-t, t\}$. Given $\Delta_t = u$, we obtain a model:

$$
\begin{aligned}
\Delta_t = u \rightarrow u+1 : P_{u,t} &= \frac{\bar{\eta}_1^a (u+t)^a}{\bar{\eta}_1^a (u+t)^a + \bar{\eta}_0^a (t-u)^a} p + (1-p)q, \\
\Delta_t = u \rightarrow u-1 : Q_{u,t} &= 1 - P_{u,t}.
\end{aligned}
$$

We now consider the continuous limit $\epsilon \rightarrow 0$,

$$
\begin{aligned}
X_\tau &= \epsilon \Delta_{[t/\epsilon]}, \\
P(x,\tau) &= \epsilon P(\Delta_t/\epsilon, t/\epsilon),
\end{aligned}
\tag{9.9}
$$

where $\tau = t/\epsilon$ and $x = \Delta_t/\epsilon$. Approaching the continuous limit, we obtain the stochastic differential equation:

$$\mathrm{d}X_\tau = [(1-p)(2q-1) + p\frac{\bar{\eta}_1^a (X_\tau + \tau)^a - \bar{\eta}_0^a (\tau - X_\tau)^a}{\bar{\eta}_1^a (X_\tau + \tau)^a + \bar{\eta}_0^a (\tau - X_\tau)^a}]\mathrm{d}\tau + \sqrt{\epsilon}\mathrm{d}B_\tau. \quad (9.10)$$

When $a = 1$ and $\bar{\eta}_1 = \bar{\eta}_0 = 1/2$ we can obtain

$$\mathrm{d}X_\tau = [(1-p)(2q-1) + p\frac{X_\tau}{\tau}]\mathrm{d}\tau + \sqrt{\epsilon}\mathrm{d}B_\tau. \quad (9.11)$$

The first term is the contribution of the dull ants, and the second term is the ants.

We can assume a stationary solution as

$$X_\infty = \bar{v}\tau, \quad (9.12)$$

where $\bar{v}$ is constant. Substituting Eq. (9.12) into Eq. (9.10), **we obtain**

$$\bar{v} = (1-p)(2q-1) + p\frac{\bar{\eta}_1^a (\bar{v}+1)^a - \bar{\eta}_0^a (1-\bar{v})^a}{\bar{\eta}_1^a (\bar{v}+1)^a + \bar{\eta}_0^a (1-\bar{v})^a}, \quad (9.13)$$

which is the self-consistent equation.

We can confirm the phase transition when we change the parameter a. In the case $q = 1/2$, $p = 1$, and $a > 1$, the solutions of Eq. (9.13) are three solutions, $\bar{v} = \pm 1$ and

$$\bar{v} = \frac{\bar{\eta}_0^{\frac{a}{a-1}} - \bar{\eta}_1^{\frac{a}{a-1}}}{\bar{\eta}_0^{\frac{a}{a-1}} + \bar{\eta}_1^{\frac{a}{a-1}}}. \quad (9.14)$$

When $\bar{\eta}_1 > \bar{\eta}_0 (\bar{\eta}_0 > \bar{\eta}_1)$, $\bar{v} = \pm 1$ is stable. We show the self-consistent equation, Eq. (9.13) in Fig. 9.2. $\bar{v} = 1(-1)$ corresponds to the good(bad) candidate. The middle solution Eq. (9.14) is unstable. In this case, the total votes sometimes select the bad candidate C_0.

Next, we consider the analog case $a = 1$ and $p = 1$. The former corresponds to the analog herder and the latter corresponds to no independents. When $q = 1/2$, $\bar{v}$ has two solutions $\bar{v} = \pm 1$. When $\bar{\eta}_1 > \bar{\eta}_0 (\bar{\eta}_0 > \bar{\eta}_1)$, $\bar{v} = -1 (\bar{v} = 1)$ is unstable. Hence, the process converges to the stable solution $\bar{v} = 1 (\bar{v} = -1)$. This means that the process converges into the correct candidate C_1 in the voting model. In ACO, parameter a is often set to $a = 1$. We show the self-consistent equation, Eq. (9.13) which has only one stable solution in Fig. 9.2. If we set $a > 1$, the process sometimes converges to an incorrect solution. In contrast, $a = 1$, the process converges to the good solution only. When $a < 1$, there is only one station Eq. (9.14) which is stable. The process converges to the solution between the correct and incorrect solutions as Fig. 9.2. In this case, we can not expect the optimization.

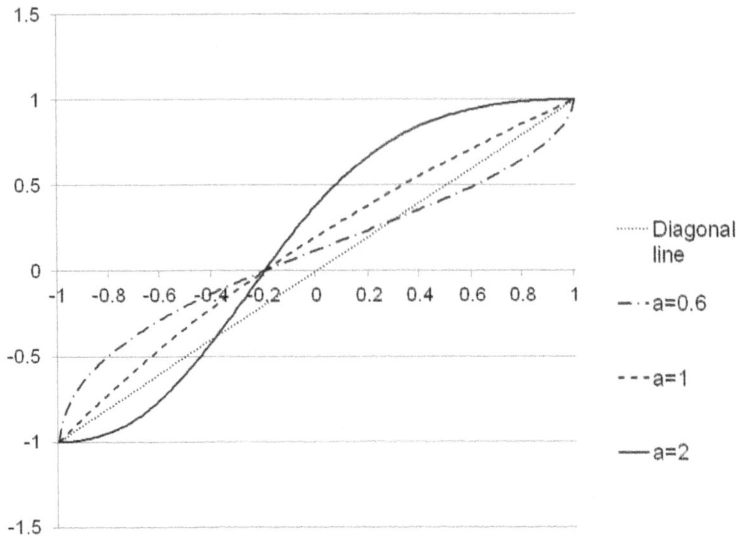

Fig. 9.2 Mean field approximation $a = 2, a = 1$, and $a = 0.8$. Here, we set $p = 1$. We use $\bar{\eta_1} = 0.6$ and $\bar{\eta_0} = 0.4$. The number of stable solutions is two for $a = 2$ and one for $a = 1$ and $a = 0.8$

9.2.2 Symmetric Case

Here, we consider a symmetric model $\bar{\eta_0} = \bar{\eta_1} = 1/2$ and $q = 1/2$. The self-consistent equation is

$$\bar{v} = p\frac{(\bar{v}+1)^a - (1-\bar{v})^a}{(\bar{v}+1)^a + (1-\bar{v})^a} = f(\bar{v}), \tag{9.15}$$

When $a > 1$, there are two stable solutions and one unstable solution $\bar{v} = 0$ which corresponds to Eq. (9.15) above p_c. This is the two-peak phase. There is a phase transition when $a > 1$. Here, we consider the phase transition about the parameter p. The critical point is $p_c = 1/a$, where the gradient of the RHS of Eq. (9.15) at $\bar{v} = 0$ is 1. In the case $a \leq 1$, there is no phase transition.

Next, we consider the critical exponent. $f(x)$ is the odd function because of the symmetry and $f(0) = 0$ and $f'(0) = 1$ at the critical point $p = p_c$. Therefore, we can set

$$\begin{aligned} f(\bar{v}) &= pa\bar{v} - \frac{a^3 - a}{3}\bar{v}^3 \cdots, \\ &= \frac{p}{p_c}\bar{v} - \frac{1-p_c^2}{3p_c^3}p\bar{v}^3 + \cdots, \end{aligned} \tag{9.16}$$

where $p_c = 1/a$. We obtain

$$\bar{v} = \pm\sqrt{\frac{p - p_c}{p_c}\frac{3p_c^3}{(1 - p_c^3)p}} \sim \pm|p - p_c|^{\beta}. \tag{9.17}$$

Therefore, the critical exponent is $\beta = 1/2$ when n is finite. This is same as the Tanh-type model discussed in Chap. 4.

9.2.3 Super-Normal Phase Transition

We expand X_τ around the solution $\bar{v}\tau$,

$$X_\tau = \bar{v}\tau + (1 - p)(2q - 1) + W_\tau. \tag{9.18}$$

Here, we set $X_\tau \gg W_\tau$, that is $\tau \gg 1$. We rewrite Eq. (9.10) using Eq. (9.18) and obtain as follows:

$$\begin{aligned} \mathrm{d}W_\tau = {} & p\frac{4a\bar{\eta}_0^a\bar{\eta}_1^a(1 + \bar{v} + (1 - p)(2q - 1))^{a-1}(1 - \bar{v} - (1 - p)(2q - 1))^{a-1}}{(\bar{\eta}_1^a(1 + \bar{v} + (1 - p)(2q - 1))^a + \bar{\eta}_0^a(1 - \bar{v} - (1 - p)(2q - 1))^a)^2} \\ & \times \frac{W_\tau}{\tau}\mathrm{d}\tau + \sqrt{\epsilon}\mathrm{d}B_\tau. \end{aligned} \tag{9.19}$$

We obtain super-normal phase transition of convergence.

Here, we consider the case $p = 1$, where there is no dull ant, and independents in the voting model. In the case, $a > 1$ the process converges to ± 1, the first term of Eq. (9.19) disappears. Hence, there is no super-normal phase transition.

In the case $a = 1$, we can obtain

$$\mathrm{d}W_\tau = \frac{\bar{\eta}_0}{\bar{\eta}_1}\frac{W_\tau}{\tau}\mathrm{d}\tau + \sqrt{\epsilon}\mathrm{d}B_\tau. \tag{9.20}$$

When $\bar{\eta}_1 > 2/3$ ($\leq 2/3$), the process converges normally(slowly). The speed of the convergence depends on $\bar{\eta}_1$. As the correct ratio increases, the convergence speed becomes the normal.

In the case $a < 1$ and $p = 0$, Eq. (9.19) becomes

$$\mathrm{d}W_\tau = a\frac{W_\tau}{\tau}\mathrm{d}\tau + \sqrt{\epsilon}\mathrm{d}B_\tau. \tag{9.21}$$

In this region, the process converges to Eq. (9.14). The critical point is $a_{vc} = 1/2$. When $1 > a \geq 1/2$, convergence becomes slow. In the case $1/2 > a$, the process converges normally.

We have to note that in this section, we set $\hat{\rho} = 0$ which corresponds to the absence evaporation of the pheromone. In next section, we consider the case that the pheromone evaporates.

9.3 Ising Model and Ant Colony Optimization

We consider the case that voter refers to previous r, which corresponds to finite memory. In other words, the parameter of evaporation is $\hat{\rho} \neq 0$. The voter refers to the r previous votes. In this section, we consider the relation between Ising model and ant colony optimization. In the previous section, we used the power response function in the voting model to obtain the ACO. In this section, we change the response function to the tanh-type. When the response function is the tanh-type, we rewrite Eq. (9.7) which is for ACO as

$$\begin{aligned} c_1(t) &= k \to k+1 : P_{k,t:l,t-r} = p\frac{e^{\lambda\bar{\eta}_1(k-l)/r}}{e^{\lambda\bar{\eta}_1(k-l)/r} + e^{\lambda\bar{\eta}_0(r-(k-l))/r}} + (1-p)q, \\ c_1(t) &= k \to k : Q_{k,t:lt-r} = 1 - P_{k,t}, \end{aligned} \tag{9.22}$$

where $c_1(t-r) = l$. This means that at time $(t-r)$, the number of votes for C_1 is $c_1(t-r) = l$. λ is a parameter.

For convenience, we define a new variable Δ_t such that

$$\Delta_t = 2c_1(t) - t = c_1(t) - c_0(t) = 2k - t. \tag{9.23}$$

We change the notation from k and l to Δ_t and Δ_{t-r} for convenience. The transition is

$$\begin{aligned} \Delta_t &= u \to u+1 : P_{u,t:s,t-r,t} = \frac{p}{2}[1 + \tanh\{\frac{\lambda(\bar{\eta}_0 + \bar{\eta}_1)(u-s)}{4r} + \frac{\lambda(\bar{\eta}_1 - \bar{\eta}_0)}{4}\}] \\ &\quad +(1-p)q, \\ \Delta_t &= u \to u-1 : Q_{u,t:s,t-r,t} = \frac{p}{2}[1 - \tanh\{\frac{\lambda(\bar{\eta}_0 + \bar{\eta}_1)(u-s)}{4r} + \frac{\lambda(\bar{\eta}_1 - \bar{\eta}_0)}{4}\}] \\ &\quad +(1-p)(1-q), \end{aligned} \tag{9.24}$$

where $\bar{\eta}_0 + \bar{\eta}_1 = 1$. Here, we change the system of coordinate from Δ_t to

$$\hat{\Delta}_t = \Delta_t/p - (1-p)(2q-1)t/p, \tag{9.25}$$

and normalize the range of the transition. We obtain.

$$\begin{aligned} &\hat{\Delta}_t = \hat{u} \to \hat{u}+1 \\ &: P_{\hat{u},t:\hat{s},t-r,t} = \frac{1}{2}[1 + \tanh\{\frac{p\lambda}{4r}(\hat{u}-\hat{s}) + \frac{(2q-1)(1-p)\lambda}{4} + \frac{p\lambda(\bar{\eta}_1 - \bar{\eta}_0)}{4}\}], \\ &\hat{\Delta}_t = \hat{u} \to \hat{u}-1 \\ &: Q_{u,t:s,t-r,t} = \frac{1}{2}[1 - \tanh\{\frac{p\lambda}{4r}(\hat{u}-\hat{s}) + \frac{(2q-1)(1-p)\lambda}{4} + \frac{p\lambda(\bar{\eta}_1 - \bar{\eta}_0)}{4}\}]. \end{aligned} \tag{9.26}$$

If we set $q = 1/2$, the last term (feedback term) corresponds to the outer field. **Then, we can obtain the equivalent to the Ising model as**

$$\hat{h} = \frac{p\lambda(\bar{\eta}_1 - \bar{\eta}_0)}{4}, \tag{9.27}$$

where $\hat{h}$ corresponds to the outer fields in the 1D Ising model. As $\bar{\eta}_1 - \bar{\eta}_0$ increases, the outer field becomes strong. The outer field is the power of the origin of the optimization. The other parameter is $p\lambda/4r = \beta J$. The Hamiltonian becomes

$$H = -\beta(J \sum_{i \neq j}^{r+1} \sigma_i \sigma_j + \hat{h}) + \hat{h}\Sigma_i^{r+1}\sigma_i. \tag{9.28}$$

9.4 Genetic Algorithm and Its Extension

9.4.1 Stochastic Differential Equation and Phase Transition

In this section, we consider a relation between the voting model and genetic algorithm(GA). At time t t-th voter selects r voters who have already voted. When $r = 2$, the model becomes GA. t-th voter can see the total votes of the selected different 2 voters to obtain the information. We consider the case that the voter selects r votes according to the marks η_t which is the mark of the t-th ant. We set the average marks for the candidate $C_1(C_0)$ is $\bar{\eta}_{1m}(\bar{\eta}_{2m})$ in m-th quiz. Hereafter, we omit m. Voters refer to the voters who have higher marks.

The herders select r referred voters with the probabilities that are the power (a) of the weighted candidates' votes. The average mark of η_1 (η_0) is $\bar{\eta}_1(\bar{\eta}_0)$ from $t = 0$ to t. For simplicity, here $\bar{\eta}_1$ and $\bar{\eta}_0$ are constants because we assume $t >> 1$. We set the correct candidate as the candidate C_1 and $\bar{\eta}_1 > \bar{\eta}_0$.

We consider the Bernoulli random variables $X(t)$, which corresponds to the vote at time t. $X(t) = 1(0)$ is the vote to the candidate $1(0)$ at t.

Voters are permitted to see only the selected r previous votes for each candidate. r is a constant number. Hence, we obtain

$$c_0^r(t) = \sum_{refered} X(i), \tag{9.29}$$

and

$$c_1^r(t) = \sum_{refered} (1 - X(i)). \tag{9.30}$$

In the scaling limit $t = c_0(t) + c_1(t) = c_0^\infty + c_1^\infty \to \infty$, we define

$$\frac{c_1(t)}{t} \Longrightarrow Z_\infty. \tag{9.31}$$

$Z(t)$ is the ratio of voters who vote for C_1 at t.

Here, we define $\pi(Z)$ as the majority probability of binomial distributions of Z. In other words, the probability of $c_1^r(t) > 1/2$. When r is odd, we have

$$\pi(Z) = \sum_{g=\frac{r+1}{2}}^{r} \binom{r}{g} Z^g (1-Z)^{r-g} = \sum_{g=\frac{r+1}{2}}^{r} P(g, Z) \equiv \Omega_r(Z), \tag{9.32}$$

where $P(g, Z)$ is the beta binomial distribution. When r is even, since the definition of the behavior of the herder, we have

$$\begin{aligned}\pi(Z) &= \sum_{g=\frac{r}{2}+1}^{r} \binom{r}{g} Z^g (1-Z)^{r-g} + \frac{1}{2}\binom{r}{r/2} Z^g (1-Z)^{r/2} \\ &= \sum_{g=\frac{r}{2}}^{r-1} \binom{r-1}{g} Z^g (1-Z)^{r-1-g} = \Omega_{r-1}(Z).\end{aligned} \tag{9.33}$$

The majority probability π in even case becomes the odd case $r-1$. we consider only the odd case $r = 2n+1$, where $n = 0, 1, 2 \ldots$.

$\pi(Z)$ can be calculated as follows,

$$\pi(Z) = \frac{(2n+1)!}{(n!)^2} \int_0^Z x^n (1-x)^n dx = \frac{1}{B(n+1, n+1)} \int_0^Z x^n (1-x)^n dx. \tag{9.34}$$

Equation (9.34) can be applied when the referred voters are selected with overlap. In fact, the referred voters are selected without overlap. However, in this chapter, we use this approximation to study the large t limit.

Here, we introduce the roulette selection for the GA. The voter selects r referred voters by the weighted candidates' votes. They are selected proportionally to the marks. As the mark increases, the probability of the selection increases. These are called roulette selection.

The herders' votes are based on the marks. We use the following functions. If the number of votes is $c_0(t)$ and $c_1(t)$ at time t, herder votes for C_1 at time $t+1$ with the following probability:

$$f(z) = \pi\left(\frac{(\bar{\eta}_1 z)}{(\bar{\eta}_1 z) + (\bar{\eta}_0 (1-z)}\right), \tag{9.35}$$

where $z(t) = c_1(t)/(c_1(t) + c_0(t))$.

Then, we can write the voting process as

$$
\begin{aligned}
c_1(t) = k \rightarrow k+1 : P_{k,t} &= p\pi(\frac{(\bar{\eta}_1 k)}{(\bar{\eta}_1 k) + (\bar{\eta}_0(t-k))}) + (1-p)q, \\
c_1(t) = k \rightarrow k : Q_{k,t} &= 1 - P_{k,t},
\end{aligned} \tag{9.36}
$$

where $q = 1/2$. The last term is for the independents. The appearance ratio is $1 - p$. They do not refer to other voters. The difference between GA and ACO is only the response function. For ACO, the power function is the response function. For GA, the binomial distribution is the response function.

We define a new variable Δ_t such that

$$
\Delta_t = 2c_1(t) - t = c_1(t) - c_0(t). \tag{9.37}
$$

We change the notation from k to Δ_t for convenience. Then, we have $|\Delta_t| = |2k - t| < t$. Thus, Δ_t holds within $\{-t, t\}$. Given $\Delta_t = u$, we obtain the random walk model:

$$
\begin{aligned}
\Delta_t = u \rightarrow u+1 : P_{u,t} &= \pi(\frac{\bar{\eta}_1(u+t)}{\bar{\eta}_1(u+t) + \bar{\eta}_0(t-u)})p + (1-p)q, \\
\Delta_t = u \rightarrow u-1 : Q_{u,t} &= 1 - P_{u,t}.
\end{aligned}
$$

We now consider the continuous limit $\epsilon \rightarrow 0$,

$$
\begin{aligned}
X_\tau &= \epsilon \Delta_{[t/\epsilon]}, \\
P(x, \tau) &= \epsilon P(\Delta_t/\epsilon, t/\epsilon),
\end{aligned} \tag{9.38}
$$

where $\tau = t/\epsilon$ and $x = \Delta_t/\epsilon$. Approaching the continuous limit, we can obtain the stochastic differential equation:

$$
\begin{aligned}
\mathrm{d}X_\tau = [(1-p)(2q-1) - p + 2p\frac{(2n+1)!}{(n!)^2} \int_0^{\frac{\bar{\eta}_1(X_\tau+\tau)}{\bar{\eta}_1(X_\tau+\tau)+\bar{\eta}_0(\tau-X_\tau)}} \\
x^n(1-x)^n dx]\mathrm{d}\tau + \sqrt{\epsilon}\mathrm{d}B_\tau.
\end{aligned} \tag{9.39}
$$

In the cases $r = 1, 2$, the equation becomes

$$
\mathrm{d}X_\tau = [(1-p)(2q-1) + p\frac{\bar{\eta}_1(X_\tau+\tau) - \bar{\eta}_0(\tau - X_\tau)}{\bar{\eta}_1(X_\tau+\tau) + \bar{\eta}_0(\tau - X_\tau)}]\mathrm{d}\tau + \sqrt{\epsilon}\mathrm{d}B_\tau. \tag{9.40}
$$

It is the same as Eq. (9.10) *when* $a = 1$. **ACO if** $a = 1$ **corresponds to GA.**

If we set $r = 2$, in the GA case, the voter refers to two other voters. If the answers of the two voters are the same, the voter votes for this answer. If the answers of the two voters are split, the voter votes for each answer with a probability $1/2$. It is nothing but a genetic algorithm. In this case, the stochastic differential equation becomes Eq. (9.40). Hence, the mark increase as τ. In the original case, $r \geq 3$ and $q = 1/2$, the correct ratio does not increase and the behavior is the same as that of the symmetric Brownian motion. However, in this case, the feedback of the mark becomes the outer field and the correct ratio increases.

We can assume a stationary solution as

$$X_\infty = \bar{v}\tau + (1-p)(2q-1)\tau, \tag{9.41}$$

where $\bar{v}$ is constant. Substituting Eq. (9.41) into Eq. (9.39), **we obtain**

$$\bar{v} = -p + 2p\frac{(2n+1)!}{(n!)^2}\int_0^{\frac{\bar{\eta}_1(\bar{v}+1+(1-p)(2q-1))}{(\bar{\eta}_1+\bar{\eta}_0)+(\bar{\eta}_1-\bar{\eta}_0)(\bar{v}+(1-p)(2q-1))}} x^n(1-x)^n dx d\tau. \tag{9.42}$$

which is the self-consistent equation.

We consider the case $q = 1/2$ and $p = 1$. The self-consistent equation, Eq. (9.42) is

$$\bar{v} + 1 = 2\frac{(2n+1)!}{(n!)^2}\int_0^{\frac{\bar{\eta}_1(\bar{v}+1)}{(\bar{\eta}_1+\bar{\eta}_0)+(\bar{\eta}_1-\bar{\eta}_0)\bar{v}}} x^n(1-x)^n dx]d\tau. \tag{9.43}$$

We show the solutions of the mean field approximation in Fig. 9.3.

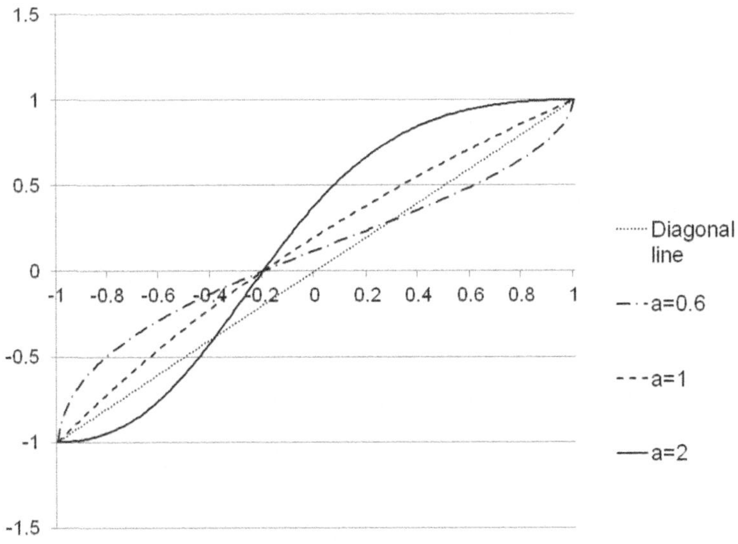

Fig. 9.3 Mean field approximation $r = 1$, $r = 3$, and $r = 5$. Here, we set $p = 1$. We use $\bar{\eta}_1 = 0.6$ and $\bar{\eta}_0 = 0.4$. The number of stable solutions is two for $r = 3$ and $r = 5$ one for $r = 1$. $r = 1$ corresponds to the standard GA, which has two parents

9.4.2 Symmetic Case

Here, we consider a symmetric model $\bar{\eta}_0 = \bar{\eta}_1 = 1/2$ and $q = 1/2$. The self-consistent equation is

$$\bar{v} = -p + 2p\frac{(2n+1)!}{(n!)^2}\int_0^{\frac{\bar{v}}{2}+\frac{1}{2}} x^n(1-x)^n dx d\tau. \quad (9.44)$$

We discuss the phase transition of this self consistent equation in Sect. 9.7 Appendix A.

9.5 Optimization

This section presents the numerical simulation results. At first 1000 independents answer the two-choice quiz randomly. In the next step, herders are introduced. A herder selects the r answers from the recent 1000 previous answers and decides the answer according to the selected answers. We have fixed the number of referred answers. Hence, this process is the equilibrium process.

We set a as the parameter for the decisions of herders. In the standard GA, the roulette selection is applied. However, here we introduce a as Eq. (9.7) in ACO,

$$\frac{(\bar{\eta}_1 k)^a}{(\bar{\eta}_1 k)^a + (\bar{\eta}_0 (t-k))^a}. \quad (9.45)$$

In the case of $a = 1$, it corresponds to the roulette selection, Eq. (9.36). a is the weight of the selection of an answer. If a is larger, the probability of the selection that has higher points increases.

Another parameter is r. Here, we perform a numerical simulation for r =2, 5, 1000. $r = 1000$ corresponds to ACO, because all previous answers are referred. The number of quizzes is 100 and the full mark is 1. We evaluate the convergence to the full mark using the number of herders t_{min}. t_{mim} is the number of herders to the first answer of the full mark. If $t_{min} < 100{,}000$, we define the process converges.

In Table 9.1, we summarize the conclusions. The convergence is fast as a and r increase. These parameters are the strength of herding and the convergence is fast

Table 9.1 Conclusions of the numerical simulations. The parameters are a and r. Most of them converge to 1

	r = 2	r = 5	r = 10	r = 1000
a = 0.5	1	No convergence	No convergence	No convergence
a = 1.0	1	1	1	1
a = 2.0	1	0.89	0.86	0.96

when the process converges. On the other hand, when $a = 2.0$ and $r = 10$, the mark is less than 1.0. This shows the convergences to the bad equilibrium for the strength of herding. In fact, here there is no independents. Therefore, it is difficult to jump from the bad equilibrium to the good equilibrium. It is called Rock in or "Rigidity" which we explained in Chap. 1.

The convergence is faster with the herding. On the other hand, strong herding causes the convergence to a bad equilibrium. It is the trade-off of social learning. The optimized parameters depend on the difficulty of the quiz. In this case, the quiz is not so difficult because of the two-choice quizzes. Therefore, the nonlinear parameters, $r = 5$, 10, and $a = 1$ show better performance.

9.6 Summary

A voting model is constructed by two types of voters—herders and independents and two candidates. The voting of independents is based on their fundamental values; on the other hand, the voting of herders is based on the number of previous votes. Herders correspond to ants that follow the pheromone trail and independents are dull ants that cannot follow the pheromone trail. There are two important parameters for ACO, power index a in Eq. (9.1) and evaporation rate $\hat{\rho}$. Although there was no evaporation at all, we show that dull ants contribute to the optimization and that this algorithm converges to the correct answers.

We show the phase transitions for the power index a. One is Ising-like phase transition. When $a > 1$, there are two stable solutions, which correspond to the correct or the incorrect answer. There are two equilibria. When $a = 1$, there is only a stable solution which corresponds to the correct answer. When $a < 1$, there is only one solution between the correct and the incorrect answer.

Another transition is super-normal transition. Here, we consider the case $p = 1$. In the case $a > 1$ there is no super-normal phase transition.

In the case $a = 1$, When $\bar{\eta}_1 > 2/3$ ($\leq 2/3$), the process converges normally (slowly). The speed of convergence depends on $\bar{\eta}_1$.

In the case $a < 1$ and $p = 0$, the process converges between the correct answer and the incorrect answers. The critical point is $a_{vc} = 1/2$. When $1 > a \geq 1/2$, the convergence is slow. In the case $1/2 > a$, the process converges normally.

Next, we show the relation among kinetic Ising model, GA and ACO. The difference is only the response function. For ACO, the power function becomes the response function. For Ising model, the tanh-type becomes the response function. For GA, the binomial distribution is the response function. The difference does not affect macro phenomena.

9.7 Appendix A Mean Field Approximation and Stochastic Differential Equation

Here we discuss the relation between the stochastic differential equation and mean field approximation. The relation between $\bar{v}$ and the voting ratio to C_1 is $\hat{v} = 2\bar{v} - 1$. We rewrite Eq. (9.13) as

$$\hat{v} = q(1-p) + p\frac{\bar{\eta}_1^a(\hat{v}+1)^a}{\bar{\eta}_1^a(\hat{v}+1)^a + \bar{\eta}_0^a(1-\hat{v})^a} \tag{9.46}$$

The first term of Eq. (9.46) is the contribution from independent voters and the second term is from herders.

References

Araújo N A M, Andrade Jr. J S, and Herrmann H J 2010 *PLOS ONE* **5** e12446

Bikhchandani S, Hirshleifer D, and Welch I 1992 *J. Polit. Econ.* **100** 992

Cont R and Bouchaud J 2000 *Macroecon. Dyn.* **4** 170

Couzin I D, Krause J, James R, Ruxton G R, and Franks N R 2002 *J. Theor. Biol.* **218** 1

Curty P and Marsili M 2006 *JSTAT* P03013

Dorigo M and Gambardella L M, Ant colonies for the travelling salesman problem. biosystems, **43(2)**, 73–81 (1997)

Dorigo M and Stutzle T *Ant Colony Optimization* (Bradford Books, 2004)

Dorigo M, Caro G D and Gambardella L M, Ant algorithms for discrete optimization, *Artificial Life*, 5(2) **5(2)** 137–172 (1999)

Eguíluz V and Zimmermann M 2000 *Phys. Rev. Lett.* **85** 5659

Galam G 1990 *Stat. Phys.* **61** 943

Hisakado M and Hino M, Between ant colony optimization and generic algorithm *J IPS Jpn* **9(3)** 8–14 (2016)

Keynes J M 1936 *General Theory of Employment Interest and Money* (Macmillan Cambridge University Press)

Maniezzo V and Colorni A, The Ant System applied to the quadratic assignment problem. *IEEE Tran on Knowledge and Data Engineering* **11(5)** 769–778 (1999)

Milgram S, Bickman L, and Berkowitz L 1969 *J. Per. Soc. Psyco.* **13** 79

Partridge B L 1982 *Sci. Am.* **245** 90

Shimomura S, Sugimoto M, Haraguchi T, Matsushita Hand Nishio Y, Ant Colony Optimization with. Intelligent and Dull Ants, *Proceedings of International Symposium on Nonlinear Theory and its Applications (NOLTA'10)* 504 (2010)

Stauffer D 2001 *Adv.Complex Syst.* **4** 19

Tarde G 1890 *Les lois de l'imitation* (Paris: Felix Alcan)

Chapter 10
Beta Binomial Distribution with Temporal Correlation

10.1 Introduction

Estimations of the probability of default (PD) and default correlation have been obtained from empirical studies on historical data from credit events. These two parameters are important for pricing financial products, such as synthetic Collateralized Debt Obligation (CDO) (Schönbucher 2003). Also called "long run PDs", these parameters are important to financial institutions for portfolio management and risk management. If the number of defaults is minimal, it is not easy to estimate these parameters (Pluto and Tasche 2011; Benjamin et al. 2001).

In this chapter, we introduce a Bayesian estimation method using the beta binomial distribution. For the usual cases, the Merton model, which incorporates the default correlation by the correlation of the asset price movements (asset correlation), is used to estimate the PD and default correlation (Merton 1974) in financial institutions. Monte Carlo simulations are necessary to estimate these parameters, except in the limit of large homogeneous portfolios, where the Merton model is used (Schönbucher 2003). In the beta binomial case, the default correlation, instead of the asset correlation, is used. Moreover, we consider a multi-term case with a temporal correlation, which refers to a time-dependent correlation (Pluto and Tasche 2011; Benjamin et al. 2001).

A phase transition occurs when the temporal correlation decays by the power-law. A power-law decay implies that PD has a long memory compared with that of exponential decay (Mantegna and Stanley 2000). When the power index is less than one, the estimator distribution of the PD does not converge to the delta function. Alternatively, when the power index is larger than one, the convergence is the same as that in the normal case. When the distribution does not converge, it is difficult to estimate the PD with limited data. The required condition for estimating PD is clarified.

The remainder of this paper is organized as follows. In Sect. 10.2, we introduce a Bayesian estimation method using the beta binomial distribution. In Sect. 10.3,

M. Hisakado, *Urn Models and Their Applications in Finance*,
https://doi.org/10.1007/978-981-96-3825-3_10

we consider the convergence of the PD estimator. In Sect. 10.4, we present the urn model for the temporal correlations. In Sect. 10.5, we study the phase transition of the Pólya urn with a discount factor using an analytical method. Finally, the summaries are presented in the last section. This chapter is based on Hisakado (2020).

10.2 Bayesian Estimation Using the Beta Binomial Distribution

We denote the PD estimation as p, and the default correlation as ρ_D, where $0 \le p \le 1$ and $0 \le \rho_D \le 1$. The distribution of p and ρ_D is $P(p, \rho_D)$. The number of obligors in the portfolio is n. p and ρ_D are estimated by a Bayesian estimation. We consider the Bernoulli random variables $X_i(s)(i = 1, 2, \ldots, n)$ that take the values 1 or 0 and s is the term. When the obligor, i, at s-th term is the default (non-default), $X_i(t) = 1(0)$ which is the Bernoulli random variables $X_i(s)$. Here, we consider the first term $s = 1$. There are n obligors in the first term. We define $N_1 = \sum_{j=1}^{n} X_j(1)$ and consider a default correlation for $X_i(1)$.

When the number of defaults is k, the Bayes formula for the posterior distribution $P(p, \rho_D|N_1 = k)$ is

$$P(p, \rho_D|N_1 = k) = \frac{P(p, \rho_D, N_1 = k)}{P(N_1 = k)} = \frac{P(N_1 = k|p, \rho_D) f(p, \rho_D)}{P(N_1 = k)}, \tag{10.1}$$

where $f(p, \rho_D)$ is the prior distribution.

We use the beta binomial distribution for $P(N_1 = k|p, \rho_D)$. The posterior distribution is given by

$$\begin{aligned} P(p, \rho_D|N_1 = k) &\propto \frac{n!}{k!(n-k)!} \frac{B(\alpha + k, n + \beta - k)}{B(\alpha, \beta)} f(p, \rho_D) \\ &\propto \frac{\Gamma(\alpha + k)}{\Gamma(\alpha)} \frac{\Gamma(n + \beta - k)}{\Gamma(\beta)} \frac{\Gamma(\alpha + \beta)}{\Gamma(\alpha + \beta + n)} f(p, \rho_D), \end{aligned} \tag{10.2}$$

where, $p = \frac{\alpha}{\alpha+\beta}$ and $\rho_D = \frac{1}{\alpha+\beta+1}$. Hence, we obtain the relations $\alpha = p\frac{1-\rho_D}{\rho_D}$ and $\beta = (1-p)\frac{1-\rho_D}{\rho_D}$. Here, we use the beta function $B(\alpha, \beta) = \Gamma(\alpha)\Gamma(\beta)/\Gamma(\alpha + \beta)$.

We consider the maximum a posteriori (MAP) estimation of Eq. (10.2). When the prior function $f(p, \rho_D)$ is a constant function, the maximum point is

$$\begin{aligned} &\frac{\partial P(p, \rho_D|N_1 = k)}{\partial p} \\ &\propto \frac{(1-\rho_D)}{\rho_D} \frac{\Gamma(\alpha + k)}{\Gamma(\alpha)} \frac{\Gamma(n + \beta - k)}{\Gamma(\beta)} (\varphi(\alpha + k) - \varphi(\alpha) - \varphi(\beta + n - k) + \varphi(\beta)) \\ &= \frac{(1-\rho_D)}{\rho_D} \frac{\Gamma(\alpha + k)}{\Gamma(\alpha)} \frac{\Gamma(n + \beta - k)}{\Gamma(\beta)} (\sum_{i=1}^{k} \frac{1}{\alpha + i - 1} - \sum_{i=1}^{n-k} \frac{1}{\beta + i - 1}) = 0, \end{aligned} \tag{10.3}$$

where $\varphi(x)$ is the digamma function, $\varphi(x) = \Gamma'(x)/\Gamma(x)$.

The summation from $i = 1$ to k is a monotonously decreasing function of p because α increases, while the second summation in Eq. (10.3) is a monotonously increasing function about p because β decreases. When $p \sim 0$, the difference between the two summations is positive. Conversely, when $p \sim 1$, the difference between the two summations becomes negative. Hence, the function $P(p|N_1 = k, \rho_D)$ has one peak in the range $0 < p < 1$. The multi-term case is presented in Sect. 10.7 Appendix A.

Next, we consider the variable ρ_D. The maximum point is

$$\frac{\partial P(p, z|N_1 = k)}{\partial z} \propto \frac{\Gamma(\alpha + k)}{\Gamma(\alpha)} \frac{\Gamma(N + \beta - k)}{\Gamma(\beta)} \frac{\Gamma(\alpha + \beta)}{\Gamma(\alpha + \beta + n)} \left(\sum_{i=1}^{k} \frac{p}{pz + i - 1} + \sum_{i=1}^{k} \frac{1 - p}{(1 - p)z + i - 1} - \sum_{i=1}^{k} \frac{1}{z + i - 1}\right) = 0,$$

where $z = (1 - \rho_D)/\rho_D$.

All summations in the last term with parenthesis are monotonously decreasing functions about z. When $z \sim 0$, the last term becomes positive. Conversely, when $z >> 1$, the last term becomes 0. When $(k - 1)/n \leq p \leq k/n$ or is sufficiently close to this condition, the last term becomes positive. $(k - 1)/p \leq n - 1$ and $(n - k - 1)/(1 - p) \leq n - 1$ become $(k - 1)/(n - 1) \leq p \leq k/(n - 1)$. In this case, the last term increases monotonously, and the peak is $z = \infty$ and $\rho_D = 0$. This implies that the optimization of ρ_D is zero for the single-term model. When p is not sufficiently close to $(k - 1)/n \leq p \leq k/n$, the last term changes from positive to negative as z increases. Therefore, a peak occurs in $P(p, z|N_1 = k)$.

We extend this method to a multi-term case. There are n_i obligors in term i and k_i defaults occur. The prior distribution for the second term is the posterior distribution, which is calculated from the first termâĽ™s data. In this way, the posterior distribution is updated in every term. We write the posterior distribution $P(p, \rho_D|k_1, k_2)$ as

$$P(p, \rho_D|N_1 = k_1, N_2 = k_2) = \frac{P(k_2|p, \rho_D, k_1)}{P(k_2)} \frac{P(k_1|p, \rho_D) f(p, \rho_D)}{P(k_1)}. \quad (10.4)$$

It is natural to assume that the number of defaults in the current term is affected by the number of defaults in previous terms, thus the defaults have a temporal correlation. When the default rate is high (low), it is reasonable to assume that the default rate will be high (low) in the next term. This is similar to volatility clustering, [1] which also has a long memory (Engle 1982; Bollerslev 1986). We confirm this using empirical data in the following sections.

We introduce the temporal correlation by adjusting α and β, and consider the j th term. The numbers of obligors and defaults in the j th term are n_j and k_j. In the same term, the correlation is ρ_D. We set the temporal correlation parameters between the i and j th terms; $\hat{d}_{i-j}$ and $j < i$. α and β are adjusted to $\alpha + \sum_{j=1}^{i-1} \hat{d}_{i-j} k_j$ and

[1] The terms with higher volatility continues in market.

$\beta + \sum_{j=1}^{i-1} \hat{d}_{i-j}(n_j - k_j)$ (Mori 2018). This implies that the previous terms' data affect the present defaults. It is easy to confirm that $\hat{d}_i = 1$ indicates that the data are correlated with ρ_D. When $\hat{d}_i = 0$, the data are independent each term.

10.3 Correlation Decay

In the previous section, $\hat{d}_i$ was introduced to represent the temporal correlation. In this section, to clarify the behavior of the parameter $\hat{d}_i$, where $i = 1, 2, \ldots, T$ and $\hat{d}_0 = 1$, the variance of the stochastic process is considered. In each term, diffusion has n_i steps and k_i defaults, where $i = 1, 2, \ldots, T$.

The adjustments related to parameters α and β are the effects of the temporal correlation from the previous conclusions. We shrink the previous terms' conclusions and add them to the initial parameters for the adjustment process. The shrinking ratio for interval i is $\hat{d}_i$.

The two term model is examined first. We consider the relation between the first and second terms. n_1 and k_1 are the number of obligors and defaults, respectively, in the first term. The second term's parameters become $\alpha + \hat{d}_1 k_1$ and $\beta + \hat{d}_1(n_1 - k_1)$. We consider the shrinking processes from α to $\alpha + \hat{d}_1 k_1$ and β to $\beta + \hat{d}_1(n_1 - k_1)$. The variance of the second term of the process is $n_1 \hat{d}_1 pq + \hat{d}_1 n_1 (n_1 - 1) pq\rho_D$, where $q = 1 - p$; that is, we approximate $\hat{d}_1 B_{\alpha,\beta}(k_1, n_1 - k_1) \sim B_{\alpha,\beta}(\hat{d}_1 k_1, \hat{d}_1 n_1 - \hat{d}_1 k_1)$ where $B_{\alpha,\beta}$ is the beta binomial distribution with parameters α and β. We approximate this variance by $n_1 \hat{d}_1 pq + \hat{d}_1 n_1 (\hat{d}_1 n_1 - 1) pq\rho_D$, and the difference becomes $n_1^2 pq\rho_D \hat{d}_1(1 - \hat{d}_1) \geq 0$. Hence, the approximation is exact when $\hat{d}_1 = 0, 1$ or $\rho_D = 0$. However, if $\hat{d}_1 \sim 0, 1$ or $\rho_D \sim 0$, this approximation can be used. In other cases, the real variance is larger than the approximation. We use this approximation to study the meaning of this process.

For the defaults of the obligors, the hypothesis $\hat{d}_1 \sim 0$ or 1 and $\rho_D \sim 0$ is given. In other words, the temporal correlation is either a high or low case, or a low correlation case. Hereafter, we use this approximation to calculate the variance of this process.

We extend the stochastic process to the multi-term case. Let $\{U_{\hat{t}}; \hat{t} \geq 1\}$ be an independent and identically distributed (i.i.d.) sequence that is uniformly distributed on [0,1]. We consider the $t + 1$-th term $m + 1$-th obligor. Here, we introduce the new Bernoulli variable $X(\hat{t})$ instead of $X_m(t)$, where

$$\hat{t} = \sum_{j=1}^{t} n_j + m. \tag{10.5}$$

$\hat{t}$ is the serial number of the process. The $\hat{t} + 1$-th discrete dynamics of the process is described by:

$$X(\hat{t} + 1) = \mathbf{1}_{U_{\hat{t}+1} \leq Z_d(\hat{t})}, \tag{10.6}$$

when $\sum_{j=1}^{t} n_j + 1 \leq \hat{t} \leq \sum_{j=1}^{t+1} n_j$.

Here $Z_d(\hat{t})$ is given by

$$Z_d(\hat{t}) \equiv \frac{\alpha + \sum_{s=\sum n_j+1}^{\hat{t}} X(s) + \sum_{j=1}^{t} \hat{d}_{t-j} k_j}{\alpha + \beta + m + \sum_{j=1}^{t} \hat{d}_{t-j} n_j}. \tag{10.7}$$

The expected value of $X(t)$ is $\mathrm{E}(X(t)) = \alpha/(\alpha + \beta)$. When $\hat{d}_i = 1$, the process is the beta binomial process.

We consider the relationship between the term i and $i + 1$. The distribution of the term i is a beta binomial distribution. Hence, the conditional variance, V_{i+1}, of term $i + 1$ can be evaluated, using the above approximation, as

$$\begin{aligned} V_{i+1} &\sim \sum_{j=1}^{i+1} n_j \hat{d}_{i+1-j} pq + (\sum_{j=1}^{i+1} n_j \hat{d}_{i+1-j})(\sum_{j=1}^{j+1} n_j \hat{d}_{i+1-j} - 1) pq \rho_D \\ &\quad - \sum_{j=1}^{i} n_j \hat{d}_{i+1-j} pq - (\sum_{j=1}^{i} n_j \hat{d}_{i+1-j})(\sum_{j=1}^{i} n_j \hat{d}_{i+1-j} - 1) pq \rho_D \\ &= pq n_{i+1} + pq n_{i+1}(n_{i+1} - 1)\rho_D + 2pq\rho_D n_{i+1} \sum_{j=1}^{i} n_j \hat{d}_{i+1-j}. \end{aligned} \tag{10.8}$$

Therefore, the difference between the summations $\sum_{j=1}^{i+1} n_j \hat{d}_{i+1-j}$ and $\sum_{j=1}^{i} n_j \hat{d}_{i+1-j}$ corresponds to the variance of $(i + 1)$-th step. Therefore, using this approximation, the correlation between the i-th and j-th terms is approximated by $\rho_D \hat{d}_{i-j}$. The term $\hat{d}_{i-j}$ plays the role of a discount factor in the correlation ρ_D. It can be seen that as time progresses, the correlation is discounted. It is reasonable to assume a monotonically decreasing function for $\hat{d}_i$ because the effects decrease as the distance between i and j increases.

The total variance of diffusion is approximated by

$$V \sim \sum_{i=1}^{T} pq n_i + \sum_{i=1}^{T} pq n_i (n_i - 1)\rho_D + 2pq\rho_D \sum_{i>j}^{T} n_i n_j \hat{d}_{i-j}. \tag{10.9}$$

The first, second, and third terms correspond to the variance of the binomial distribution, constant correlation ρ_D in the portfolio, and temporal correlation, respectively.

In summary, when $\hat{d}_i \sim 0, 1$ or $\rho_D \sim 0$, the correlation between term i and term j is approximated by

$$\Sigma_D \sim \rho_D \begin{pmatrix} 1 & \hat{d}_1 & \hat{d}_2 & \cdots & \hat{d}_T \\ \hat{d}_1 & 1 & \hat{d}_1 & \ddots & \vdots \\ \ddots & \ddots & \ddots & \ddots & \ddots \\ \vdots & \ddots & \ddots & \ddots & \hat{d}_1 \\ \hat{d}_T & \cdots & \hat{d}_2 & \hat{d}_1 & 1 \end{pmatrix}.$$

It is the correlation matrix of the default correlation. The average PD, the correlation of the Bernoulli random variables, and the temporal correlation using this approximation are p, ρ_D, and $\hat{d}_i$, respectively.

In the Bayesian estimation, if the scaled variance converges as the data increase, these parameters can be estimated correctly. Conversely, if the variance does not converge, the parameters cannot be estimated. It should also be considered whether the process has a stationary solution, which will be discussed in the following sections.

It is difficult to estimate all $\hat{d}_i$ values because of the limited data. By introducing a prior distribution for $\hat{d}_i$, the estimation becomes a hierarchical Bayesian estimation. It is reasonable to assume that the prior distribution is a monotonically decreasing function. Therefore, we considered two temporal correlations, an exponential and power decay, to have a long memory.

10.4 Urn Model with the Temporal Correlation

In this section, we consider a multi-term urn process with correlations. In this section, we consider the case, $n_j = N, j \geq 1$. In the first term, the urn contains θ_0 red balls and $n_0 - \theta_0$ white balls. Then, the balls are sequentially taken out from the urn. For example, a single ball is taken out, after which we return the ball to the urn and add ω balls of the same color to the urn. Thus, the total number of balls increases by ω a process, which is a correlation parameter in the same term. In fact, as we take out more red balls, the probability that a red ball is taken out increases because the number of red balls in the urn increases. This is the correlation in the same term. We repeat the process N times in the first term. Hence, the number of the total balls is $n_0 + N\omega$ in the end of the first term. This is the Pólya urn model, with a beta binomial distribution (BBD).

Next, we extend the process to multi-term process. We repeat the same process as the first process with the updated parameter θ_t at the $t + 1$-th term. In the $t + 1$-th term , the urn contains $\theta_t = \theta_0 + \sum_i^t k_i \hat{d}_{t-i+1}\tilde{\omega}$ red balls, and the number of white balls is $n_0 - \theta_0 + \sum_i^t (N - k_i)\hat{d}_{t-i+1}\tilde{\omega}$. Here, the total number of initial balls in the first term is n_0, $\hat{d}_i$ denotes the kernel function or discount factor that represents the decrease in the effects of self-excitation. Note that θ_t depends on the history of the number of the red balls taken out. For the normalization we set $\hat{d}_0 = 1$. Here, we introduce the variable $N(i)$ which is the number of red balls taken out in the i-th term, and its value is $N(i) = k_i$. $\tilde{\omega}$ is the parameter for temporal correlation, where

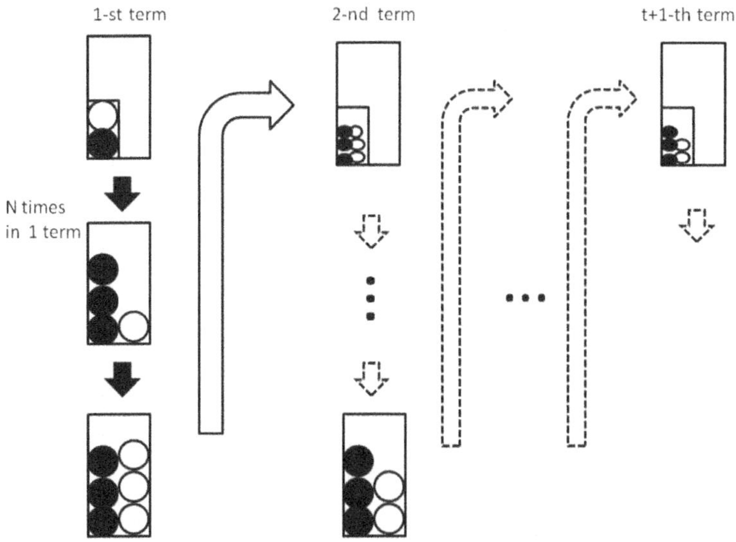

Fig. 10.1 Illustration of the multi-term urn process. The process in each term is a Pólya urn model. In each term, we repeat the process where we take out a ball and place back ω same collar balls, N times. In this figure, two balls are taken out in one term, $N = 2$. The process is presented by the black arrow. The conclusion of each term affects the initial condition of the next term as the temporal correlation. It corresponds to the white arrow. The total number of balls in the initial condition of each term is n_0 and the number of red balls in the initial condition of t-th term is $\theta_t = \theta_0 + \sum_i^t k_i \hat{d}_{t-i+1}\tilde{\omega}$ and the number of white balls is $n_0 - \theta_0 + \sum_i^t (N - k_i)\hat{d}_{t-i+1}\tilde{\omega}$. $\hat{d}_i$ denotes the kernel function or discount factor. The number of red balls taken out in the i-th is $N(i) = k_i$. $\tilde{\omega}$ is the scale factor for the adjustment for the temporal correlation

ω denotes the scale parameter of the added red balls. $\tilde{\omega}$ is the ratio of the number of balls placed back to the urn to the number of balls drawn from the urn, when $\hat{d}_0 = 1$ and $\hat{d}_i = 0, i \geq 1$. In this case only the previous term affects this term. In summary, we update only the parameter θ_t and other parameters are fixed in each term.

As the first term, we sequentially take out balls from the urn. After that, we return the taken out ball to the urn and ω balls of the same color are added to the urn. We repeat the process N times in this $t + 1$-th term. This is the definition of one term. After the process of $t + 1$ term, we set the initial condition of $t + 2$-term and continue the $t + 2$-th process. Each term is also the Pólya urn model. In this model, we use two types of correlations: the first is a correlation in the same term using the parameter ω, and the second one is the temporal correlation using the parameters $\tilde{\omega}$ which is independent from ω and $\hat{d}_i$, which is the kernel function. The temporal correlation decays as a function of time using the parameter $\hat{d}_i$ (Fig. 10.1).

10.5 Phase Transition in the Estimation of PD

In this section, we determine whether the PD in the Bayesian estimation converges. To simplify the model, we set $n_j = 1, j \geq 1$ in Eq. (10.7). This does not affect the outcome of the PD estimation. Let $\{U_t; t \geq 1\}$ be an independent and identically distributed (i.i.d.) sequence that is uniformly distributed on [0,1]. The discrete dynamics of the process are described by:

$$X(t+1) = \mathbf{1}_{U_{t+1} \leq Z_d(t)}.$$

Here, $Z_d(t)$ is the weighted sum of $X(s), s \leq t$ with the discount factor $\hat{d}_{t-s}$,

$$Z_d(t) \equiv \frac{\alpha + \sum_{s=1}^{t} X(s)\hat{d}_{t-s}}{\alpha + \beta + \sum_{s=1}^{t} \hat{d}_{t-s}}. \tag{10.10}$$

This is the Pólya urn model with discount factor $\{\hat{d}_i\}$.

The expected value of $X(t)$ is $\mathrm{E}(X(t)) = \alpha/(\alpha+\beta)$. The PD estimator is $Z(t)$,

$$Z(t) \equiv \sum_{s=1}^{t} X(s)/t.$$

The success of the PD estimation depends on the behavior of the variance of $Z(t)$. Specifically, if the variance of $Z(t)$ converges, then the PD can be estimated.

10.5.1 *Stochastic Differential Equation and Mean Field Approximation*

First, the stochastic process is rewritten using $c_1(t) = \sum_{s=1}^{t} X(s)$;

$$\begin{aligned} c_1(t) = k \rightarrow k+1 : P_{k,t} &= \frac{\alpha + \sum_{s=1}^{t} X(s)\hat{d}_{t-s}}{\alpha + \beta + \sum_{s=1}^{t} \hat{d}_{t-s}}, \\ c_1(t) = k \rightarrow k : Q_{k,t} &= 1 - P_{k,t}, \end{aligned} \tag{10.11}$$

where $P_{k,t}$ and $Q_{k,t}$ are the process probabilities. The sum of $P_{k,t}$ and $Q_{k,t}$ is 1.

For convenience, we define a new variable Δ_t such that

$$\Delta_t = 2c_1(t) - t. \tag{10.12}$$

We change the variables from k to u and $c_1(s)$ to $\Delta_s = 2X(s) - 1$. Given $\Delta_t = u$, we obtain the following random walk model:

$$\Delta_t = u \to u+1 : P_{u,t} = \frac{\alpha + \sum_{s=1}^{t} \hat{d}_{t-s}(x_s+1)/2}{\alpha + \beta + \sum_{s=1}^{t} \hat{d}_{t-s}},$$
$$\Delta_t = u \to u-1 : Q_{u,t:s,t-r} = 1 - P_{u,t}.$$

We now consider the continuous limit $\epsilon \to 0$,

$$X_\tau = \epsilon \Delta_{[t/\epsilon]},$$
$$P(x,\tau) = \epsilon P(\Delta_t/\epsilon, t/\epsilon), \tag{10.13}$$

where $\tau = t/\epsilon$ and $x = \Delta_t/\epsilon$. By approaching the continuous limit, we obtain the following stochastic partial differential equation:

$$\mathrm{d}X_\tau = \frac{\alpha - \beta + \int_{\sigma=1}^{\tau} d(\tau-\sigma)\mathrm{d}X_\sigma}{\alpha + \beta + \int_1^\tau d(\tau-\sigma)\mathrm{d}\sigma}\mathrm{d}\tau + \sqrt{\epsilon}\mathrm{d}B_\tau, \tag{10.14}$$

where $d(\tau)$ is the continuous function of $\hat{d}_t$, the discount factor, and $\mathrm{d}X_\tau = \epsilon x_{[t/\epsilon]}$.

We are interested in the behavior of X_τ in the limit $\tau \to \infty$. We assume that the stationary solution is

$$X_\infty = \bar{v}\tau, \tag{10.15}$$

where $\bar{v}$ is a constant. By substituting Eq. (10.15) into Eq. (10.14), we obtain

$$\bar{v} = \frac{\alpha - \beta + \bar{v}\hat{T}}{\alpha + \beta + \hat{T}}, \tag{10.16}$$

where $\hat{T} = \lim_{\tau\to\infty} \int_1^\tau d(\tau-\sigma)\mathrm{d}\sigma$. **Eq.** (10.16) **is a self-consistent equation.** When $\hat{T} < \infty$, Eq. (10.16) is solved when $\bar{v} = (\alpha-\beta)/(\alpha+\beta)$. The process converges to the average point. On the other hand, when $\hat{T} \to \infty$, we can obtain the identity equation $\bar{v} = \bar{v}$, meaning that the process does not converge to the delta function. The expected value of X_s is $(\alpha-\beta)/(\alpha+\beta)$. Hence, the phase transition at the point $\hat{T}$ diverges to infinity. When the distribution does not converge, we cannot estimate the parameters correctly, even if the number of data increases. This is a critical issue when using Bayesian estimation. In other words, $\hat{T} < \infty$ is a compulsory condition for parameter estimation.

This phase transition is considered the transition between the self-averaging process and the non-self-averaging process. When $\hat{T} < \infty$, it is the self-averaging process because the distribution converges to the delta function. Conversely, when $\hat{T}$ diverges to infinity, it is a non-self-averaging process (Kadanoff 2000). When the self-averaging process, we can estimate the parameters.

We show Eq. (10.16) in Fig. 10.2. (a) is the case $\hat{T} < \infty$ and (b) is the case $\hat{T} \to \infty$. In the case of (a), there is the solution, $\bar{v} = (\alpha-\beta)/(\alpha+\beta)$. On the other hand, in the case of (b), the solutions are on the line. In this case, the solutions are spread over the range.

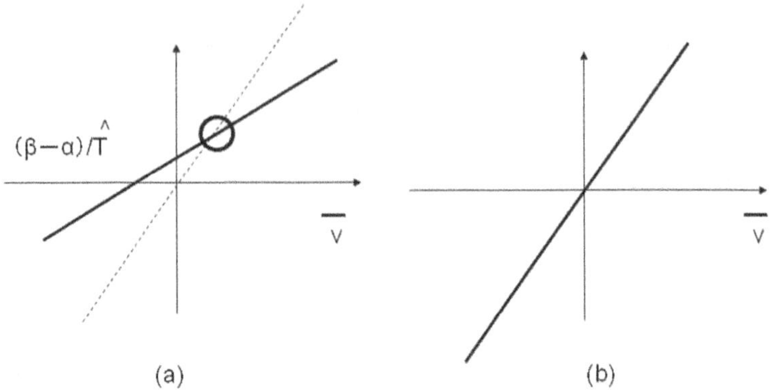

Fig. 10.2 Solutions of the self-consistent equation (10.16). (**a**) $\hat{T} < \infty$ and (**b**) $\hat{T} \to \infty$. In the case of (**a**), there is a solution, $\bar{v} = 1 - 2q$. On the other hand, in the case of (**b**), the solutions are spread on the line

10.5.2 Correlation Function

To understand the phase transition, we investigated the correlation function, $C(t)$. $C(t)$ is defined as the correlation between $X(1)$ and $X(t)$ such that

$$C(t) \equiv \mathrm{E}(X(t+1)|X(1)=1) - \mathrm{E}(X(t+1)|X(1)=0) = \frac{\mathrm{Cov}(X(1), X(t+1))}{\mathrm{V}(X(1))}. \tag{10.17}$$

The function $C(t)$ represents the propagation of the memory of $X(1)$ to the later variables $X(t+1)$.

Using $X(t+1) = \mathbf{1}_{U_{t+1} \leq Z_d(t)}$, we obtain the next relation for the conditional expectation value of $X(t+1)$ with the condition $X(1) = x$, $\mathrm{E}(X(t+1)|X(1)=x)$, as

$$\mathrm{E}(X(t+1)|X(1)=x) = \frac{\alpha + \sum_{s=1}^{t} \mathrm{E}(X(s)|X(1)=x)\hat{d}_{t-s}}{\alpha + \beta + \sum_{s=1}^{t} \hat{d}_{t-s}}.$$

As $C(t) = \mathrm{E}(X(t+1)|1) - \mathrm{E}(X(t+1)|0)$, we obtain the following recursive relation for $C(t)$ as

$$C(t) = \frac{\sum_{s=1}^{t} C(s-1)\hat{d}_{t-s}}{\alpha + \beta + \sum_{s=1}^{t} \hat{d}_{t-s}}. \tag{10.18}$$

This recursive relation contains all the information about the asymptotic behavior of $C(t)$. If we assume a functional form for $\hat{d}_i$ with the initial condition $C(0) = 1$, we can estimate $C(t)$ for $t \geq 1$.

Exponential Decay Case

We consider the exponential decay case, $\hat{d}_i = r^i, r \leq 1$. $\hat{T}$ is finite, and there is no phase transition. We decompose the numerator of Eq. (10.18) as $C(t-1) + \sum_{s=1}^{t-1} C(s-1)r^{t-s}$. We rewrite the second term using Eq. (10.18) for $t-1$ as

$$\sum_{s=1}^{t-1} C(s-1)r^{t-s} = r\sum_{s=1}^{t-1} C(s-1)r^{t-1-s} = r \cdot C(t-1)(\alpha + \beta + \sum_{s=1}^{t-1} r^{t-1-s}).$$

We then obtain the next recursive relation for $C(t)$:

$$C(t) = \frac{1 + r(\alpha + \beta + \sum_{s=1}^{t-1} r^{t-1-s})}{\alpha + \beta + \sum_{s=1}^{t} r^{t-s}} C(t-1). \tag{10.19}$$

As we are interested in the asymptotic behavior of $C(t)$, we estimate the decay rate, r_{eff}, with $C(t) \sim r_{eff}^t$, which gives

$$r_{eff} \equiv \lim_{t\to\infty} C(t)/C(t-1) = r + \frac{1-r}{(\alpha+\beta)(1-r)+1} < 1, \tag{10.20}$$

where $r_{eff} < 1$ for $r < 1$, and $C(t)$ decays exponentially.

Numerical studies of the system were conducted. To estimate $C(t)$, the recursive relation of Eq. (10.18) is solved for $t \leq 2 \times 10^5$. A Monte Carlo sampling procedure is adopted for the variance of $Z(t)$. We obtained 10^4 sample sequences for $\{X(t)\}, t = 1, \ldots, 2 \times 10^5$ and estimated the variance of $Z(t)$. Figure 10.3(a) represents $C(t)$ versus t. It is clearly shown that $C(t)$ decays exponentially. Figure 10.3(b) shows the plot of V$(Z(t))$ versus t. For all $r < 1 \in \{0.8, 0.9, 0.99\}$, $V(Z(t))$ decays as $1/t$. When $r = 1$, the $Z(t)$ distribution converges to the beta distribution. Hence, there is no phase transition for $r < 1$.

Power-Law Decay Case

For the case of power-law decay, namely $\hat{d}_i = \frac{1}{(1+i)^{\gamma}}$, when $\gamma > 1$ and $\hat{T} < \infty$, the process converges to the delta function. In contrast, when $\gamma \leq 1$ and $\hat{T}$ goes to infinity, the process does not converge.

The behaviors of $C(t)$ and $V(Z(t))$ were investigated using the numerical method, in the same manner as in the exponential decay case. Figure 10.4(a) shows the double logarithmic plot of $C(t)$ versus t. It can be seen that $C(t)$ decays in the power-law form for $\gamma \in \{1.5, 2, 3\}$. For small γ, such as $\gamma = 0.5, 0.1$, the slope is extremely small. Figure 10.4(b) shows the double logarithmic plot of $V(Z(t))$ versus t. For $r = 3.0, 2.0, 1.5$, $V(Z(t))$ decays as $1/t$. At $\gamma = 1$, the slope of the decay is less than one. For $r < 1$, the curve is concave down. These results indicate the validity of the self-consistent equation analysis.

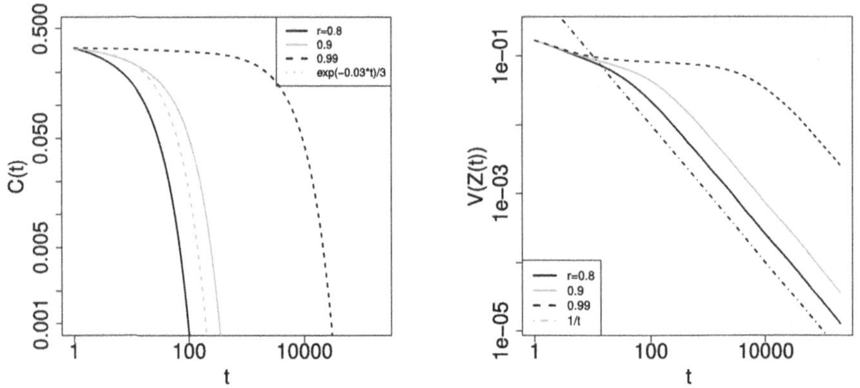

(a) (b)

Fig. 10.3 Plots of (**a**) $C(t)$ and (**b**) $V(Z(t))$ versus t, for $r \in \{0.8, 0.9, 0.99\}$. For comparison, $\exp(-0.03t)/3$ and $1/t$ are given in (**a**) and in (**b**), respectively

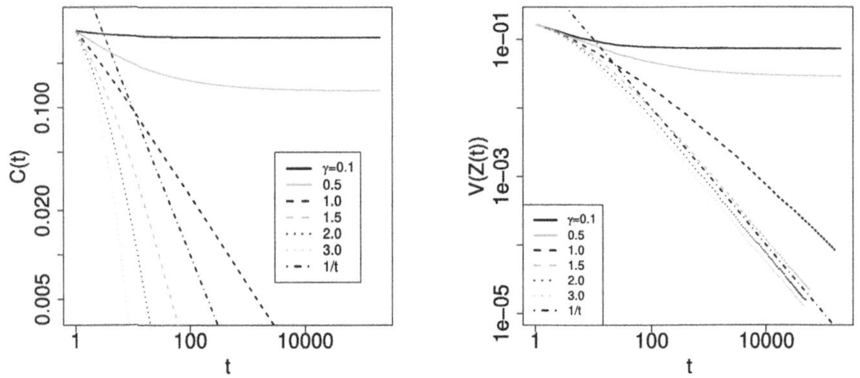

(a) (b)

Fig. 10.4 Plots of (**a**) $C(t)$ and (**b**) $V(Z(t))$ versus t, for $\gamma \in \{3.0, 2.0, 1.5, 1.0, 0.5, 0.1\}$

10.5.3 Impact Analysis

In Fig. 10.5, we show the conditional distributions, $P(X/t|X(1) = 0)$ and$P(X/t|X(1) = 1)$ for the power decay case. The initial condition affects in $\gamma \leq 1$. On the other hand $\gamma > 1$, these conditional distributions are the same. This is the impact analysis of the initial condition and shows the phase transition from self-average to non-self-average process at $\gamma = 1$.

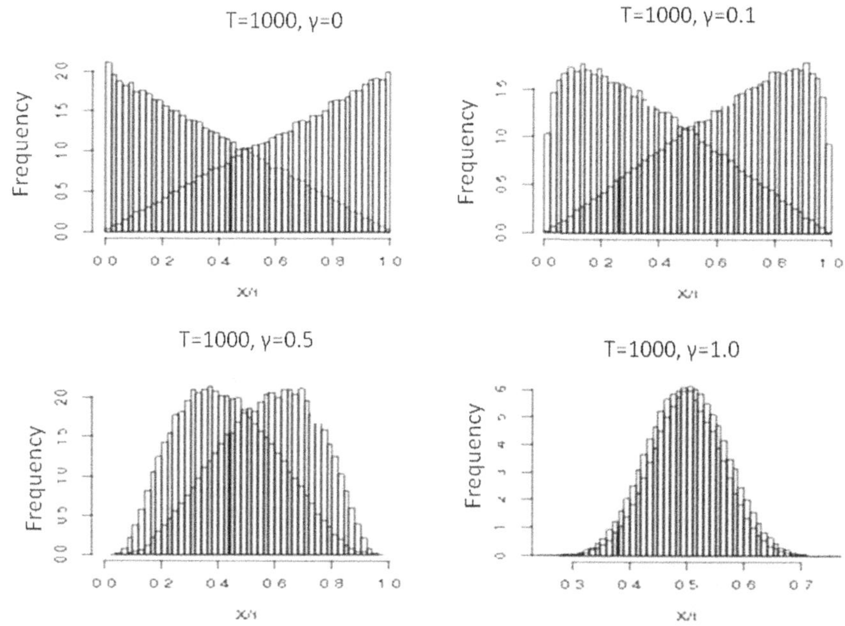

Fig. 10.5 The conditional distributions, $P(X/t|X(1) = 0)$ and $P(X/t|X(1) = 1)$. The initial condition affects the distribution in the case $\gamma \leq 1$

10.6 Summary

In this chapter, we introduced a hierarchical Bayesian estimation method using the beta binomial distribution (BBD) to estimate the parameters, probability of default (PD), and default correlation. Moreover, we considered a multi-term case with the temporal correlation. We confirmed the phase transitions when the temporal correlation decayed by a power curve, indicating that the correlation had a long memory. Conversely, in the case of exponential decay, there was no phase transition. When the power index, γ, was above or equal to one, the estimator distribution of the PD converged. Conversely, when the power index was less than 1, the distribution did not converge. We call this phase transition a "short memory-long memory transition". In summary, the condition for the estimating of parameters is $\hat{T} = \lim_{\tau \to \infty} \int_1^\tau d(\tau - s)\mathrm{d}s < \infty$. By sufficient data collection, these parameters can be estimated correctly.

10.7 Appendix A Extension of Eq. (10.2)

We extend the maximum a posteriori (MAP) estimation, Eq. (10.2), for the multi-term case. The numbers of obligors and defaults in the j-th term are n_j and k_j, respectively. When a prior function, $f(p, \rho_D)$, is a constant function, the maximum point is

$$\frac{\partial P(p, \rho_D|n_1, \ldots, n_T, k_1, \ldots, k_T)}{\partial p} \propto \frac{(1-\rho_D)}{\rho_D} \frac{\prod_{j=1}^{T} \Gamma(\alpha_j + k_j)}{\prod_{j=1}^{T} \Gamma(\alpha_j)} \frac{\prod_{j=1}^{T} \Gamma(n_j + \beta_j - k_j)}{\prod_{j=1}^{T} \Gamma(\beta_j)}$$
$$\times (\sum_{j=1}^{T} \{\varphi(\alpha_j + k_j) - \varphi(\alpha_j) - \varphi(\beta_j + n_j - k_j) + \varphi(\beta_j)\})$$
$$= \frac{(1-\rho_D)}{\rho_D} \frac{\prod_{j=1}^{T} \Gamma(\alpha_j + k_j)}{\prod_{j=1}^{T} \Gamma(\alpha_j)} \frac{\prod_{j=1}^{T} \Gamma(n_j + \beta_j - k_j)}{\prod_{j=1}^{T} \Gamma(\beta_j)}$$
$$\times \{\sum_{j=1}^{T} (\sum_{i=1}^{k_j} \frac{1}{\alpha_j + i - 1} - \sum_{i=1}^{n_j - k_j} \frac{1}{\beta_j + i - 1})\} = 0, \quad (10.21)$$

where $\varphi(x)$ is the digamma function.

α_j and β_j are the adjusted α and β. $\alpha_j = \alpha + \sum_{l=1}^{j-1} \hat{d}_{j-l} k_l$ and $\beta_j = \beta + \sum_{l=1}^{j-1} \hat{d}_{j-l}(n_l - k_l)$. The first term in the last set of parentheses in Eq. (10.21) is a monotonously decreasing function of p because α increases. The second term in the last set of parentheses increases monotonously about p because β decreases. When $p \sim 0$, the difference between the two terms is positive because $\alpha_1 = \alpha$. In contrast, when $p \sim 1$, the difference between the two terms becomes negative because $\beta_1 = \beta$. Hence, the function, $P(p|N_1 = k, \rho_D)$, has one peak in the range $0 < p < 1$.

References

Benjamin N, Cathcart A, and Ryan K, *Low Default Portfolios: A Proposal for Conservative Estimation of Default Probabilities* (Financial Services Authority, 2006).

Bollerslev T, Generalized autoregressive conditional heteroskedasticity, J. Econometrics **31(3)**, 307327 (1986).

Engle R F, Autoregressive Conditional Heteroscedasticity with Estimates of the Variance of United Kingdom Inflation, Econometarica **50(4)**, 987–1007 (1982).

Hisakado M and Mori S, Phase transition in the Bayesian estimation of the default portfolio, Physica A: Statistical Mechanics and its Applications, **544**, 123480 (2020).

Kadanoff L, Statistical Physics, World Scientific Publishing (2000).

Mantegna N M and Stanley H E, *Introduction to Econophysics: Correlations and Complexity in Finance* (Cambridge University Press, 2000).

Merton R C, On the Pricing of Corporate Debt: The Risk Structure of Interest Rates, J. Fin. **29(2)**, 449–470 (1974).

Mori S, Hisakado M, and Nakayama K, Mean field vector model on networks and multi-variate beta distribution, Phys. Rev E, **99** 52307 (2018).

Pluto K and Tasche D, *Estimating Probabilities of Default for Low Default Portfolios* In: Engelmann B., Rauhmeier R. (eds) The Basel II Risk Parameters. Springer, Berlin, Heidelberg (2011).

Schönbucher P J, *Cresit Derivatives Pricing Models:Models, Pricing, and Inplementation* (John Wiley & Sons, Ltd. 2003).

Chapter 11
Merton Model with Temporal Correlation

11.1 Introduction

In this chapter, we study a Bayesian estimation method using the Merton model. Under normal circumstances, the Merton model incorporates the default correlation by the correlation of asset price movements (asset correlation), which is used for estimating the PD and the correlation. A Monte Carlo simulation is an appropriate tool to estimate parameters, except under the limit of large homogeneous portfolios (Schönbucher 2003). In this case, the distribution becomes a Vasicek distribution that can be calculated analytically (Vasicek 2002).

In Chap. 10, we discussed parameter estimation using a beta binomial distribution with the default correlation and considered a multi-term case with a temporal correlation. A non-equilibrium phase transition, similar to that of the Ising model, occurs when the temporal correlation decays by a power law. In this study, we discuss the phase transition when using the Merton model. When the power index is less than one, the estimator distribution of the PD converges slowly to the delta function. In contrast, when the power index is greater than one, the convergence is the same as that in the normal case. When the distribution converges slowly, estimation of the long-run PD with limited data becomes time-consuming.

To confirm the decay form of the temporal correlation, we investigate the empirical default data. We verify the estimation of the power index in the slow-convergence range. This demonstrates that even if adequate historical data are available, accurate estimation of PD parameters, asset correlation, and temporal correlation is time-consuming.

The remainder of this paper is organized as follows. In Sect. 11.2, we introduce the stochastic process of the Merton model and consider the convergence of the PD estimator. In Sect. 11.3, we describe the application of the Bayesian estimation approach to the empirical data of the default history and confirm its parameters. We use Merton model and BBD as the models. The estimated parameter is in the slow-convergence phase. Finally, the summary is presented in Sect. 11.4. This chapter is based on Hisakado et al. (2022).

M. Hisakado, *Urn Models and Their Applications in Finance*,
https://doi.org/10.1007/978-981-96-3825-3_11

11.2 Asset and Default Correlations

11.2.1 Varince and Asset Correaltion

In this section, we consider whether the time series of a stochastic process using the Merton model converges (Merton 1974). We demonstrate that convergence is intimately related to the phase transition. Using this conclusion, we discuss the possibility of estimating parameters.

Normal random variables, S_t, are hidden variables that explain the status of economics, and S_t affects all obligors in the t-th term. To introduce the temporal correlation of the defaults from different terms, let $\{S_t, 1 \le t \le T\}$ be the time series of the stochastic variables of the correlated normal distribution with the following correlation matrix:

$$\Sigma_A \equiv \begin{pmatrix} 1 & \hat{d}_1 & \hat{d}_2 & \cdots & \hat{d}_{T-1} \\ \hat{d}_1 & 1 & \hat{d}_1 & \ddots & \vdots \\ \ddots & \ddots & \ddots & \ddots & \ddots \\ \vdots & \ddots & \ddots & \ddots & \hat{d}_1 \\ \hat{d}_{T-1} & \cdots & \hat{d}_2 & \hat{d}_1 & 1 \end{pmatrix}, \tag{11.1}$$

where $(S_1, \cdots, S_T)^T \sim \mathrm{N}_T(0, \Sigma_A)$.

We consider two cases of temporal correlation: exponential decay, $\hat{d}_i = r^i, 0 \le r \le 1$, and power decay, $\hat{d}_i = 1/(i+1)^{-\gamma}, \gamma \ge 0$. The exponential decay corresponds to a short memory and the power decay corresponds to intermediate and long memories (Florescu et al. 2016). Without loss of generality, we assume that the number of obligors in the t-th term is constant and denote it as n.

The asset correlation, ρ_A, is the parameter that describes the correlation between the value of the assets of the obligors in the same term. We consider the i-th asset value, $\hat{U}_{it}$, at time t, to be

$$\hat{U}_{it} = \sqrt{\rho_A} S_t + \sqrt{1-\rho_A} \epsilon_{it}, \tag{11.2}$$

where $\epsilon_{it} \sim \mathrm{N}(0, 1)$ is i.i.d. In this formulation, the equal-time correlation of U_{it} is ρ_A. It is the one factor-Gaussian Copula.

We consider the assets of two obligors, $\hat{U}_{1t}$ and $\hat{U}_{2t}$, in term t to confirm the temporal correlation. The assets of the two obligors have correlation ρ_A. S_t is the global economic factor that affects the two obligors at t:

$$\begin{aligned} \hat{U}_{1t} &= \sqrt{\rho_A} S_t + \sqrt{1-\rho_A} \epsilon_{1t}, \\ \hat{U}_{2t} &= \sqrt{\rho_A} S_t + \sqrt{1-\rho_A} \epsilon_{2t}, \end{aligned} \tag{11.3}$$

where ϵ_1 and ϵ_2 are the individual factors for the obligors. Here, there is no correlation among ϵ_1, ϵ_2, and S_t because they are independent of each other. This is one of the methods to create the correlated normal distribution using Cholesky decomposition. In the following term, $t+1$, the assets of the two obligors are $\hat{U}_{1t+1}$ and $\hat{U}_{2t+1}$. Assets have the same correlation, ρ_A, through global factor S_{t+1}. We can write this as:

$$\begin{aligned}\hat{U}_{1t+1} &= \sqrt{\rho_A} S_{t+1} + \sqrt{1-\rho_A}\epsilon_{1t+1},\\ \hat{U}_{2t+1} &= \sqrt{\rho_A} S_{t+1} + \sqrt{1-\rho_A}\epsilon_{2t+1}.\end{aligned} \tag{11.4}$$

The temporal correlation between t and $t+1$ is $\hat{d}_1$. The correlation between $\hat{U}_{1t}$ and $\hat{U}_{2t+1}$ is $\hat{d}_1\rho_A$. Similarly, we obtain the temporal correlation matrix, Eq. (11.1).

The discrete dynamics of the process are described by

$$X_i(t) = 1_{\hat{U}_{it} \le Y}, \tag{11.5}$$

where Y is the threshold and $1 \le i \le n$. They are the Bernoulli random variables. When $X_i(t) = 1(0)$, the i-th obligor in the t-th term is default (non-default). Equation (11.5) corresponds to the conditional default probability for $S_t = S$ as

$$G(S) \equiv \mathrm{P}(X_i(t) = 1|S_t = S) = \Phi\left(\frac{Y - \sqrt{\rho_A}S}{\sqrt{1-\rho_A}}\right), \tag{11.6}$$

where $\Phi(x)$ is the standard normal distribution, $G(S_t)$ is the distribution of the default probability during the t-th term in the portfolio, and the average PD is $p = \Phi(Y)$, which corresponds to "long run PDs."

The default correlation, ρ_D, is

$$\rho_D = f(\rho_A) \equiv \frac{\mathrm{P}(X_i(t) = 1 \cap X_j(t) = 1) - p^2}{p(1-p)} = \frac{\Phi_2((\Phi^{-1}(p), \Phi^{-1}(p)), \rho_A) - p^2}{p(1-p)},$$

where Φ_2 denotes the bivariate normal distribution with standardized marginals. We define the mapping function, $\rho_D = f(\rho_A)$, between the default correlation ρ_D and asset correlation ρ_A. Note that the mapping function, $\rho_D = f(\rho_A)$, depends on p. By the temporal correlation of S_t, we obtain the asset correlation of the asset values at different times as

$$\mathrm{Cor}(U_{it}, U_{jt'}) = \rho_A \Sigma_{t,t'}, \tag{11.7}$$

where ρ_A is the asset correlation and Σ is the correlation matrix for $S_t, t = 1, \cdots, T$. In Appendix A, we explain the calculation of Eq. (11.7). The default correlation between $X_i(t)$ and $X_j(t')$ is given by

$$\frac{\mathrm{P}(X_i(t) = 1 \cap X_j(t') = 1) - p^2}{p(1-p)} = \frac{\Phi_2((\Phi^{-1}(p), \Phi^{-1}(p)), \rho_A \hat{d}_{|t-t'|}) - p^2}{p(1-p)} = f(\rho_A \hat{d}_{|t-t'|}).$$

We are interested in the unbiased estimator of the PD, $Z(T) \equiv \sum_{t,i} X_i(t)/(nT)$, and limit $\lim_{T\to\infty} Z(T)$. Because the covariance of $X_i(t)$ and $X_j(t')$ is $p(1-p)f(\rho_A \hat{d}_{|t-t'|})$, the variance of $Z(T)$ is

$$\mathrm{V}(Z(T)) = p(1-p)\frac{1}{nT} + p(1-p)\frac{(n-1)}{nT} f(\rho_A) + 2p(1-p)\frac{1}{T^2}\sum_{i=1}^{T-1} f(\rho_A \hat{d}_i)(T-i).$$

The first term is from the binomial distribution, the second term is from the default correlation in the same term, and the third term is from the temporal correlation. In limit $T \to \infty$, the first two terms disappear, and the convergence of estimator $Z(T)$ is governed by the third term.

$f(\rho_A)$ is explicitly expressed as

$$\begin{aligned} f(\rho_A) &= \frac{\Phi_2((\Phi^{-1}(p), \Phi^{-1}(p)), \rho_A) - p^2}{p(1-p)} \\ &= \frac{1}{p(1-p)} \frac{1}{2\pi(1-\rho_A^2)} \\ &\times \left(\int_{-\infty}^{\Phi^{-1}(p)} dx \int_{-\infty}^{\Phi^{-1}(p)} dy \exp(-\frac{1}{2(1-\rho_A^2)}(x^2+y^2-2\rho_A xy)) - p^2 \right). \end{aligned} \quad (11.8)$$

We expand $f(\rho_A)$ at $\rho_A = 0$ as

$$f(\rho_A) = \frac{\rho_A}{2\pi p(1-p)} (\int_{-\infty}^{\Phi^{-1}(p)} x \exp(-\frac{1}{2}x^2) dx)^2 + O(\rho_A^2).$$

We denote the coefficient as A and $f(\rho_A) \simeq A\rho_A$. A is defined as

$$A \equiv \frac{1}{2\pi p(1-p)} \left(\int_{-\infty}^{\Phi^{-1}(p)} x \exp(-\frac{1}{2}x^2) dx \right)^2 > 0. \quad (11.9)$$

Hence, we can confirm $A > 0$.

For large i and $\hat{d}_i \to 0$, we have the asymptotic behavior of the default correlation as

$$C(t) \equiv \mathrm{Cor}(X_i(s), X_j(s+t)) = f(\rho_A \hat{d}_t) \simeq A\rho_A \hat{d}_t.$$

We note that default correlation $f(\rho_A \hat{d}_i)$ obeys the same decay law as that of $\hat{d}_i$ for large i and $\hat{d}_i \to 0$.

We plot the mapping function $\rho_D = f(\rho_A)$ in the (ρ_A, ρ_D)-plane in Fig. 11.1 under the conditions $f(0) = 0$ and $f(1) = 1$. Straight lines $\rho_D = A\rho_A$ with A, as expressed in Eq. (11.9), are plotted in Fig. 11.1b. We confirm that A in Eq. (11.9) is the same as the slope of the tangent line at point $(\rho_A, \rho_D) = (0, 0)$. We plot the relation between p and A in Fig. 11.1c. From the convexity of $f(\rho_A)$ in Fig. 11.1a, we find the following inequality:

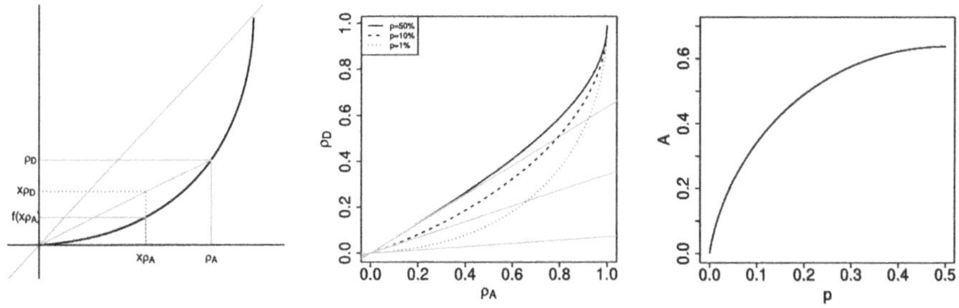

Fig. 11.1 **a** Plot of the mapping function $f(\rho_A)$ versus ρ_A, demonstrates the relation between the asset correlation ρ_A and default correlation ρ_D. $f(\rho_A)$ is downward convex and $f(x\rho_A) \leq xf(\rho_A)$ for $0 \leq x \leq 1$. **b** We set $p = 0.5$ (solid), 0.1(broken), and 0.01 (dotted). The gray lines represent the lines with a slope of A in Eq. (11.9). **c** Plot of A versus p

$$A\rho_A\hat{d}_i < f(\rho_A\hat{d}_i) < f(\rho_A)\hat{d}_i = \rho_D\hat{d}_i \ , \ \text{for} \ \ \rho_A, \rho_D > 0, 0 < \hat{d}_i < 1$$

Using this inequality, we form the upper bound for $V(Z(T))$ as:

$$\begin{aligned} \mathrm{V}(Z(T)) &= p(1-p)\frac{1}{nT} + p(1-p)\frac{(n-1)}{nT}f(\rho_A) + 2p(1-p)\frac{1}{T^2}\sum_{i=1}^{T-1} f(\rho_A\hat{d}_i(T-i)) \\ &\leq p(1-p)\frac{1}{nT} + p(1-p)\frac{(n-1)}{nT}f(\rho_A) + 2p(1-p)\rho_D\frac{1}{T^2}\sum_{i=1}^{T-1}\hat{d}_i(T-i). \end{aligned}$$

The lower bound is then

$$V(Z(T)) \geq p(1-p)\frac{1}{nT} + p(1-p)\frac{(n-1)}{nT}f(\rho_A) + 2p(1-p)A\rho_A\frac{1}{T^2}\sum_{i=1}^{T-1}\hat{d}_i(T-i).$$

In both the upper and lower bounds, the third term is proportional to $\frac{1}{T^2}\sum_{i=1}^{T-1}\hat{d}_i(T-i)$. Thus, we can estimate the asymptotic behavior of $V(Z(T))$ using the following expression:

$$\mathrm{V}(Z(T)) = p(1-p)\frac{1}{nT} + p(1-p)\frac{(n-1)}{nT}\rho_D + 2p(1-p)c\frac{1}{T^2}\sum_{i=1}^{T-1}\hat{d}_i(T-i), \tag{11.10}$$

where c is a positive constant and $\rho_D > c > A\rho_A$.

11.2.2 Exponential Decay Case

In this subsection, we discuss the convergence of $Z(T)$ for the exponential decay model $\hat{d}_i = r^i, r \leq 1$:

$$V(Z(T)) \simeq p(1-p)\frac{1}{nT} + p(1-p)\frac{(n-1)}{nT}f(\rho_A) + 2p(1-p)c\frac{1}{T^2}\sum_{i=1}^{T-1} r^i(T-i). \quad (11.11)$$

The first two terms on the right-hand side (RHS) behave as $\propto 1/T$ and, thus, converge to 0 in limit $T \to \infty$. For $r \neq 1$, the third term is

$$2p(1-p)c\frac{1}{T^2}[T\frac{1-r^{T-1}}{1-r} + \frac{(T-1)r^{T-1}(1-r) - (1-r^{T-1})r}{(1-r)^2}] \propto 1/T$$

and it converges to 0 in limit $T \to \infty$. In addition, $C(t) \simeq A\rho_A r^t$ for large t. We conclude that as the number of data samples increases, the distribution of $Z(T)$ converges to a delta function, and therefore, the PD can be estimated empirically.

Thus, we calculate $C(t) = f(\rho_A r^t)$ and $V(Z(t))$ numerically for $t \leq 10^5$. We set $\rho_A = 0.5, n = 10^4, p = 0.5, 0.1, 0.01$ and $r = 0.8, 0.9, 0.99, 0.999$. Figure 11.2(a)–(c) shows the double logarithmic plot of $C(t)$ versus t. Here, it is clearly seen that $C(t)$ decays exponentially. Figure 11.2(d)–(f) shows the plot of V($Z(t)$) versus t. For all $r < 1 \in \{0.8, 0.9, 0.99, 0.999\}$, $V(Z(t))$ decays as $1/t$. When $r = 1$, there is no temporal correlation decay case and all obligors are correlated ρ_A. Hence, there is no phase transition for $r < 1$.

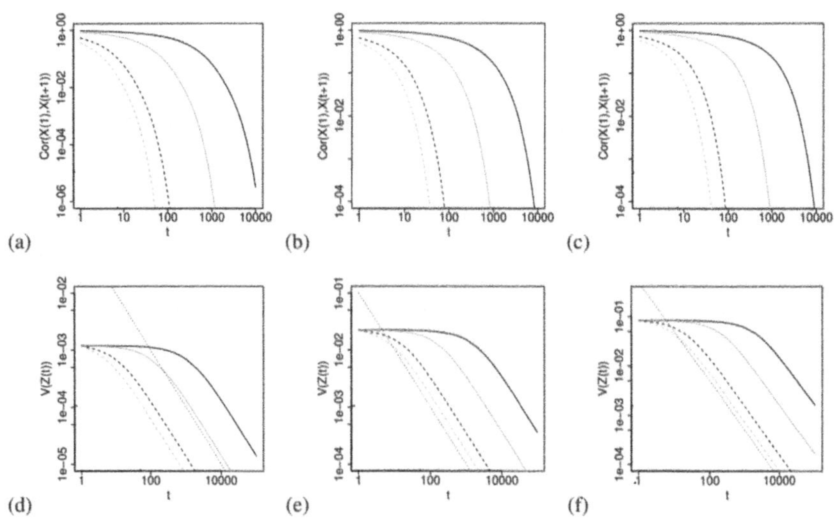

Fig. 11.2 Plots of (**a**), (**b**), (**c**) $C(t)$ and (**d**), (**e**), (**f**) $V(Z(t))$ versus t, for $r \in \{0.8, 0.9, 0.99, 0.999\}$; exponential decay case. The PDs are 1% for (**a**) and (**d**), 10% for (**b**) and (**e**), and 50% for (**c**) and (**f**)

11.2.3 Power Decay Case

In this subsection, we consider the power decay case $\hat{d}_i = 1/(i+1)^\gamma, i = 1, 2, \cdots,$ where $\gamma \geq 0$ is the power index. The power correlation affects the number of defaults over long periods of time. Ranges $\gamma \leq 1$ and $\gamma > 1$ are called long memory and intermediate memory, respectively (Florescu et al. 2016). In contrast, the exponential decay affects short periods of time and is called short memory. The asymptotic behavior of $V(Z(T))$ is given as:

$$\mathrm{V}(Z(T)) \simeq p(1-p)\frac{1}{nT} + p(1-p)\frac{(n-1)}{nT} f(\rho_A) + 2p(1-p)c\frac{1}{T^2}$$
$$\sum_{i=1}^{T-1}(i+1)^{-\gamma}(T-i).$$

$\gamma > 1$ Case

We obtain

$$\begin{aligned} V(Z(T)) &\simeq \frac{p(1-p)}{nT} + \frac{p(1-p)(n-1)f(\rho_A)}{nT} + \frac{2p(1-p)c}{T^2}\sum_{i=1}^{T-1}(T-i)/(i+1)^\gamma \\ &\simeq \frac{p(1-p)}{nT} + \frac{p(1-p)(n-1)\rho_D}{nT} + 2pqcT^{-\gamma}/(\gamma-1). \end{aligned} \tag{11.12}$$

The first two terms decrease as $1/T$, and the third term decreases as $1/T^\gamma$, where $\gamma > 1$. Hence, the significant terms are the first two terms, and the convergence of $V(Z(T))$ behaves as $\sim 1/T$. The convergence speed is the same as that of the independent binomial case.

$\gamma = 1$ Case

$V(Z(T))$ behaves as

$$V(Z(T)) \simeq \frac{p(1-p)}{nT} + \frac{p(1-p)(n-1)f(\rho_A)}{nT} + \frac{2p(1-p)c}{T^2}\sum_{i=1}^{T-1}(T-i)/(i+1). \tag{11.13}$$

The RHS of Eq. (11.13) is evaluated as

$$RHS \simeq \frac{p(1-p)}{nT} + \frac{p(1-p)(n-1)f(\rho_A)}{nT} + 2p(1-p)c[(T+1)\log T - T + 2] \sim \log T/T. \tag{11.14}$$

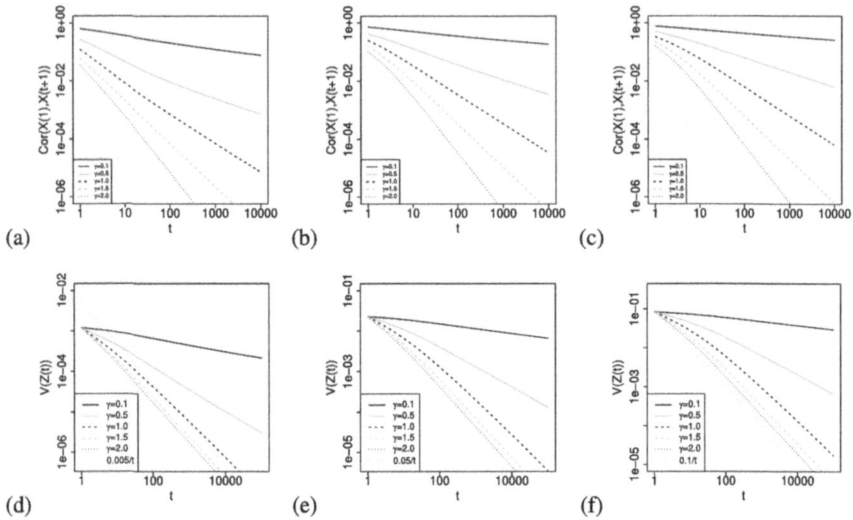

Fig. 11.3 Plots of (**a**), (**b**), (**c**) $C(t)$ and (**d**), (**e**), (**f**) $V(Z(t))$ versus t, for $\gamma \in \{3.0, 2.0, 1.5, 1.0, 0.5, 0.1\}$. Power decay case. The PDs are 1% for (**a**) and (**d**), 10% for (**b**) and (**e**), and 50% for (**c**) and (**f**)

In conclusion, $V(Z(T))$ behaves asymptotically as

$$V(Z(T)) \sim \log T/T \tag{11.15}$$

and estimator $Z(T)$ converges to p more slowly than in the normal case.

$\gamma < 1$ Case

$V(Z(T))$ is calculated as

$$V(Z(T)) \simeq \frac{p(1-p)}{nT} + \frac{p(1-p)(n-1)f(\rho_A)}{nT} + 2p(1-p)c[\frac{1}{(1-\gamma)(2-\gamma)T^{\gamma}}] \sim T^{-\gamma}. \tag{11.16}$$

Thus, we can conclude that $V(Z(T))$ behaves as

$$V(Z(T)) \sim T^{-\gamma}. \tag{11.17}$$

In conclusion, a phase transition occurs when the temporal correlation decays according to the power law. When the power index, γ, is less than one, the PD estimator, $Z(T)$, slowly converges to p. Conversely, when the power index, γ, is greater than one, the convergence behavior is the same as that of the binomial distribution. This phase transition is called a "super-normal transition" discussed in Chap. 3, which is the transition between long memory and intermediate memory. This transition is different from the phase transition found when we used the beta binomial distribution in Chap. 10. In that study, when the power index was less than one, the PD estimator, $Z(T)$, did not converge to p when the beta binomial model was used (Hisakado et al. 2020).

To confirm the above discussion, we calculate $C(t) = f(\rho_A(t+1)^{-\gamma})$ and $V(Z(T))$. Figure 11.3(a)–(c) shows the double logarithmic plot of $C(t)$ versus t. $C(t)$ decays according the power law for $\gamma \in \{0.1, 0.5, 1.0, 1.5, 2.0\}$. For small γ, such as $0.5, 0.1$, the slope is extremely small. Figure 11.3(d)–(f) shows the double logarithmic plot of $V(Z(t))$ versus t. For $\gamma \in \{0.1, 0.5, 1.0, 1.5, 2.0\}$, $V(Z(t))$ decays as $1/t$. At $\gamma \leq 1$, the slope of the decay becomes less than one. In this case, the convergence becomes slower than in the normal case.

We introduced the power law decay for asset correlation. It is the correlation of the hidden time series for the asset value. We evaluated the default correlation when the asset correlation is the power law decay and investigated the convergence of the variance related to the default correlation.

When $\gamma > 1$, $Z(T)$ converges to p, as in the normal case. However, when $\gamma \leq 1$, the convergence is slower than that in the normal case. Hence, there is a phase transition at $\gamma = 1$. The transition does not depend on the average PD, p. The phase transition depends on the relation between the asset and the default correlations, $f(\rho_A)$.

In conclusion if $\sum_{i=1}^{T} \hat{d}_i(T-i)$ in the limit $T \to \infty$ is infinite, we can observe the "super phase", where the convergence behavior is slower than the binomial distribution. If $\sum_{i=1}^{T} \hat{d}_i(T-i)$ changes from finite to infinite with the parameter, we can find the phase transition. One example of a phase transition between "normal phase" and "super phase" is the case that the asset correlation is power law decay. They correspond to intermediate memory and long memory, respectively (Florescu et al. 2016). Long memory is observed in the volatility of financial data. On the other hand, when the asset correlation is an exponential decay, we cannot find "super phase" because $\sum_{i=1}^{T} \hat{d}_i(T-i)$ is finite. It corresponds to short memory.

11.3 Estimation of the Parameters

As discussed in the previous section, whether the temporal correlation obeys the exponential decay or the power decay is an important issue because there exists a super-normal transition in the latter case. Furthermore, the appearance of a transition affects whether we can estimate the PD.

11.3.1 Bayesian Estimation Using the Merton Model

We explain the estimation of the parameters using the Merton model (Tasche 2013). There is a prior belief in the possible value of the PD. The prior belief is updated by observations using the prior distribution as a weighting function. Here, we use the prior function, which is a uniform prior distribution.

To calculate the unconditional probability $P(X_1 = k_1, \cdots, X_T = k_T)$, we approximate the solution using Monte Carlo simulations and numerical integration. Here, the number of obligors and defaults in the t-th term is n_t and k_t, respectively, and they are observable variables. The likelihood is

$$P(X_1 = k_1, \cdots, X_T = k_T) \sim \sum_{i=1}^{n} \prod_{t=1}^{T} \frac{n_t!}{k_t!(n_t - k_t)!} G(S_t^i)^{k_t} (1 - G(S_t^i))^{(n_t - k_t)}, \tag{11.18}$$

where $G(S_t^i)$ is defined as the probability that an obligor will default in term t, which is conditional to the i-th path realization of all global factors, such that

$$G(S_t^i) = \Phi(\frac{\Phi^{-1}(p) - S_t^i \sqrt{\rho_A}}{\sqrt{1 - \rho_A}}), \tag{11.19}$$

where ρ_A is the asset correlation among obligors within a one term window and Φ is the cumulative normal distribution. S_t^i is the correlated multi-dimensional normal distribution, and we use the MAP estimation to estimate the parameters. We use Normal Copula (Schönbucher 2003).

We estimated parameters using the beta binomial distribution provided in Chap. 10. **The differences between using the Merton model and the beta binomial distribution are the default correlation and the asset correlation. The default correlation is defined by the binary variables. In contrast, asset correlation is defined by continuous variables.** The other difference is that the parameters can be calculated analytically when using the beta binomial distribution. Hence, it is easier to estimate parameters using the beta binomial distribution than using the Merton model.

11.3.2 Data Analysis

First, the S&P default data from 1981 to 2018 (S&P 2019) are used. The average PD is 1.51 % for all ratings, 3.90 % for speculative grade (SG) ratings, and 0.09% for investment grade (IG) ratings. The SG rating represents ratings under BBB-(Baa3) and IG represents rating above BBB-(Baa3). In Fig. 11.4(a), we show the historical default rate of the S&P. The solid and dotted lines correspond to the speculative grade and investment grade samples, respectively. We use Moody's default data from 1920 to 2018 for 99 terms (Moody's 2019). This includes the Great Depression in 1929 and the Great Recession in 2008. The average default rate is 1.50% for all ratings, 3.70% for speculative ratings, and 0.14% for investment grade. In Fig. 11.4(b), we show the historical default rate of Moody's.

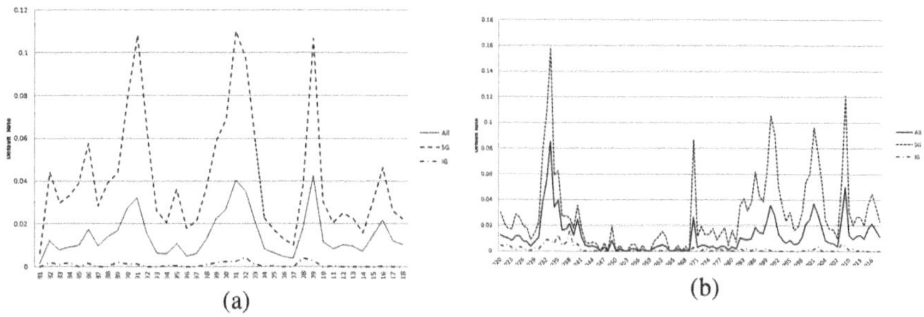

Fig. 11.4 (a) S&P default rate in 1981–2018. (b) Moody's default rate in 1920–2018. The solid and dotted lines correspond to the speculative grade (SG) and investment grade (IG) of the samples

Table 11.1 MAP estimation of parameters for the exponential and power decay models using the Merton model

		Exponential decay			Power decay		
Number	Model	p (%)	ρ_A (%)	r (%)	p (%)	ρ_A (%)	γ
1	Moody's 1920–2018	1.43	17.8	0.858	7.40	38.7	0.090
2	S&P 1981–2018	1.43	6.4	0.597	1.85	12.0	0.610
3	Moody's 1981–2018	1.61	7.1	0.613	1.92	12.4	0.622
4	S&P 1990–2018	1.72	7.5	0.616	2.97	12.5	0.616
5	Moody's 1990–2018	1.92	10.0	0.678	2.40	12.1	0.624
6	Moody's 1920–2018 SG	3.00	18.9	0.838	6.15	32.2	0.146
7	S&P 1981–2018 SG	4.53	8.7	0.588	4.42	11.7	0.628
8	Moody's 1981–2018 SG	4.28	9.4	0.603	3.97	11.5	0.619
9	S&P 1990–2018 SG	4.93	11.2	0.639	5.40	13.9	0.626
10	Moody's 1990–2018 SG	4.51	11.1	0.648	6.09	14.6	0.619
11	Moody's 1920–2018 IG	0.04	35.3	0.891	3.40	51.4	0.102
12	S&P 1981–2018 IG	0.02	25.8	0.483	0.02	20.3	9.189
13	Moody's 1981–2018 IG	0.01	21.9	0.672	1.84	33.8	0.618
14	S&P 1990–2018 IG	0.01	37.4	0.712	1.63	46.7	0.630
15	Moody's 1990–2018 IG	0.01	33.0	0.794	3.44	51.1	0.003

We estimate the parameters p, ρ_A, r and γ of the Merton model using the Bayesian method. We summarize the results in Table 11.1. We show ρ_D instead of ρ_A, as we must compare it with that of the beta binomial distribution model. The estimation of parameters is the maximum a posteriori (MAP) estimation. We notice that the power index γ is smaller than one for all cases, and the values are smaller than the phase transition point, $\gamma = 1$.

Table 11.2 Most likelihood estimate of parameters for the exponential and power decay models by beta binomial distribution

		Exponential decay			Power decay		
Number	Model	p (%)	ρ_D (%)	r (%)	p (%)	ρ_D (%)	γ
1	Moody's 1920–2018	0.96	1.9	0.044	0.94	2.0	4.7
2	S&P 1981–2018	1.53	0.8	0.026	1.54	0.8	5.7
3	Moody's 1981–2018	1.53	0.8	0.022	1.52	0.7	5.9
4	S&P 1990–2018	1.66	0.9	0.023	1.64	0.9	5.7
5	Moody's 1990–2018	1.67	0.9	0.019	1.61	0.8	6.0
6	Moody's 1920–2018 SG	2.36	3.8	0.044	2.34	4.1	4.7
7	S&P 1981–2018 SG	4.16	2.0	0.026	4.20	2.0	5.7
8	Moody's 1981–2018 SG	4.18	2.0	0.022	4.35	1.9	6.0
9	S&P 1990–2018 SG	4.42	2.5	0.024	4.43	2.6	5.6
10	Moody's 1990–2018 SG	4.33	2.3	0.020	4.31	2.2	5.9
11	Moody's 1920–2018 IG	0.13	0.8	0.17	0.11	0.9	3.0
12	S&P 1981–2018 IG	0.11	0.4	0.12	0.09	0.3	3.6
13	Moody's 1981–2018 IG	0.10	0.6	0.05	0.09	0.3	4.6
14	S&P 1990–2018 IG	0.09	0.4	0.12	0.09	0.4	3.7
15	Moody's 1990–2018 IG	0.09	0.4	0.06	0.07	0.7	4.2

Here we introduce the MAP estimation using the beta binomial distribution which used the default correlation directly to compare the MAP estimation using the Merton model. The estimation method is in Chap. 9. The beta binomial model is known as the correlated binomial distribution and we apply it to the defaults in the portfolio with the correlation. The conclusions are shown in Table 11.2 for the exponential and power decay models. We confirmed small r and large γ values, which represent the small temporal correlation. Parameter γ for the power decay is larger than the phase transition point, $\gamma = 1$. The PD and default correlation are almost the same as the estimations by the exponential and power decay models. The reason behind this is that the power exponent γ is adequately large and there is only a small difference between the exponential and power decay models.

We confirm that r and γ both have large differences between the values estimated by the beta binomial distribution and the Merton model. The reason for this is shown in Fig. 11.1(a). We set $\hat{d}_{1A}$ and $\hat{d}_{1D}$ for ρ_A and ρ_D, respectively. From this, we obtain the inequality

$$\hat{d}_{1A} = \frac{\hat{d}_{1A}\rho_A}{\rho_A} >> \frac{f(\hat{d}_{1A}\rho_A)}{f(\rho_A)} = \frac{f(\hat{d}_{1A}\rho_A)}{\rho_D} = \hat{d}_{1D}.$$

Table 11.3 WAIC and WBIC for exponential and power decay using the Merton model

		Exponential decay		Power decay	
Number	Model	WAIC	WBIC	WAIC	WBIC
1	Moody's 1920–2018	572.9	746.7	568.6	745.9
2	S&P 1981–2018	271.5	332.9	272.1	334.8
3	Moody's 1981–2018	277.6	339.0	277.5	341.4
4	S&P 1990–2018	214.3	256.4	214.6	258.3
5	Moody's 1990–2018	219.5	262.1	219.6	264.7
6	Moody's 1920–2018 SG	564.3	731.9	560.2	733.8
7	S&P 1981–2018 SG	268.7	328.1	268.9	330.5
8	Moody's 1981–2018 SG	274.1	333.6	274.4	337.1
9	S&P 1990–2018 SG	212.0	253.9	212.8	255.2
10	Moody's 1990–2018 SG	217.3	260.4	218.3	262.8
11	Moody's 1920–2018 IG	247.6	351.2	244.9	351.2
12	S&P 1981–2018 IG	116.0	153.3	115.0	160.3
13	Moody's 1981–2018 IG	110.3	156.3	108.8	156.7
14	S&P 1990–2018 IG	87.6	114.3	86.8	117.7
15	Moody's 1990–2018 IG	81.7	111.4	82.5	114.5

Hence, the difference in the estimated parameter between the Merton model and the beta binomial model becomes large. We find a large convexity for the mapping function f. Hence, r and γ for a default correlation are much smaller than those for asset correlation.

Next, we discuss whether the correlation has long memory. In Table 11.3, we calculated the WAIC and WBIC for each model that uses the Merton model. Using Moody's data from 1920, the power decay model is found to be superior to the exponential decay model. Therefore, it seems that the default rate has a long memory. As γ is less than 1 for long history data, the phase converges slowly. In other words, parameter estimation becomes difficult because the convergence speed decreases when the temporal correlation is the power decay.

In Table 11.4, we show the AIC and BIC for each model using the beta binomial distribution and compare them to the estimation using the Merton model. We obtain the same conclusion using Moody's data from 1920: the power decay model is superior to the exponential decay model. The parameter γ is not less than 1 in the power decay case when the beta binomial distribution is used.

Table 11.4 AIC and BIC for exponential and power decay using the beta binomial distribution

		Exponential decay		Power decay	
Number	Model	AIC	BIC	AIC	BIC
1	Moody's 1920–2018	746.8	780.8	746.8	780.9
2	S&P 1981–2018	352.0	382.9	353.6	384.5
3	Moody's 1981–2018	362.2	395.0	363.4	396.2
4	S&P 1990–2018	285.0	315.5	285.8	316.3
5	Moody's 1990–2018	293.8	326.3	298.6	331.1
6	Moody's 1920–2018 SG	730.8	762.0	730.6	761.8
7	S&P 1981–2018 SG	346.8	366.7	348.8	368.7
8	Moody's 1981–2018 SG	356.0	385.9	360.4	390.3
9	S&P 1990–2018 SG	283.0	302.6	283.8	303.4
10	Moody's 1990–2018 SG	291.0	320.7	291.6	321.3
11	Moody's 1920–2018 IG	302.2	334.9	300.4	333.1
12	S&P 1981–2018 IG	134.9	164.5	136.1	165.7
13	Moody's 1981–2018 IG	142.3	173.7	140.2	171.7
14	S&P 1990–2018 IG	106.0	135.3	107.2	136.5
15	Moody's 1990–2018 IG	111.8	142.8	112.4	143.4

11.4 Summary

In this chapter, we introduced the Merton model with the temporal asset correlation and discussed the convergence of the PD estimator. We adopted a Bayesian estimation method to estimate the model's parameters and discussed its implications for the estimation of PD. We found a phase transition when the temporal correlation decayed according to the power curve, indicating that the correlation had a long memory. When the power index, γ, was larger than one, the estimator distribution of the PD converged normally. When the power index was less than or equal to one, the distribution converged slowly. This phase transition is called the "super-normal transition." In the case of exponential decay, there was no phase transition.

In Chap. 10, we studied the beta binomial distribution model with a temporal default correlation. The PD estimator also showed a phase transition in the power decay case. The transition depends on whether the distribution converges or not. This was different from the phase transition observed in this study.

The main difference between the Merton and the beta binomial distribution models is the incorporation of the default correlation. In the latter model, the default correlation is defined using binary variables. In the former model, the default correlation is incorporated into the asset correlation, which is defined using a continuous variable. The implication of this study is the difficulty in the PD estimation when we adopt the former model. The estimated power index is in the slow-convergence region of the PD. Even with empirical data over a long period of time, the PD is difficult to estimate when we adopt the Merton model with temporal correlation.

11.5 Appendix A Double t Copula Model

In this Appendix, we consider the double t copula model instead of the one-factor Gaussian Copula. We consider the i-th asset value, $\hat{U}_{it}$, at time t, to be

$$\hat{U}_{it} = \sqrt{\rho_A}(\frac{\nu-2}{\nu})^{1/2}\hat{S}_t + \sqrt{1-\rho_A}(\frac{\nu-2}{\nu})^{1/2}\epsilon_{it}, \tag{11.20}$$

where ϵ_{it} is the independent random variable following the Student distribution with ν degree of freedom. Let $\{\hat{S}_t, 1 \le t \le T\}$ be the time series of the stochastic variables of the correlated Student distribution with ν degree of freedom as shown Eq. (11.1). This model is a simple extension of the one-factor Gaussian Copula and is known as the double t copula (Hull 2006).

In this formulation, the equal-time correlation of U_{it} is ρ_A. The discrete dynamics of the process are described by

$$X_i(t) = 1_{\hat{U}_{it} \le Y}, \tag{11.21}$$

where Y is the threshold and $1 \le i \le n$. When $X_i(t) = 1(0)$, the i-th obligor in the t-th term is default (non-default). Equation (11.21) corresponds to the conditional default probability for $\hat{S}_t = \hat{S}$ as

$$G(\hat{S}) \equiv \mathrm{P}(X_i(t) = 1|\hat{S}_t = \hat{S}) = t_\nu\left((\frac{\nu-2}{\nu})^{1/2}\frac{Y - \sqrt{\rho_A}(\frac{\nu-2}{\nu})^{1/2}\hat{S}}{\sqrt{1-\rho_A}}\right), \tag{11.22}$$

where t_ν is the standard univariate Student t, $G(S_t)$ is the distribution of the default probability during the t-th term in the portfolio, and the average PD is $p = t_\nu(Y)$. $\rho_D = f(\rho_A)$ is explicitly expressed as

$$\begin{aligned} f(\rho_A) &= \frac{t_{2,\nu}((t_\nu^{-1}(p), t_\nu^{-1}(p)), \rho_A) - p^2}{p(1-p)} \\ &\le \frac{1}{p(1-p)} \\ &\times \left(\frac{1}{2\pi(1-\rho_A^2)}\int_{-\infty}^{t_\nu^{-1}(p)} dx \int_{-\infty}^{t_\nu^{-1}(p)} dy \exp(-\frac{1}{2(1-\rho_A^2)}(x^2+y^2-2\rho_A xy)) - p^2\right). \end{aligned}$$

The inequality depends on the variance of χ square distribution, which is independent of the correlated normal distribution. When the degree of freedom is infinity, the model becomes the Gaussian copula model.

We expand $f(\rho_A)$ at $\rho_A = 0$ as

$$f(\rho_A) \le \frac{\rho_A}{2\pi p(1-p)}\left(\int_{-\infty}^{t_\nu^{-1}(p)} x \exp(-\frac{1}{2}x^2)dx\right)^2 + O(\rho_A^2).$$

We denote the coefficient as A and $f(\rho_A) \simeq A\rho_A$. A is defined as

$$A \equiv \frac{1}{2\pi p(1-p)} \left(\int_{-\infty}^{\Phi^{-1}(p)} x \exp(-\frac{1}{2}x^2) dx \right)^2 > 0. \qquad (11.23)$$

Hence, we can confirm $A > 0$. Then, we can see the "super-normal" phase transition when the double t copula model as the normal copula model in the same story we discussed in Sect. 11.2.

References

2018 Annual Global Corporate Default Study and Rating Transitions (S& P Global Ratings, 2019).

Florescu I, Mariani M C, Stanley H E, and Viens F G (Eds.) *Handbook of High-Frequency Trading and Modeling in Finance* John Wiley & Sons (2016).

Hisakado Mand Mori S, Physica A, Phase transition in the Bayesian estimation of the default portfolio **544** 123480 (2020).

Hisakado M, and Mori S, Parameter estimation of default portfolios using the Merton model and phase transition. Physica A: Statistical Mechanics and its Applications, **563**, 125435 (2022).

Hull J and White H, Valuing credit derivatives using an implied copula approach, *J of Derivatives* **14(2)** 8–28 (2006).

Merton R C, On the Pricing of Corporate Debt: The Risk Structure of Interest Rates, J. Fin. **29**2), 449–470 (1974).

Moody's Annual Default Study: Corporate default and recovery dates, 1920–2018 (Moody's 2019).

Schönbucher P J, *Cresit Derivatives Pricing Models: Models, Pricing, and Implementation* (John Wiley & Sons, Ltd. 2003).

Tasche D, Bayesian estimation of probabilities of default for low default portfolios, J. Risk Management in Financial Institutions, **6(3)** 302–326 (2013).

Vasicek O, The distribution of loan portfolio value, *Risk* **15(12)**, 160–162, 2002.

Chapter 12
Self-Exciting Negative Binomial Distribution and the Hawkes Process

12.1 Introduction

Recently, the Hawkes process was used to represent the clustering defaults of the time series (Hawkes 1971; Kirchner 2017; Blanc et al. 2017; Errais et al. 2010; Kanazawa et al. 2020a, b). Clustering defaults mean that as the number of events increases, the probability of an event increases. This phenomenon corresponds to self-excitation and constitutes a temporal correlation. From a physical viewpoint, this process is characterized by a phase transition from the steady state to the non-steady state. This is one kind of non-equilibrium phase transition and the relation between the Hawkes process and the branching process is shown in Hawkes et al. (1974). The extinction phase of the branching process corresponds to the steady state in the Hawkess process. Confirmation of the steady state is important for finance and risk management to hedge risks because it is not possible to manage the non-steady state.

In Chaps. 10 and 11, we discussed the parameter estimation of the urn process, which has a correlation in the same term, and considered a multi-year case with a temporal correlation. In this chapter, we introduce an extended multi-term urn process and discuss the relationship between the new urn process and the Hawkes process. In the continuous time limit, one can obtain the self-exciting NBD (SE-NBD) process and Hawkes process as the parameters approach certain limits.

We study the properties of the phase transitions of these processes. The simultaneous and temporal effects of the correlation were confirmed by analyzing empirical default data. The results confirmed that the urn process fits more accurately than the Hawkes process. The reason is the variance of the intensity function is nonzero. In fact some firms affect many companies and some firms do not affect the companies. This is an effect of the network because a network with hubs has a large variance of the degree distributions.

M. Hisakado, *Urn Models and Their Applications in Finance*,
https://doi.org/10.1007/978-981-96-3825-3_12

The remainder of this article is organized as follows. In Sect. 12.2, we introduce the multi-term urn process. Here, we discuss the relation between the Hawkes process and phase transition. In Sect. 12.3, we present our study of the phase transition of this model. In Sect. 12.4, we present the power law of the distribution function at the critical point. In Sect. 12.5, we discuss the application of this process to empirical data of historical defaults and confirm its parameters. Finally, the conclusions are presented in Sect. 12.6. This chapter is based on Hisakado et al. (2022).

12.2 From the Multi-term Urn Process to the Hawkes Process

In this section, we consider a multi-term urn process with correlations. In the first term, the urn contains θ_0 red balls and $n_0 - \theta_0$ white balls. Then, the balls are sequentially taken out from the urn. For example, a single ball is taken out, after which we return the ball taken out to the urn and add ω balls of the same color to the urn. Thus, the total number of balls increases by ω a process, which is a correlation parameter in the same term. In fact, as we take out more red balls, the probability that a red ball is taken out increases, because the number of red balls in the urn increases. This is the correlation in the same term. We repeat the process N times in the first term. Hence, the number of total balls is $n_0 + N\omega$ in the end of the first term. This is nothing but the Pólya urn model, which has a beta binomial distribution (BBD).

Next, we extend the process to a multi-term process. We repeat the same process as the first process with the updated parameter θ_t at the $t+1$-th term. In the $t+1$-th term, the urn contains $\theta_t = \theta_0 + \sum_i^t k_i \hat{d}_{t-i+1}\tilde{\omega}$ red balls, and $n_0 - \theta_t$ white balls. Here, the total number of initial balls is n_0, $\hat{d}_i$ denotes the kernel function or discount factor that represents the decrease in the effects of self-excitation. Note that θ_t depends on the history of number of the red balls taken out. For normalization, we set $\hat{d}_0 = 1$.

12.2.1 Urn Model

We consider the Bernoulli random variables $X_i(s)(i = 1, 2, \ldots, n)$ that take the values 1 or 0 and s is the term. $X_i(s) = 1(0)$ means that a red (white) ball is taken out in the i-th operation at t-th term. Here, we introduce the variable N_i which is the number of red balls taken out in the i-th term and its value is $N_i = k_i = \sum_{s=1}^N X_i(s)$.

$\tilde{\omega}$ is a parameter for temporal correlation, where ω denotes the scale parameter of the added red balls. $\tilde{\omega}$ is the ratio of the number of balls placed back to the urn to the number of balls drawn from the urn, when $\hat{d}_0 = 1$ and $\hat{d}_i = 0, i \geq 1$. In this case only the previous term affects the this term. In summary, we update only the parameter θ_t and other parameters are fixed in each term.

As the first term, we sequentially take out balls from the urn. After that, we return the taken out ball to the urn and ω balls of the same color are added to the urn. We repeat the process N times in this $t+1$-th term. This is the definition of the one term. After the process of $t + 1$ term, we set the initial condition of $t + 2$-term and continue the $t + 2$-th process. Each term is also the Pólya urn model. In this model, we use two types of correlations: the first is a correlation in the same term using the parameter ω, and the second is the temporal correlation using the parameters $\tilde{\omega}$ which is independent from ω and $\hat{d}_i$, which is the kernel function. The temporal correlation decays as a function of time using the parameter $\hat{d}_i$. We show multi-term urn process in Fig. 10.1.

In Chap. 10, we introduced a similar urn process with a correlation in the same term and temporal correlation. The difference is the initial condition of the $t + 1$-th term. The number of red balls is the same, but the number of white balls is $n_0 - \theta_0 + \sum_i^t (N - k_i)\hat{d}_{t-i+1}\tilde{\omega}$, where $N_i = k_i$, and the total number of balls in the initial condition of each term is not constant, n_0.

When k_1 red balls are taken out in the first term, $N_1 = k_1$, the probability in the first term can be calculated as

$$
\begin{aligned}
P(N_1 = k_1) &= \frac{N!}{k_1!(N-k_1)!} \\
&\times\left[\frac{\theta_0}{n_0}\cdots\frac{\theta_0+(k_1-1)\omega}{n_0+(k_1-1)\omega}\right]\left[\frac{n_0-\theta_0}{n_0+k_1\omega}\cdots\frac{n_0-\theta_0+(N-k_1-1)\omega}{n_0+(N-1)\omega}\right] \\
&= \frac{N!}{k_1!(N-k_1)!}\,\frac{\Pi_{i=0}^{k_1-1}(\theta_0+i\omega)\Pi_{j=0}^{N-k_1-1}(n_0-\theta_0+j\omega)}{\Pi_{l=0}^{N-1}(n_0+l\omega)} \\
&= \frac{N!}{k_1!(N-k_1)!}\,\frac{\Pi_{i=0}^{k_1-1}(p+i\rho/(1-\rho))\Pi_{j=0}^{N-k_1-1}(q+j\rho/(1-\rho))}{\Pi_{l=0}^{N-1}(1+l\rho/(1-\rho))},
\end{aligned}
\tag{12.1}
$$

where $p = \theta_0/n_0$, $q = 1 - \theta_0/n_0$, and $\omega/n_0 = \rho/(1-\rho)$. This is known as the beta binomial distribution BBD(α, β,N) where $p = \alpha/(\alpha+\beta)$ and $q = \beta/(\alpha+\beta)$. Here, α and β correspond to the parameters of the beta distribution in the continuous limit of the BBD. Note that $\rho = 1/(1+\alpha+\beta)$ plays the role of a correlation in this term.[1] The variance of the BBD is $Npq + N(N-1)pq\rho$ and the second term corresponds to the correlation in the same term. Hence, ω is the correlation parameter.

Here, we consider the temporal correlation by adjusting the initial conditions of the red balls, θ_t in each term. The temporal correlation is the decay of the correlation at different times, $\mathrm{Cor}(N_t, N_{t+\tau})$ as τ increases. The total number of balls in the initial condition of each term is n_0. The expected value of the ratio of the taken out red balls in $t + 1$-th term is θ_t/n_0, where $\theta_t = \theta_0 + \sum_i^t k_i\hat{d}_{t-i+1}\tilde{\omega}$. This implies that the events of the previous terms affect the present events. Here, the number of the events is the number of red balls taken out. As the number of events increases, the expected value of the number of events in the next term increases. Hence, we can introduce the temporal correlation by adjusting the initial condition of each term. When $\theta_t = \theta_0$, there is no temporal correlation, because each term is independent. In Table 12.1, we summarize the parameters of the urn process. We study the absorption process of this multi-term urn model in Sect. 12.8 Appendix B.

[1] See Chap. 2.

Table 12.1 Summary of the parameters for the urn process

Parameter and definition	
θ_0	Number of red balls at the first term
n_0	Number of total balls in the initial condition for each term
N	Number of balls taken out in each term
ω	Increase of the balls when we take out a ball It is related to the correlation in the same term
$\hat{d}_i$	Weight of the red balls taken out at the previous i-th terms (discount factor or the kernel function)
k_i	Number of red balls taken out in i-th term
$\tilde{\omega}$	Scale parameter for the initial condition in each term One of the parameters for the temporal correlation

12.2.2 Double Scaling Limit

Here, we set the first double scaling limit $N/n_0 = \Delta$, $N \to \infty$ and $n_0 \to \infty$ [2]. We obtain

$$P(N_1 = k_1) \sim \mathrm{NBD}(N_1 = k_1|K_0, M_0/K_0) = \frac{(K_0 + k_1 - 1)!}{k_1!(K_0 - 1)!}(\frac{K_0}{K_0 + M_0})^{K_0}(\frac{M_0}{K_0 + M_0})^{k_1}, \tag{12.2}$$

where $M_0 = \theta_0 N/n_0 = \theta_0\Delta$ and $K_0 = \theta_0/\omega$. In Sect. 12.9 Appendix C, we show the process. Here, $\sim$ means that the stochastic variable on the LHS obeys the probability distribution on the RHS. This is a negative binomial distribution (NBD), $\mathrm{NBD}(N_1 = k_1|K_0, M_0/K_0)$. The parameter $M_0/K_0 = \omega N/n_0 = \omega\Delta$ is related to the correlation in this term. The mean and variance of NBD are given by M_0 and $M_0 + M_0^2/K_0$, respectively.

Negative binomial distribution $\mathrm{NBD}(N_1 = k_1|K_0, M_0/K_0)$ has another face:

$$\begin{aligned}\mathrm{NBD}(N_1 = k_1|K_0, M_0/K_0) &= \int_0^\infty \mathrm{Poisson}(k_1|\lambda) \cdot \mathrm{Gamma}(\lambda|K_0, M_0/K_0)d\lambda, \\ &= \int_0^\infty \frac{\lambda^{k_1}e^{-\lambda}}{k_1!} \frac{\lambda^{K_0-1}}{\Gamma(K_0)(M_0/K_0)^{K_0}} e^{-\lambda K_0/M_0} d\lambda, \end{aligned} \tag{12.3}$$

where $\mathrm{Poisson}(k_1|\lambda)$ is the Poisson process and $\mathrm{Gamma}(\lambda|K_0, M_0/K_0)$ is the gamma distribution. We show the proof Eq. (12.3) in Sect. 12.10 Appendix D. From this result, the NBD is a Poisson process with the intensity function λ which obeys the gamma distribution. In the multi-term model, we can decompose the process in the same manner. Hence, λ_t is the intensity function of the t-th term and obeys the gamma distribution. $\mathrm{Gamma}(\lambda|K_0, M_0/K_0)$ has an average of M_0 and a variance of M_0^2/K_0.

[2] We show the limit of Moody's correlated binomial model instead of BBD in Sect. 12.7 Appendix A.

This means that the urn process in the first term corresponds to a Poisson process with intensity function λ, which has a gamma distribution in the double scaling limit. This intensity function yields the variance of the Poisson process, which has a constant intensity function. We refer to this as the NBD process. The average of N_1 converges to

$$\lim_{\Delta \to \infty} \frac{\mathrm{E}(N_1)}{\Delta} = \theta_0. \tag{12.4}$$

Then, the NBD process is the self-averaging process. Note that the BBD process is the non-self-averaging process. In the application of the double scaling limit, the non-self-averaging process becomes the self-averaging process. Therefore, we can apply the NBD process to the time series. The reason of the difference is that the BBD is the conditional probability which is the feedback, but the NBD process is not the conditional probability.

12.2.3 Discrete Multi-term Process

Next, we define the $t+1$-th term with the temporal correlation using the conditional probability. We define the conditional probability of $t+1$-th term,

$$P(N_{t+1} = k_{t+1}|N_0 = k_0, \ldots, N_t = k_t) = \mathrm{NBD}(N_{t+1} = k_{t+1}|K_t, M_t/K_t), \tag{12.5}$$

where k_i is the history of the number of red balls taken out. The conditional probability is defined by updating the parameters K_t and M_t. The only difference between the first term and the $t+1$-th term is the number of white balls in the initial condition of the term. The other conditions are the same as those in the first term. We can obtain the parameters at the $t+1$-th term of the intensity function:

$$\begin{aligned} M_t = \theta_t N/n_0 &= \frac{\theta_0 + \sum_i^t N_i \hat{d}_{t+1-i}\tilde{\omega}}{n_0} N = (\theta_0 + \sum_i^t N_i \hat{d}_{t+1-i}\tilde{\omega})\Delta \\ &= M_0 + M_0/L_0 \sum_i^t N_i \hat{d}_{t+1-i}, \end{aligned} \tag{12.6}$$

where $M_0 = \theta_0 N/n_0 = \theta_0 \Delta$, $K_0 = \theta_0/\omega$, $L_0 = \theta_0/\tilde{\omega}$ and $\omega N/n_0 = \omega\Delta = M_0/K_0$. The other parameters are obtained as follows:

$$K_t = \theta_t/\omega = \frac{\theta_0 + \sum_i^t N_i \hat{d}_{t+1-i}\tilde{\omega}}{\omega} = K_0 + K_0/L_0 \sum_i^t k_i \hat{d}_{t+1-i}, \tag{12.7}$$

and

$$M_t/K_t = \omega N/n_0 = \omega\Delta = M_0/K_0. \tag{12.8}$$

We refer to this process as the discrete self-exciting negative binomial distribution (SE-NBD) process. The self-exciting is introduced by the conditional probability, Eq. (12.5). Note that for all process parameters, M_t/K_t is a constant M_0/K_0. This signifies that the correlation in the same term does not depend on the term t. Under this condition, the process exhibits the reproductive property of the NBD. The mean of the intensity function is M_t, and the variance is M_t^2/K_t. As ω increases, K_t decreases, and the variance of the intensity function increases. The correlation in the same term affects the variance of the intensity function. In the limit $K_0 \to \infty (\omega \to 0)$ with Δ fixed, the intensity function has a vanishing variance and thus its distribution converges to the delta function. This is known as the discrete Hawkes process (Kirchner 2017). L_0 acts as a parameter for the temporal correlation as $\tilde{\omega}$.

In summary, we obtained that the discrete SE-NBD process N_t obeys NBD for M_t from the urn process,

$$N_{t+1} \sim \mathrm{NBD}\left(K_t, M_0/K_0\right), t \geq 0, \tag{12.9}$$

where

$$M_t = M_0 + M_0/L_0 \sum_{s=1}^{t} N_s \hat{d}_{t+1-s}, t \geq 1, \tag{12.10}$$

and

$$K_t = K_0 + K_0/L_0 \sum_{s=1}^{t} N_s \hat{d}_{t+1-s}, t \geq 1. \tag{12.11}$$

In the limit $K_0 \to \infty(\omega = 0)$, the discrete Hawkes process, N_t, is shown to obey the Poisson process for M_t from the urn process,

$$N_{t+1} \sim \mathrm{Poisson}\left(M_t\right), t \geq 0, \tag{12.12}$$

where

$$M_t = M_0 + M_0/L_0 \sum_{s=1}^{t} N_s \hat{d}_{t+1-s}, t \geq 1. \tag{12.13}$$

In the limit $L_0 \to \infty(\tilde{\omega} = 0)$, the process becomes an NBD process, that does not have self-excitation capabilities. Δ corresponds to the number of balls taken out in a term, and represents the interval between the terms.

12.2.4 Continuous Limit

SE-NBD Process

Here, we introduce the counting process, $\tilde{N}_t = \sum_i N_i$. We set the second double scaling limit $\Delta = N/n_0 \rightarrow 0$, $\omega \rightarrow \infty$ with $\omega\Delta = \omega'$, as the continuous limit of the process $\tilde{N}_t$. Note that in this limit $\omega\Delta = \omega'$ is constant, and the process has the reproductive property of a discrete SE-NBD process. We obtain the mean of the intensity function at t, λ_t

$$\mathrm{E}(\lambda_t|F_t) = \lim_{\Delta\rightarrow 0} \frac{\mathrm{E}[\tilde{N}_{t+\Delta} - \tilde{N}_t|F_t]}{\Delta} = \lim_{\Delta\rightarrow 0} \frac{M_t}{\Delta} = (\theta_0 + \tilde{\omega}\sum_{i<t} N_i \hat{d}_{t-i}), \quad (12.14)$$

where F_t is the history of the number of red balls and $N_1 = k_1, \ldots, N_t = k_t$ and it corresponds to the Hazard function. The variance of the intensity of the distribution at time t is

$$\mathrm{V}(\lambda_t|F_t) = \lim_{\Delta\rightarrow 0} \frac{M_t^2/K_t}{\Delta} = \omega'(\theta_0 + \tilde{\omega}\sum_{i<t} \hat{d}_{t-i} N_i). \quad (12.15)$$

In the continuous SE-NBD process, the intensity function exhibits a gamma distribution as a discrete SE-NBD process.

Hawkes Process

Next, we consider the continuous limit of the discrete Hawked process, Eq. (12.12). Here, we employ the counting process, $\tilde{N}_t = \sum_i N_i$. The limit $\Delta = N/n_0 \rightarrow 0$ is set, as the continuous limit of process $\tilde{N}_t$. The intensity function λ_t,

$$\lambda_t = \lim_{\Delta\rightarrow 0} \frac{E[\tilde{N}_{t+\Delta} - \tilde{N}_t|F_t]}{\Delta} = \lim_{\Delta\rightarrow 0} \frac{M_t}{\Delta} = (\theta_0 + \tilde{\omega}\sum_{i<t} k_i \hat{d}_{t-i}), \quad (12.16)$$

We can then obtain the Hawkes process,

$$\tilde{N}_{t+\Delta} - \tilde{N}_t \sim \text{Poisson}\left(\theta_t \Delta\right), t \geq 0, \quad (12.17)$$

where

$$\theta_t = \theta_0 + \tilde{\omega}\sum_{s<t} N_s \hat{d}_{t-s}, t \geq 0. \quad (12.18)$$

We can confirm that the intensity function becomes a delta function in the limit $\omega' \rightarrow 0$ of Eq. (12.15), which corresponds to the continuous Hawkes process.

12.3 Phase Transition of SE-NBD Process

12.3.1 Mean Field Approximation

In this section, we consider the phase transition of the SE-NBD process. Here we apply the mean field approximation that we set $\bar{v} = N_\infty$. We consider the increase in Δ,

$$E[\tilde{N}_{t+\Delta} - \tilde{N}_t] = E[\lambda_t | F_t] = [\theta_0 + \bar{v}\tilde{\omega} \sum_i \hat{d}_i]\Delta, \tag{12.19}$$

where we use Eq. (12.18). We set the average, $\bar{v}$, of the intensity function and LHS of Eq. (12.19) is $\bar{v}\Delta$. Then we obtain,

$$\bar{v} = \theta_0 / (1 - \tilde{\omega}\hat{T}), \tag{12.20}$$

where $\hat{T} = \sum^\infty \hat{d}_i$.

In the limit $\tilde{\omega} \to 0$, the temporal correlation is zero, and the process is the NBD process, where the phase transition disappears.

The SE-NBD includes the Hawkes process as a branching process.[3] The branching ratio is

$$\nu = \tilde{\omega}\hat{T}, \tag{12.21}$$

and the condition for the steady state is

$$\nu = \tilde{\omega}\hat{T} < 1. \tag{12.22}$$

The phase transition between the steady state and non-steady state occurs at $\nu = 1$, which is the critical point. The transition point is the same as that in the Hawkes process.

The parameter ν is termed the effective reproduction number for an infectious disease. It is the number of patients infected by one patient in the infection model. If the effective reproduction number is above 1, the number of patients increases to infinity indicating the non-steady state. We will discuss this phase transition in detail in Chap. 13.

In the SE-NBD process, the distribution of the intensity function has a variance. In contrast, in the Hawkes process, the intensity function is a delta function. The variance of the intensity function is the origin of the variance of the branching ratios. This means that the SE-NBD process has the variance of the branching ratios, because the intensity function has a variance. For example, the mixture of branching ratios affects not only the expected value of the intensity function but also the variance of the intensity function. Hence, the SE-NBD process, which uses the gamma distribution as the intensity function, may be useful. We confirm this in Sect. 12.5.

[3] We discuss in Chap. 13.

We consider the exponential and power decay cases as the kernel functions. These correspond to short and long memories (Florescu et al. 2016) as the kernel function. When we consider the exponential decay case, $\hat{d} = e^{-\hat{\beta}t} = r^t$, the condition for the steady state is $M_0/L_0 < \hat{\beta}$. When we consider the power decay case, $\hat{d} = 1/(1+t)^{\gamma}$, the condition for the steady state is $M_0/L_0 < \gamma - 1$. In Sect. 12.5, we use the exponential kernel for the empirical default data.

12.4 Power Law Distribution of the Intensity Function at the Critical Point

12.4.1 Poisson Noise

We begin with a discrete SE-NBD process $\{N_t\}, t = 1, \ldots$. Here, $N_t \in \{0, 1, \ldots\}$ represents the size of an event at time t. This is the process we obtained from the urn process. This event corresponds to the taken out of the red ball from the urn. N_t obeys the NBD with M_t,

$$N_{t+1} \sim \text{NBD}\left(\frac{M_t}{\omega} = K_t, \omega\Delta\right), t \geq 0,$$

$$M_t = M_0 + n \sum_{i=1}^{t} N_s h(t-i), t \geq 1,$$

where $n = M_0/(1-r)L_0$. We adopt the exponential decay kernel function, $h(t) = (1-r)\hat{d} = (1-r)r^t, 0 \leq r < 1$. In addition, we replace the normalization factor $(1-r)$ of $h(t)$ with $1/\tau$ to ensure that $\int_0^{\infty} h(t)dt = \frac{1}{\tau}\int_0^{\infty} e^{-t/\tau}dt = 1$.

The stochastic process $\{N_t\}, t = 1, \ldots$ is non-Markovian. We focus on the time evolution of the intensity function M_t, which satisfies the following recursive relation.

$$M_{t+1} = r(M_t - M_0) + M_0 + nh(0)N_{t+1}.$$

Here, we use the relationship $\sum_{i=1}^{t+1} N_i h(t+1-i) = N_{t+1}h(0) + r\sum_{i=1}^{t} N_s h(t-i)$. The stochastic difference equation for the excess intensity, $\hat{z}_t \equiv M_t - M_0$ is

$$\hat{z}_{t+1} - \hat{z}_t = (r-1)\hat{z}_t + nh(0)N_{t+1}. \tag{12.23}$$

We take the continuous time limit as in Sect. 12.2. We divided the unit time interval by infinitesimal time intervals with width $dt = \Delta$. The decreasing factor r^t during the interval dt is replaced with $r^{dt} = e^{-dt/\tau} \simeq 1 - dt/\tau + o(dt/\tau)$.

N_{t+1} is the noise for time interval $[t, t+1]$, and it is necessary to prepare the noise for infinitesimal interval $[t, t+dt)$. For this purpose, we use the reproductive property of the NBD. If N_{t+1}, the noise in the interval $[t, t+1)$ obeys $\text{NBD}(\theta_t/\omega, \omega)$.

The noise for $[t, t+dt)$ is denoted by $d\hat{\xi}^{NB}_{(\theta_t/\omega',\omega')}(t)$. Here, the double scaling limit is applied to change the parameter from ω to ω'. As Eq. (12.17) the stochastic difference equation (SDE) then becomes

$$d\hat{z}_t = \hat{z}_{t+dt} - \hat{z}_t = -\frac{1}{\tau}\hat{z}_t dt + \frac{n}{\tau} d\hat{\xi}^{NBD}_{(\frac{\theta_0+\hat{z}_t}{\omega'},\omega')}(t). \tag{12.24}$$

The state-dependent NBD noise $\hat{\xi}^{NBD}_{\theta_t/\omega,\omega}$ defines the noise in infinitesimal time interval $[t, t+dt)$ with the following probabilistic rules:

$$d\hat{\xi}^{NBD}_{(\theta_t/\omega',\omega')}(t) \equiv \hat{\xi}^{NBD}_{(\theta_t/\omega',\omega')}(t+dt) - \hat{\xi}^{NBD}_{(\theta_t/\omega',\omega')}(t) \sim \text{NBD}(\theta_t dt/\omega', \omega').$$

The characteristic function for $d\hat{\xi}^{NBD}_{(\theta_t/\omega',\omega')}$ is $\phi(s) = (\frac{1}{1+\omega'-\omega' e^{is}})^{\theta_t dt/\omega'}$. The infinite divisibility is clear from the functional form. In addition, $d\hat{\xi}^{NBD}_{(\theta_t/\omega',\omega')}$ can be written as the superposition of the Poisson noise with the state-dependent random intensity λ_t,

$$d\xi^{NBD}_{(\theta_t/\omega',\omega')} \sim \text{Poisson}(\lambda_t dt),\ \lambda_t \sim \text{Gamma}(\theta_t/\omega', \omega').$$

as we discussed in Sect. 12.2. When we denote the timing and size of the i-th event as t_i and k_i, and the number of events before t as $\hat{N}(t)$, we can rewrite the state-dependent NBD noise $\hat{\xi}^{NBD}_{(\frac{\theta_0+\hat{z}_t}{\omega'},\omega')}(t)$ as

$$\hat{\xi}^{NBD}_{(\frac{\theta_0+\hat{z}_t}{\omega'},\omega')}(t) = \sum_{i=1}^{\hat{N}(t)} k_i \delta(t - t_i).$$

Probability of the occurrence and non-occurrence of an event of size k during time interval dt is given by,

$$P\left(d\hat{\xi}^{NBD}_{(\frac{\theta_0+\hat{z}_t}{\omega'},\omega')} = k\right) = \begin{cases} 1 - \frac{\hat{z}_t+\theta_0}{\omega'}\ln(\omega'+1)dt & k=0 \\ \frac{1}{k}\left(\frac{\hat{z}_t+\theta_0}{\omega'}\right)\left(\frac{\omega'}{\omega'+1}\right)^k dt & k \geq 1. \end{cases} \tag{12.25}$$

In the limit $\omega' \to 0$, the probabilities become

$$\lim_{\omega'\to 0} P\left(d\hat{\xi}^{NBD}_{(\frac{\theta_0+\hat{z}_t}{\omega'},\omega')} = k\right) = \begin{cases} 1 - (\hat{z}_t + \theta_0)dt & k=0 \\ (\hat{z}_t+\theta_0)dt & k=1 \\ 0 & k \geq 2. \end{cases}$$

k_i is restricted to one or zero, and the state-dependent noise becomes a Poisson noise.

12.4.2 Power Law

The SDE (12.24) is interpreted as

$$\hat{z}(t+dt)-\hat{z}(t)=\begin{cases} -\frac{1}{\tau}\hat{z}_t & \text{Prob.} = 1-\frac{\hat{z}_t+\theta_0}{\omega'}\ln(\omega'+1)dt \\ \frac{nk}{\tau} & \text{Prob.} = \frac{1}{k}\left(\frac{\hat{z}_t+\theta_0}{\omega'}\right)\left(\frac{\omega'}{\omega'+1}\right)^k dt, k=1,\ldots. \end{cases}$$

The same procedure is adopted to derive the master equation for the probability density function (PDF) of $\hat{z}_t$ in Kanazawa et al. (2020b, a). This yields

$$\frac{\partial}{\partial t}P_t(z)=\frac{1}{\tau}\frac{\partial}{\partial z}zP_t(z)+\sum_{k=1}^{\infty}\frac{1}{\omega' k}\left(\frac{\omega'}{\omega'+1}\right)^k\left\{(\theta_0+z-\frac{nk}{\tau})P_t(z-\frac{nk}{\tau})-(\theta_0+z)P_t(z)\right\}. \tag{12.26}$$

The Laplacian representation of the steady state is denoted as $P_{SS}(z)$ as $\tilde{P}_{SS}(s)$.

$$\tilde{P}_{SS}(s)\equiv\int_0^{\infty}P_{SS}(z)e^{-sz}dz.$$

The master equation for $\tilde{P}_{SS}(s)$ is

$$\left[\sum_{k=1}^{\infty}\frac{1}{\omega' k}\left(\frac{\omega'}{\omega'+1}\right)^k\left(e^{-\frac{nk}{\tau}s}-1\right)+\frac{s}{\tau}\right]\frac{d}{ds}\tilde{P}_{SS}(s)$$
$$=\sum_{k=1}^{n}\frac{1}{\omega k'}\left(\frac{\omega'}{\omega'+1}\right)^k\theta_0\left(e^{-\frac{nk}{\tau}s}-1\right)\tilde{P}_{SS}(s). \tag{12.27}$$

Thus,

$$\frac{d}{ds}\ln\tilde{P}_{SS}(s)=\theta_0-\frac{\theta_0 s/\tau}{\sum_{k=1}^{\infty}\frac{1}{\omega' k}\left(\frac{\omega'}{\omega'+1}\right)^k(e^{-\frac{nk}{\tau}s}-1)+\frac{s}{\tau}}.$$

We integrate the equation with the initial condition $\tilde{P}_{SS}(0)=1$.

$$\ln\tilde{P}_{SS}(s)=\theta_0 s-\int_0^s\frac{\theta_0 s'/\tau}{\sum_{k=1}^{\infty}\frac{1}{\omega' k}\left(\frac{\omega'}{\omega'+1}\right)^k(e^{-\frac{nk}{\tau}s'}-1)+\frac{s'}{\tau}}ds'.$$

Here, the large z behavior of $P_{SS}(z)$ near the critical point $n=1$ is of interest, and we study the integral at $s\simeq 0$ and $n=1-\epsilon, \epsilon << 1$. We expand $e^{\frac{nk}{\tau}s}=1-\frac{nks}{\tau}+\frac{1}{2}(\frac{nks}{\tau})^2+o(s^2)$ and calculate the summation over k in the denominator of the integral. Therefore, we have

$$\ln\tilde{P}_{SS}(s)\simeq\theta_0 s-\int_0^s\frac{\frac{2\theta_0\tau}{\omega'+1}}{\frac{2\tau}{\omega'+1}\epsilon+s'}ds'.$$

Near the critical point, the excess intensity distribution exhibits power-law behavior with a non-universal exponent, up to an exponential truncation:

$$P_{SS}(z) \propto z^{-1+2\frac{\theta_0\tau}{\omega'+1}} e^{-\frac{2\tau\epsilon}{\omega'+1}z}. \tag{12.28}$$

The power law exponent of the PDF of the excess intensity is $1-\frac{2\theta_0\tau}{\omega'+1}$, and depends on ω', which is the correlation simultaneously. In the limit $\omega' \to 0$, the result coincides with Hawkes process in Kanazawa et al. (2020b, a). The power law exponent increases with ω' and converges to 1 in the limit $\omega' \to \infty$. $\omega' \to 0$ is no correlation at the same time and the intensity function becomes the delta function. $\omega' \to \infty$ is the strong correlation limit at the same time and the intensity function has a large variance. The correlation in the same term alters critical behavior. In addition, the length scale of the exponential decay for the off-critical case is $(\omega'+1)/(2\tau\epsilon)$, which is also an increasing function of ω'.

12.5 Parameter Estimation Using Default Data

We use empirical data on financial defaults to estimate the parameters. First, the S&P default data from 1981 to 2020 are used. A speculative grade (SG) rating represents ratings below BBB-(Baa3), whereas an investment grade (IG) rating indicates ratings above BBB-(Baa3). We also use Moody's default data from 1920 to 2020, a period of 100 years that includes the Great Depression of 1929 and the Great Recession of 2008.

We estimate the parameters using Bayes' formula

$$\begin{aligned} P(K_0, M_0, L_0, \hat{\beta}|k_0, k_1, \ldots, k_T) &= \frac{P(r_T|K_0, M_0, L_0, \hat{\beta},))}{P(k_T)} \cdots \frac{P(k_0|K_0, M_0, L_0, \hat{\beta})}{P(k_0)} \\ &\times f(K_0, M_0, L_0, \hat{\beta}), \end{aligned} \tag{12.29}$$

where $f(K_0, M_0, L_0, \hat{\beta})$ is a prior distribution, for which we used a uniform distribution. We apply the maximum a posteriori (MAP) estimation expressed by Eq. (12.29). The use of the NBD to determine the distribution P would be the parameter estimation for the discrete SE-NBD process introduced in Sect. 12.2. The use of the Poisson distribution instead of the NBD would be parameter estimation of the discrete Hawkes process. We present the results of the optimization using the discrete SE-NBD, discrete Hawkes, and NBD processes in Tables 12.2, 12.3, and 12.4. NBD was employed to confirm the existence of self-excitation. For the Hawkes process, $K_0 \to \infty(\omega = 0)$, and for the NBD process, $L_0 \to \infty(\tilde{\omega} = 0)$, as discussed in Sect. 12.2.

When K_0 is large, the process is nearly a Hawkes process. When L_0 is large, it is nearly an NBD process with no self-excitation. The estimated K_0 is small for the SE-NBD process, in particular, for IG. As shown in Table 12.4, we obtain a much

Table 12.2 MAP estimation of the parameters for the discrete SE-NBD and discrete Hawkes processes

		SE-NBD					Hawkes			
No.	Model	K_0	L_0	M_0/K_0	$\hat{\beta}$	$\bar{v}$	M_0	L_0	$\hat{\beta}$	$\bar{v}$
1	Moody's 1920–2020	0.28	6.17	18.89	2.94	58.35	3.4	3.55	15.98	86.85
2	S&P 1981–2020	1.06	27.80	18.95	16.08	71.22	13.3	15.76	18.40	83.65
3	Moody's 1981–2020	1.03	32.12	22.55	15.97	82.79	17.2	21.09	16.69	91.72
4	S&P 1990–2020	1.51	62.52	23.04	16.23	78.64	29.2	45.37	13.01	81.82
5	Moody's 1990–2020	1.58	86.55	27.92	14.40	90.00	40.6	73.03	19.19	91.38
6	Moody's 1920–2020 SG	0.29	5.91	17.81	3.03	56.04	2.9	3.00	13.57	105.32
7	S&P 1981–2020 SG	1.05	25.66	17.90	16.22	69.90	12.0	13.93	17.57	85.29
8	Moody's 1981–2020 SG	1.02	30.57	21.65	15.99	80.65	15.4	18.53	15.71	92.38
9	S&P 1990–2020 SG	1.54	60.05	21.82	16.02	76.39	28.2	43.74	18.97	79.59
10	Moody's 1990–2020 SG	1.62	86.79	26.76	15.44	87.05	39.6	71.53	14.57	88.57
11	Moody's 1920–2020 IG	0.13	1.10	4.06	0.99	2.13	1.14	0.40	0.98	1.24
12	S&P 1981–2020 IG	0.39	2.87	3.06	15.32	2.05	1.2	2.67	14.68	2.05
13	Moody's 1981–2020 IG	0.28	6.07	5.37	1.26	2.36	1.7	6.02	14.63	2.34
14	S&P 1990–2020 IG	0.33	2.73	3.75	16.80	2.32	1.2	2.65	17.46	2.32
15	Moody's 1990–2020 IG	0.28	4.13	5.80	13.27	2.69	1.8	5.58	16.26	2.68

better AIC for the SE-NBD process. This implies an intensity function is not a delta function, as in the Hawkes process. In fact, certain defaulted obligors affect other obligors. An obligor connected to many obligors affects many other obligors. This corresponds to the chain bankruptcy situation and this is called network effects. In fact, the AIC for the SE-NBD process was smaller than that for the NBD process. Hence, we can confirm the existence of self-excitation in this historical credit dataset.

12.6 Summary

In this chapter, we considered a multi-term urn process with a correlation in the same term and temporal correlation. Each term is the Pólya urn model, representing the correlation in the same term. The temporal correlation represents the correlation effects from the previous terms. When the number of red balls is much smaller, we obtain the Poisson process with the gamma distribution intensity function, the NBD process. We introduced the temporal correlation as the conditional distribution of the intensity function. This is equivalent to a self-exciting negative binomial distribution (SE-NBD) with conditional parameters. We referred to this process as the discrete SE-NBD process. This process becomes a discrete Hawkes process without correlation in the same term but with a temporal correlation between two different terms. In the standard continuous limit of the discrete SE-NBD process, we obtained the Hawkes process. On the other hand, taking the double-scaling limit obtained a continuous SE-NBD process. The difference between the continuous SE-NBD and Hawkes processes is the variance in the intensity function. In other words, at the limit where

Table 12.3 MAP estimation of the parameters for the NBD process

No.	Model	K_0	M_0/K_0	$\bar{\upsilon}$
1	Moody's 1920–2020	0.47	80.64	37.86
2	S&P 1981–2020	1.54	38.61	59.62
3	Moody's 1981–2020	1.52	45.76	69.78
4	S&P 1990–2020	2.07	34.37	71.29
5	Moody's 1990–2020	2.14	38.86	83.23
6	Moody's 1920–2020 SG	0.47	76.28	35.81
7	S&P 1981–2020 SG	1.55	37.14	57.57
8	Moody's 1981–2020 SG	1.53	44.00	67.45
9	S&P 1990–2020 SG	2.12	32.46	68.97
10	Moody's 1990–2020 SG	2.20	36.65	80.58
11	Moody's 1920–2020 IG	0.29	7.18	2.05
12	S&P 1981–2020 IG	0.46	4.42	2.05
13	Moody's 1981–2020 IG	0.37	6.30	2.33
14	S&P 1990–2020 IG	0.41	5.63	2.32
15	Moody's 1990–2020 IG	0.36	7.32	2.65

Table 12.4 AIC for the discrete SE-NBD, discrete Hawkes, and NBD processes

No.	Model	SE-NBD process	Hawkes process	NBD process
1	Moody's 1920–2020	791.9	2193.1	904.0
2	S&P 1981–2020	386.7	1010.6	407.9
3	Moody's 1981–2020	399.3	1186.1	420.6
4	S&P 1990–2020	316.3	923.0	323.5
5	Moody's 1990–2020	327.1	1098.6	332.4
6	Moody's 1920–2020 SG	781.5	2060.1	893.0
7	S&P 1981–2020 SG	383.2	975.8	405.1
8	Moody's 1981–2020 SG	396.5	1140.0	417.8
9	S&P 1990–2020 SG	313.9	894.0	321.0
10	Moody's 1990–2020 SG	325.1	1062.4	329.9
11	Moody's 1920–2020 IG	321.7	490.1	360.3
12	S&P 1981–2020 IG	150.0	197.6	153.2
13	Moody's 1981–2020 IG	156.8	257.1	156.8
14	S&P 1990–2020 IG	121.3	168.3	124.2
15	Moody's 1990–2020 IG	127.4	219.7	127.9

the intensity function becomes the delta function, the continuous SE-NBD process becomes the Hawkes process. The continuous SE-NBD process is a marked point process.

We observed a phase transition from the steady to non-steady state, which is the same type of phase transition as that in the Hawkes process. We observe a difference in the distribution of the intensity function at the critical point. The distribution functions of both models obey the power law at the critical point and have different indexes. We applied this process to the default data to estimate the parameters. According to our observation, the urn process is more effective for a default portfolio because of network effects.

12.7 Appendix A Moody's Correlated Binomial and Negative Binomial Distribution

The distribution of the Moody's correlated binomial distribution is

$$P(N_1 = k_1) = \frac{N!}{k_1!(N-k_1)!}\sum_{j=0}^{N-k_1}[(-1)^j \frac{(N-k_1)!}{j!(N-k_1-j)!}\Pi_{i=0}^{j+k_1-1} p_{i0}], \quad (12.30)$$

where $p_{i0} = 1-(1-p)(1-\rho)^i$. We can confirm that the distribution is the sum of the products $\Pi_{i=1}^{j+k} p_{i0}$.

Here we consider the product

$$\begin{aligned} N^{k_1}\Pi_{i=1}^{k_1} p_{i0} &\sim N^{k_1} p\cdot(p+\rho)\cdots(p+(k_1-1)\rho), \\ &= M_0\cdot(M_0+\frac{M_0}{K_0})\cdots(M_0+(k_1-1)\frac{M_0}{K_0}) \\ &= (\frac{M_0}{K_0})^{k_1}\frac{\Gamma(K_0+k_1)}{\Gamma(K_0)}. \end{aligned} \quad (12.31)$$

We take the double scaling limit $p \to \infty$, $\rho \to 0$, and $N \to \infty$ with the condition, $pN = M_0$, $\rho N = M_0/K_0$.

For the BBD, we can calculate in the same way,

$$\begin{aligned} N^{k_1}\Pi_{i=1}^{k_1} p_{i0} &= N^{k_1}\frac{\Pi_{i=0}^{k_1-1}(\theta_0+i\omega)}{\Pi_{l=0}^{k_1-1}(n_0+l\omega)} = \Pi_i^{k_1-1}(\theta_0\frac{N}{n_0}+i\omega\frac{N}{n_0}) \\ &= \frac{1}{k_1!}(\frac{M_0}{K_0})^{k_1}\frac{\Gamma(K_0+k_1)}{\Gamma(K_0)}. \end{aligned} \quad (12.32)$$

Then, in the limit, we can obtain the same $N^{k_1}\Pi_{i=1}^{k_1} p_{i0}$.

The distribution is in the double scaling limit,

$$
\begin{aligned}
P(N_1 = k_1) &= \frac{N!}{k_1!(N-k_1)!}\sum_{j=0}^{N-k_1}[(-1)^j \frac{(N-k_1)!}{j!(N-k_1-j)!}\Pi_{i=0}^{j+k_1-1} p_{i0}] \\
&\sim \frac{1}{k_1!}[N^{k_1}\Pi_{i=0}^{k_1} p_{i0} - \frac{1}{2!}N^{k_1+1}\Pi_{i=0}^{k_1+1} p_{i0} + \frac{1}{3!}N^{k_1+2}\Pi_{i=0}^{k_1+1} p_{i0}\cdots] \\
&= \frac{1}{k_1!}\sum_{i=0}(-1)^i \frac{1}{i!}(\frac{M}{K})^{k_1+i}\frac{\Gamma(K_0+k_1+i)}{\Gamma(K_0)} \qquad (12.33)
\end{aligned}
$$

Hence, in the double scaling limit of the Moody's correlated binomial model, we can obtain the NBD.

12.8 Appendix B Absorption of the Urn Model Before the Poisson Process Limit

In this Appendix, we present the urn model before the Poisson Process limit. Let $\{U_{\hat{t}}; \hat{t} \geq 1\}$ be an independent and identically distributed (i.i.d.) sequence that is uniformly distributed on [0,1]. We consider the $t+1$-th term $m+1$-th ball. Here, we introduce the variable $X(\hat{t})$ instead of $X_m(t)$, where

$$\hat{t} = Nt + m. \qquad (12.34)$$

$\hat{t}$ is the serial number of the process. The $\hat{t}+1$-th discrete dynamics of the process is described by:

$$X(\hat{t}+1) = \mathbf{1}_{U_{\hat{t}+1} \leq Z_d(\hat{t})}. \qquad (12.35)$$

Here, $Z_d(\hat{t})$ is given by

$$Z_d(\hat{t}) \equiv \frac{\alpha + \sum_{s=Nt+1}^{\hat{t}} X(s) + \tilde{\omega}\sum_{j=1}^{t} \hat{d}_{t-j}k_j}{\alpha+\beta+m}, \qquad (12.36)$$

where $\alpha = \theta_0/\omega = \theta_0$ and $\beta = (n_0 - \theta_0)/\omega = n_0 - \theta_0$. Here, we set $\omega = 1$.

To simplify the model, we set $N = 1$, $j \geq 1$ in Eq. (12.36). Let $\{U_t; t \geq 1\}$ be an independent and identically distributed (i.i.d.) sequence that is uniformly distributed on [0,1]. The discrete dynamics of the process are described by:

$$X(t+1) = \mathbf{1}_{U_{t+1} \leq Z_d(t)}.$$

Here, $Z_d(t)$ is the weighted sum of $X(s)$, $s \leq t$ with the discount factor $\hat{d}_{t-s}$,

$$Z_d(t) \equiv \frac{\alpha + \tilde{\omega}\sum_{s=1}^{t} X(s)\hat{d}_{t-s}}{\alpha+\beta}. \qquad (12.37)$$

Here, we apply the mean field approximation. We set $\bar{v}$ is the equilibrium ratio of the red ball, $\bar{v} = \sum X(t)/t$ in the large t limit. We obtain the self-consistent equation,

$$\bar{v} \equiv \frac{\alpha + \tilde{\omega}\hat{T}\bar{v}}{\alpha + \beta}. \tag{12.38}$$

where $\hat{T} = \sum_{s=1}^{t} \hat{d}_{t-s}$. When $\bar{v} = 1$, after that only red ball is taken out. This corresponds to Eq. (12.20) for Hawkes peocess. Hence, the process becomes an absorption process. The condition of the absorption is

$$\tilde{\omega}\hat{T}_c = \beta, \tag{12.39}$$

where $\hat{T}_c$ is the transition point. When $\hat{T} > \hat{T}_c$, it is the state of absorption. On the other hand, $\hat{T} < \hat{T}_c$, we can take out not only the red ball but also white ball.

12.9 Appendix C Proof of Eq. (12.2)

We show the relation between the beta binomial and the negative binomial distributions. First, we begin with the beta binomial distribution.

$$\begin{aligned} P(N_1 = k_1) &= \frac{N!}{k_1!(N-k_1)!} \\ &\times [\frac{\theta_0}{n_0} \cdots \frac{\theta_0 + (k_1 - 1)\omega}{n_0 + (k_1 - 1)\omega}][\frac{n_0 - \theta_0}{n_0 + k_1\omega} \cdots \frac{n_0 - \theta_0 + (N - k_1 - 1)\omega}{n_0 + (N-1)\omega}] \\ &= \frac{N!}{k_1!(N-k_1)!} \frac{\Pi_{i=0}^{k_1-1}(\theta_0 + i\omega)\Pi_{j=0}^{N-k_1-1}(n_0 - \theta_0 + j\omega)}{\Pi_{l=0}^{N-1}(n_0 + l\omega)} \\ &= \frac{N!}{k_1!(N-k_1)!} \frac{\Pi_{i=0}^{k_1-1}(\theta_0 + i\omega)}{\Pi_{l=0}^{k_1-1}(n_0 + l\omega)} \cdot \frac{\Pi_{j=0}^{N-k_1-1}(n_0 - \theta_0 + j\omega)}{\Pi_{m=0}^{N-k_1-1}(n_0 + m\omega)} \end{aligned} \tag{12.40}$$

First, we consider the first part. In the double scaling limit, we have

$$\begin{aligned} &\frac{N!}{k_1!(N-k_1)!} \frac{\Pi_{i=0}^{k_1-1}(\theta_0 + i\omega)}{\Pi_{l=0}^{k_1-1}(n_0 + l\omega)} = \frac{1}{k_1!}\Pi_i^{k_1-1}(\theta_0 \frac{N}{n_0} + i\omega\frac{N}{n_0}) \\ &= \frac{1}{k_1!}(\frac{M}{K})^{k_1} \frac{\Gamma(K + k_1)}{\Gamma(K)}. \end{aligned} \tag{12.41}$$

Here, we set the first double scaling limit $N/n_0 = \Delta$, $N \to \infty$ and $n_0 \to \infty$. Next, we consider the second part,

$$\frac{\Pi_{j=0}^{N-k_1-1}(n_0-\theta_0+j\omega)}{\Pi_{m=0}^{N-k_1-1}(n_0+m\omega)} = \frac{\Pi_{j=0}^{K+k_1-1}(n_0-K\omega+j\omega)}{\Pi_{j=0}^{K+k_1-1}(n_0+N\omega-k_1\omega-K\omega+j\omega)}$$
$$=(\frac{1}{1+M/K})^{K+k_1}. \tag{12.42}$$

We obtain

$$P(N_1=k_1)\sim \mathrm{NBD}(N_1=k_1|K_0,M_0/K_0)=\frac{(K_0+k_1-1)!}{k_1!(K_0-1)!}(\frac{K_0}{K_0+M_0})^{K_0}(\frac{M_0}{K_0+M_0})^{k_1}, \tag{12.43}$$

where $M_0=\theta_0 N/n_0=\theta_0\Delta$ and $K_0=\theta_0/\omega$. Here, $\sim$ means that the stochastic variable on the LHS obeys the probability distribution on the RHS. This is a negative binomial distribution (NBD), $\mathrm{NBD}(N_1=k_1|K_0,M_0/K_0)$. Parameter $M_0/K_0=\omega N/n_0=\omega\Delta$ is related to the correlation in this term. The mean and the variance of the NBD are given M_0 and $M_0+M_0^2/K_0$, respectively.

12.10 Appendix D Proof of Eq. (12.3)

$$\int_0^\infty \mathrm{Poisson}(k_1|\lambda)\cdot \mathrm{Gamma}(\lambda|K_0,M_0/K_0)d\lambda$$
$$=\int_0^\infty \frac{\lambda^{k_1}e^{-\lambda}}{k_1!}\frac{\lambda^{K_0-1}}{\Gamma(K_0)(M_0/K_0)^{K_0}}e^{-\lambda K_0/M_0}d\lambda,$$
$$=\frac{(M_0/K_0)^{-K_0}}{k_1!\Gamma(K_0)}\int_0^\infty \frac{\Gamma(K_0+k_1)}{\Gamma(K_0+k_1)}\frac{(M_0/(M_0+K_0))^{K_0+k_1}}{(M_0/(M_0+K_0))^{K_0+k_1}}\lambda^{K_0+k_1-1}e^{-\lambda/(M_0/(M_0+K_0))}d\lambda$$
$$=\frac{(M_0/K_0)^{-K_0}}{k_1!\Gamma(K_0)}\Gamma(K_0+k_1)(\frac{M_0}{M_0+K_0})^{K_0+k_1}$$
$$=\frac{\Gamma(K_0+k_1)}{k_1!\Gamma(K_0)}(\frac{M_0}{M_0+K_0})^{k_1}(\frac{K_0}{M_0+K_0})^{K_0}=\mathrm{NBD}(N_1=k_1|K_0,M_0/K_0), \tag{12.44}$$

where we use the relation at the third equal,

$$\int_0^\infty \frac{1}{\Gamma(K_0+k_1)}\frac{1}{(M_0/(M_0+K_0))^{K_0+k_1}}\lambda^{K_0+k_1-1}e^{-\lambda/(M_0/(M_0+K_0))}d\lambda=1, \tag{12.45}$$

because it is the integral of $\mathrm{Gamma}(\lambda|K_0+k_1,M_0/(M_0+K_0))$.

References

Blanc P, Donier J, and Bouchard J -P, Quadratic Hawkes processes for financial prices *Quant. Fin.* **17** 171 (2017).

Errais E, Giesecke K, Goldberg L R, Affine point processes and portfolio credit risk *SIAM J. Fin Math* **1** 642–665 (2010).

Florescu I, Mariani M C, Stanley H E, Viens, F. G. (Eds.) *Handbook of High-Frequency Trading and Modeling in Finance* (John Wiley & Sons, 2016).

Hawkes A G, Spectra of some self-exciting and mutually exciting point processes. Biometrika, **58(1)**, 83–90 (1971).

Hawkes A G and Oakes D, A Cluster Process Representation of a Self-Exciting Process *J App Prob* **11** 493–503 (1974).

Hisakado M, Hattori K, and Mori S, From the multiterm urn model to the self-exciting negative binomial distribution and Hawkes processes, *Phys Rev E.* **106** 034106 (2022).

Kanazawa K and Sornett D, Field master equation theory of the self-excited Hawkes process *Phys. Rev. Research* **2** 033442 (2020a).

Kanazawa K and Sornett D, Non-universal power law distribution of intensities of the self-excited Hawkes process: a field-theoretical approach*Phys. Rev. Lett.* **125** 138301 (2020b).

Kirchner M, An estimation procedure for the Hawkes process *Quant. Fin.* **17(7)** 999–1020 (2017).

Chapter 13
Multi-dimensional Self-Exciting NBD Process and Hawkes Graph

13.1 Introduction

In Chap. 12, we introduced the self-exciting negative binomial (SE-NBD) process, which has a correlation in the same term and temporal correlation. The correlation in the same term is the same as that considered in the Pólya urn model, while the temporal correlation is incorporated by introducing the conditional initial condition. The SE-NBD process is a Poisson process with a gamma-distributed intensity function. In the SE-NBD process with a conditional intensity function, as the number of events increases, the probability of the events also increases, similar to the Hawkes process with a conditional intensity function.

The SE-NBD process is within the double-scaling limit of the urn process. In the standard continuous limit, such an urn process becomes the Hawkes process with no correlations in the same term. Both processes have the same phase transition point from the steady state to the non-steady state. The difference between these processes is the variance in the intensity function. The variance in the intensity function of the SE-NBD is non-zero, and the intensity function is the gamma distribution.

For the default portfolio, the results obtained by the urn process are superior to those obtained by the Hawkes process. This result is due to network effects that correspond to the variance of the intensity function. In fact, some firms affect many other firms, whereas others do not affect other firms. This is among the network effects because a network with hubs has a large variance in the degree distributions. In this study, we used 13 sector portfolios to extend this research. We use the Hawkes graph to demonstrate the interaction among sectors (Embrechts et al. 2017).

The remainder of this paper is organized as follows. In Sect. 13.2, we review the SE-NBD process. In Sect. 13.3, we introduce the multidimensional SE-NBD process. In Sect. 13.4, We discuss the relation to Branching process. In Sect. 13.5, we describe the application of this process to the default history empirical data and discuss the network. Finally, we draw a summary in section. This chapter is based on Hisakado et al. (2022).

M. Hisakado, *Urn Models and Their Applications in Finance*,
https://doi.org/10.1007/978-981-96-3825-3_13

13.2 Review of the Negative Binomial Distribution Process

In this section, we review the discrete self-exciting negative binomial distribution (SE-NBD) process which was discussed in Chap. 12 as the preparation of this chapter. N_t denotes the number of events at time t, and $\hat{d}_i$ denotes the discount factor for the temporal correlation or the kernel function. In addition, we considered the exponential decay $\hat{d}_i = r^{i-1}$ in this study. This corresponds to a short memory (Florescu et al. 2016). The number of events at time t corresponds to the negative binomial distribution (NBD),

$$P(N_{t+1} = k_{t+1}) \sim \text{NBD}(N_{t+1} = k_{t+1}|K_t, M_t/K_t = M_0/K_0)$$
$$= \frac{(K_t + k_{t+1} - 1)!}{k_{t+1}!(K_t - 1)!}\left(\frac{K_t}{K_t + M_t}\right)^{K_t}\left(\frac{M_t}{K_t + M_t}\right)^{k_{t+1}}, \quad (13.1)$$

where

$$M_t = M_0 + M_0/L_0 \sum_i^t k_i \hat{d}_{t+1-i}, \quad (13.2)$$

and

$$K_t = K_0 + K_0/L_0 \sum_i^t k_i \hat{d}_{t+1-i}, \quad (13.3)$$

and

$$M_t/K_t = M_0/K_0. \quad (13.4)$$

Conditional probability is defined by updating parameters K_t and M_t. Note that for all process parameters M_t/K_t, M_0/K_0 is constant, which simultaneously corresponds to the correlation at the same time. Under these conditions, the process exhibits the reproductive properties of an NBD. L_0 and $\hat{d}_i$ are the parameters for the temporal correlation. The negative binomial distribution $\text{NBD}(N_{t+1} = k_{t+1}|K_t, M_t/K_t = M_0/K_0)$ has the following form:

$$\text{NBD}(N_{t+1} = k_{t+1}) = \int_0^\infty \text{Poisson}(k_{t+1}|\lambda_{t+1}) \cdot \text{Gamma}(\lambda_{t+1}|K_t, M_0/K_0) d\lambda_{t+1},$$
$$= \int_0^\infty \frac{\lambda^{k_{t+1}} e^{-\lambda_{t+1}}}{k_{t+1}!} \frac{\lambda_{t+1}^{K_t - 1}}{\Gamma(K_t)(M_0/K_0)^{K_t}} e^{-\lambda K_0/M_0} d\lambda_{t+1}, \quad (13.5)$$

where $\text{Poisson}(k_{t+1}|\lambda_{t+1})$ is a Poisson process, and $\text{Gamma}(\lambda_{t+1}|K_t, M_0/K_0)$ is a gamma distribution. The intensity function λ_{t+1} follows a gamma distribution. Note that $\text{Gamma}(\lambda|K_t, M_0/K_0)$ has an average M_t and variance M_t^2/K_t. The intensity function has a variance compared to the Poisson process, which has a zero-variance intensity function, namely, the Hawkes process (Hawkes 1971). We call this the discrete self-exciting negative binomial distribution (SE-NBD) process. Self-excitation is introduced by the conditional probability through K_t and M_t. When $M_t = M_0$ and $K_t = K_0$, the non-self-exciting case is referred to as the NBD process. In summary, we obtain a discrete SE-NBD process N_t that obeys the NBD for M_t as,

$$N_{t+1} \sim \text{NBD}\left(K_t, M_0/K_0\right), t \geq 0, \tag{13.6}$$

where

$$M_t = M_0 + M_0/L_0 \sum_{s=1}^{t} N_s \hat{d}_{t+1-s}, t \geq 1, \tag{13.7}$$

and

$$K_t = K_0 + K_0/L_0 \sum_{s=1}^{t} N_s \hat{d}_{t+1-s}, t \geq 1. \tag{13.8}$$

In the limit $K_0 \to \infty$, we obtain a discrete Hawkes process. N_t follows a Poisson process for M_t,

$$N_{t+1} \sim \text{Poisson}\left(M_t\right), t \geq 0, \tag{13.9}$$

where

$$M_t = M_0 + M_0/L_0 \sum_{s=1}^{t} N_s \hat{d}_{t+1-s}, t \geq 1. \tag{13.10}$$

Here, we set the average $\bar{v}$ of the intensity function. **The mean field approximation in Eq.** (13.7) **is as follows:**

$$\bar{v} = M_0/(1 - (M_0/L_0)\hat{T}), \tag{13.11}$$

where $\hat{T} = \sum^{\infty} \hat{d}_i = 1/(1-r)$. When $M_0/L_0\hat{T} < 1$, we obtain a steady state. However, when $M_0/L_0\hat{T} \geq 1$, a non-steady state is observed. **This denotes the phase transition between the steady and non-steady states and the transition point is** $M_0/L_0\hat{T} = 1$. This self-consistent equation and the transition point are the same as those for the discrete Hawkes process, as described in Eq. (13.11).

13.3 Multi-dimensional SE-NBD Process

13.3.1 Introduction of Multi-dimensional SE-NBD Process

We extended the discrete SE-NBD process to a multidimensional model (MD-SE-NBD). $N_t^{(i)}$ obeys NBD for $M_t^{(i)}$, where $i = 1, 2, \ldots, D$,

$$N_{t+1}^{(i)} \sim \text{NBD}\left(K_t^{(i)}, M_0^{(i)}/K_0^{(i)}\right), t \geq 0, \tag{13.12}$$

where

$$M_t^{(i)} = M_0^{(i)} + \sum_{j=1}^{D} M_0^{(i)}/L_0^{(ij)} \sum_{s=1}^{t} N_s^{(j)} \hat{d}_{t+1-s}^{(i)}, t \geq 1, \tag{13.13}$$

and

$$K_t^{(i)} = K_0^{(i)} + \sum_{j=1}^{D} K_0^{(i)}/L_0^{(ij)} \sum_{s=1}^{t} N_s^{(j)} \hat{d}_{t+1-s}^{(i)}, t \geq 1, \tag{13.14}$$

where $\hat{d}_s^{(i)} = r_i^{s-1}$ denotes the exponential decay. In this model, the interaction from i to j is $M_0^{(j)}/L_0^{(ji)}$ and from j to i is $M_0^{(i)}/L_0^{(ij)}$. Note that, in this model, the interaction is asymmetrical.

Accordingly, we also consider the discrete multidimensional Hawkes process (MD-Hawkes) $N_t^{(i)}$ that follows a Poisson distribution for $M_t^{(i)}$, where $i = 1, 2, \ldots, D$,

$$N_{t+1}^{(i)} \sim \text{Poisson}\left(M_t^{(i)}\right), t \geq 0, \tag{13.15}$$

and where

$$M_t^{(i)} = M_0^{(i)} + \sum_{j=1}^{D} M_0^{(i)}/L_0^{(ij)} \sum_{s=1}^{t} N_s^{(j)} \hat{d}_{t+1-s}^{(i)}, t \geq 1, \tag{13.16}$$

and additionally, where $\hat{d}_s^{(i)} = r_i^{s-1}$ denotes the exponential decay.

Here, we consider the interaction between lines j and i as S_{ij},

$$S_{ij} = (M_0^{(i)}/L_0^{(ij)})T^{(i)}, \tag{13.17}$$

where $T^{(i)} = \sum_{s=1}^{\infty} \hat{d}_s^{(i)}$. In the matrix form, we obtain the following expression:

$$\mathbf{S} \equiv \begin{pmatrix} S_{11} & S_{12} & \cdots & S_{1D} \\ S_{21} & S_{22} & \ddots & \vdots \\ \vdots & \ddots & \ddots & \vdots \\ S_{D1} & \cdots & \cdots & S_{DD} \end{pmatrix}. \tag{13.18}$$

In a single case, we considered an event as the impact. The total expected number of events affected by an event is

$$\frac{M_0}{L_0} \sum_{s=1}^{\infty} \hat{d}_s.$$

In the case $\hat{d}_i = r^{i-1}$, it becomes $M_0/L_0(1-r)$. S_{ij} is the multidimensional extension of this formula and denotes the number of affected events in the i-th line by an event in the j-th line. This is the direct effect of the event. In the next subsection, we discuss the indirect effects of an event through the network.

That is, parameter S_{ij} is the effective reproduction number for an infectious disease from lines j to i. This denotes the number of patients infected by one patient in the infection model. Therefore, we can study the interaction between lines i and j using matrix **S**.

Here, we set the average $\bar{v}^{(i)}$ of the i-th line and vector $\bar{\boldsymbol{v}} = \bar{v}^{(i)}$; then, we obtain the matrix-form equations for $\bar{v}^{(i)}$ using Eq. (13.13),

$$\bar{\boldsymbol{v}} = \mathbf{M_0} + \mathbf{S}\bar{\boldsymbol{v}}, \tag{13.19}$$

where $\mathbf{M_0}$ is the vector $\mathbf{M_0} = M_0^{(i)}$. Solving Eq. (13.19), we obtain the equilibrium solution for $\bar{\boldsymbol{v}}$,

$$\bar{\boldsymbol{v}} = (\mathbf{E} - \mathbf{S})^{-1}\mathbf{M_0}, \tag{13.20}$$

where **E** is the unit matrix. Equation (13.20) corresponds to Eq. (13.11) in the single-line case, **while Eq.** (13.19) **denotes the mean field approximation of the MD-SE-NBD process. We can confirm the transition point between the steady and non-steady states, that is, E = S described in Eq.** (13.20) **for a multidimensional case.** In the same way, we can obtain the equilibrium solution for the hybrid case that includes both the SE-NBD and Hawkes processes. This means that some lines are MD-SE-NBD and the rest are MD-Hawkes processes that interact with each other.

13.3.2 Impact Analysis of the MD-SE-NBD Process

In this subsection, we consider the impact sensitivity. In other words, we study the increase in future events when an event is added through the network. This denotes the sensitivity to adding an event.

To study the impact sensitivity, we considered the first shock for the i-th line at time 0. This is to compare sectors using a network such as Google's page rank (Langville et al. 2006). The shock vector is $\mathbf{v}_0^{(i)} = (0, \ldots, 1, \ldots, 0)$, in which only i-th element is 1, and the others are 0. This corresponds to an event on the i-th line. We considered the increase in events due to this impact. In the single-line case, the added number of events for the shock event is

$$v_\infty = \frac{M_0}{L_0}\sum_{i=1}^{\infty}(\frac{M_0}{L_0} + r)^i = \frac{M_0}{L_0}\frac{1}{1 - r - M_0/L_0}, \tag{13.21}$$

where $\hat{d}_i = r^{i-1}$. The proof is given in Sect. 13.7 Appendix A. **Note that** $M_0/(1 - r)L_0 = (M_0/L_0)\hat{T} = 1$ **is the transition point.** We extend this to a multidimensional case (Fig. 13.1).

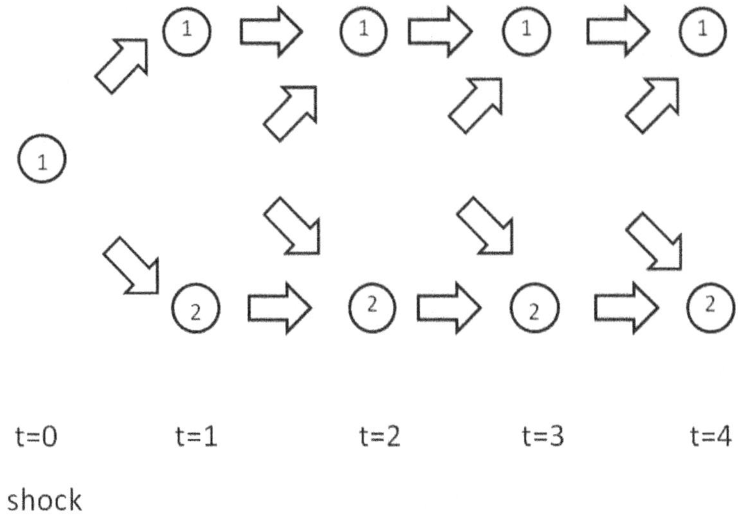

Fig. 13.1 Contagion of shock for the two-dimensional case. The impact decays exponentially

Here, we introduce two matrices $\hat{S}$

$$\hat{S} = \hat{S}^{(ij)} = M_0^{(i)}/L_0^{(ij)}, \tag{13.22}$$

and the diagonal matrix, $\hat{\boldsymbol{T}}$,

$$\hat{\boldsymbol{T}} = \hat{T}^{(ii)} = r_{(i)}. \tag{13.23}$$

Here, we consider the parameter vector

$$\mathbf{M}_t = (M_t^{(i)}). \tag{13.24}$$

$\mathbf{M}_t^+$ includes effects of the impact $\mathbf{v}_0^{(i)}$, while $\mathbf{M}_t$ does not include the impact. The effect of the impact at time t, $\mathbf{v}_t^{(i)}$, is defined as the difference

$$\mathbf{v}_t^{(i)} = \sum_{s=1}^{t} \mathrm{E}[M_s^{(i)+} - M_s^{(i)}], \tag{13.25}$$

where $\mathrm{E}(x)$ denotes the expected value of x. This is the sum of the increases in the expected value of the intensity function for the shock at time 0. In other words, it signifies the expected number of events for a shock in the future. We can obtain $\mathbf{v}_t^{(i)}$,

$$\mathbf{v}_t^{(i)} = \hat{S} \sum_{j=1}^{t} (\hat{\boldsymbol{T}} + \hat{S})^j \mathbf{v}_0^{(i)}, \tag{13.26}$$

where $j = t$ is the first contagion and $j = 0, \ldots, t-1$ denotes the temporal decay of the impacts.

In the limit $t \to \infty$, we calculate $\mathbf{v}_{\infty}^{(i)}$,

$$\mathbf{v}_{\infty}^{(i)} = \hat{S}\sum_{j=1}^{\infty}(\hat{T}+\hat{S})^{j}\mathbf{v}_{0}^{(i)} = (\mathbf{E}-\mathbf{S})^{-1}\mathbf{S}\mathbf{v}_{0}^{(i)}. \tag{13.27}$$

It is an extension of Eq. (13.21). $\mathbf{v}_{\infty}^{(i)}$ denotes the increase in the number of events in each line for the shock of $\mathbf{v}_{0}^{(i)}$. Sum of the elements of the vector $\mathbf{v}_{\infty} = \sum_i \mathbf{v}_{\infty}^{(i)}/D$ denotes the total increase in the number of events for the average shock, that is, $\sum_i \mathbf{v}_{0}^{(i)}/D$.

The shock of the upstream region has a larger impact than that of the downstream region. We discuss the relationship between the impact analysis and the branching process in Sect. 13.8 Appendix B. **Note that we can confirm the transition point $\mathbf{E} = \mathbf{S}$ depicted in Eq.** (13.27). **it is consistent with the conclusion of the mean field approximation, Eq.** (13.19).

13.3.3 Continuous Limit

In this section, we discuss the continuous limit of the discrete MD-SE-NBD model. Here, we introduce the counting process $\tilde{N}_t^{(i)} = \sum_s N_s^{(i)}$. $M_t^{(i)}$ is the average of the i-th line, as defined in Eq. (13.13). We set $M_t^{(i)} = \theta_t^{(i)}\Delta$. We can confirm that $M_t^{(i)}$ increases as Δ increases, and Δ plays the role of time. Moreover, we set $M_t^{(i)}/K_t^{(i)} = \omega_i'$, which is independent of t. Hence, the process has the reproductive properties of an NBD.

We obtain the mean of the intensity function in the continuous limit at t, $\lambda_t^{(i)}$, which is the intensity function at t, as shown in Eq. (13.5) for the single-line case,

$$\mathrm{E}(\lambda_t^{(i)}|F_t) = \lim_{\Delta\to 0}\frac{\mathrm{E}[\tilde{N}_{t+\Delta}^{(i)} - \tilde{N}_t^{(i)}|F_t]}{\Delta} = \lim_{\Delta\to 0}\frac{M_t^{(i)}}{\Delta} = (\theta_0^{(i)} + \sum_{j=1}^{D}\sum_{s<t}\tilde{\omega}_{ij}N_s^{(j)}\hat{d}_{t-s}^{(i)}), \tag{13.28}$$

where F_t is the history of the number of events and $\tilde{\omega}_{ij} = \theta^{(i)}/L_0^{(ij)}$. Equation (13.28) corresponds to the hazard function (Schönbucher 2003).

The variance in the intensity of the distribution at time t is

$$\mathrm{V}(\lambda_t^{(i)}|F_t) = \lim_{\Delta\to 0}\frac{(M_t^{(i)})^2/K_t^{(i)}}{\Delta} = \omega_i'(\theta_0^{(i)} + \sum_{j=1}^{D}\sum_{s<t}\tilde{\omega}_{ij}\hat{d}_{t-s}^{(i)}N_s^{(i)}). \tag{13.29}$$

In the continuous MD-SE-NBD process, the intensity function follows a gamma distribution as a discrete MD-SE-NBD process. In summary, we can obtain, in the continuous limit,

$$\tilde{N}_{t+\Delta}^{(i)} - \tilde{N}_t^{(i)} \sim \text{NBD}\left(\theta_t^{(i)}\Delta/\omega_i', \omega_i'\right), t \geq 0, \tag{13.30}$$

where

$$\theta_t^{(i)} = \theta_0^{(i)} + \sum_{j=1}^{D}\sum_{s<t} \tilde{\omega}_{ij} N_s^{(j)} \hat{d}_{t-s}^{(i)}, t \geq 0. \tag{13.31}$$

In the limit $\omega_i' \to 0$, the variance of the intensity function vanishes. Subsequently, the continuous SE-NBD process becomes the Hawkes process.

$$\tilde{X}_{t+\Delta}^{(i)} - \tilde{X}_t^{(i)} \sim \text{Poisson}\left(\theta_t^{(i)}\Delta\right), t \geq 0, \tag{13.32}$$

where

$$\theta_t^{(i)} = \theta_0^{(i)} + \sum_{j=1}^{D}\sum_{s<t} \tilde{\omega}_{ij} N_s^{(j)} \hat{d}_{t-s}^{(i)}, t \geq 0. \tag{13.33}$$

Next, we consider the correlation functions of the continuous MD-SE-NBD process. The calculations are presented in detail in Sect. 13.9 Appendix C. We can obtain the covariance as,

$$\text{E}[N_t^{(i)}, N_{t+\tau}^{(j)}] = [C_{ij}(\tau) + \bar{v}_i\bar{v}_j + (1+\omega_i')\bar{v}_i\delta_{ij}\delta(\tau)]\Delta^2, \tag{13.34}$$

where $C_{ij}(\tau)$ denotes the covariance density function for $\tau > 0$, δ_{ij} denotes the Kronecker's delta, and $\delta(x)$ denotes the delta function. There is no correlation between the i-th and j-th lines at the same time. The SE-NBD process has a larger correlation at the same time than the Hawkes process.

Here, we define $g_{ij}(x) = \tilde{\omega}_{ij}\hat{d}_x^{(i)}$ and set $\tau > 0$. We can obtain the integral equation for the correlation function using Eq. (13.20)

$$C_{ik}(\tau) = (\omega_i' + 1)\bar{v}_i g_{ii}(\tau) + \sum_{j=1}^{D}\int_0^{\infty} g_{kj}(w)C_{ij}(\tau - w)dw. \tag{13.35}$$

We show Eq. (13.35) in Sect. 13.8 Appendix B. The MD-Hawkes process is the case where $\omega_i' = 0$. The difference between the SE-NBD and Hawkes processes is the correlation in the same term, which is the first term on the RHS.

13.4 Branching Process

13.4.1 Impact and Branching Process

In this section, we consider the relationships between the impact and branching process to clarify the mean of the impact analysis based on the single-line case. We consider the case of $\hat{d}_1 = 1$ and $\hat{d}_s = 0, s = 2, 3, \ldots$ for Eqs. (13.6) and (13.9), which denote the discrete NBD and Hawkes processes, respectively. In this case, we obtain $\hat{T} = 1$. This corresponds to an event that affects only the next term.

In Eqs. (13.6) and (13.9), M_0 denotes the birth probability of the 0-th generation, which corresponds to the parent events, and the second term corresponds to the birth of the offspring events.

We focus on a single parent's event and its offspring events. N_s denotes the number of offspring events. The parent event occurs at t, and the offspring events occur in the s-th generation at $t + s$. Here, we set the parent event to the 0-th generation.

We define p_k as the offspring distribution of k children. With probability p_0, there are no offspring events. Using Eqs. (13.6) and (13.9), the offspring distribution becomes NBD$(K_0/L_0, M_0/K_0L_0)$, while the Poisson distribution becomes Poisson (M_0/L_0). Because these distributions have a reproductive function, the sum of the distributions becomes Eqs. (13.6) and (13.9). This corresponds to the decomposition of the SE-NBD and Hawkes processes, focusing on a single parent. This is a discrete-time branching process. An image of the process is shown in Fig. 13.2.

In summary, the process has the following properties:

- $N_0 = 1$ this is the parent event.
- When $N_n = i$, where $i = 0, 1, 2, \ldots$, if $i = 0$, $N_{n+1} = 0$. This corresponds to the extinction of the offspring event.
- If $i > 0$, new offspring events occur independently with the probability NBD$(K_0/L_0, M_0/K_0)$ or Poisson(M_0/L_0) at the next term.

Note that the impact is considered as the sum of the offspring events $\sum_k N_k - 1$. This impact excludes the parent event.

Here, we consider the extinction probability ϵ and the probability of $N_n = 0$ for any n, considering the phase transition. After this term, there are no offspring events for the parent event. We can obtain the self-consistent equation

$$x = \sum_{k=0}^{\infty} p_k x^k, \tag{13.36}$$

and its solution is $x = \epsilon$. Here, we introduce the moment-generating function

$$g(x) = \sum_{k=0}^{\infty} p_k x^k. \tag{13.37}$$

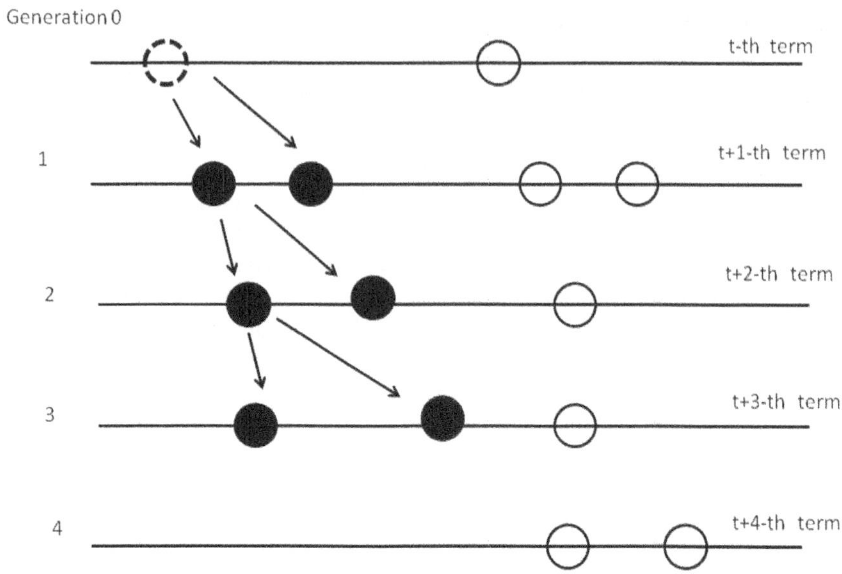

Fig. 13.2 Parent and offspring events. We focus on a single's parent event and its offspring events. The dotted circle denotes the parent event, and the black and white circles denote the offspring and other events, respectively

We rewrite Eq. (13.36) using the generating function,

$$x = g(x). \tag{13.38}$$

Equation (13.38) has the solution $x = 1$ because $g(1) = \sum_k p(k) = 1$. If $g'(1) > 1$, then there is only one solution, $x = 1$; accordingly, the probability is 1. If $g'(x) < 1$, then there are two solutions, where the smaller solution denotes the extinction probability ϵ. We show the image in Fig. 13.3. The transition point is $g'(1) = 1$. This is known as the branching process phase transition. Note that the order parameter of the phase transition is the survival probability $1 - x = 1 - \epsilon$. **Hence, we change the variable from** x to $y = 1 - x$ **which is the order parameter.** We rewrite Eq. (13.38) around $y \sim 0$,

$$1 - y = g(1 - y) = 1 - g'(1)y + g''(1)y^2/2 - \cdots .$$

Then, we can obtain,

$$y \sim \frac{2(g'(1) - 1)}{g''(1)} = \frac{2(M_0/L_0 - 1)^\beta}{g''(1)}, \tag{13.39}$$

where the critical exponent, $\beta = 1$ and $g''(1) > 0$.

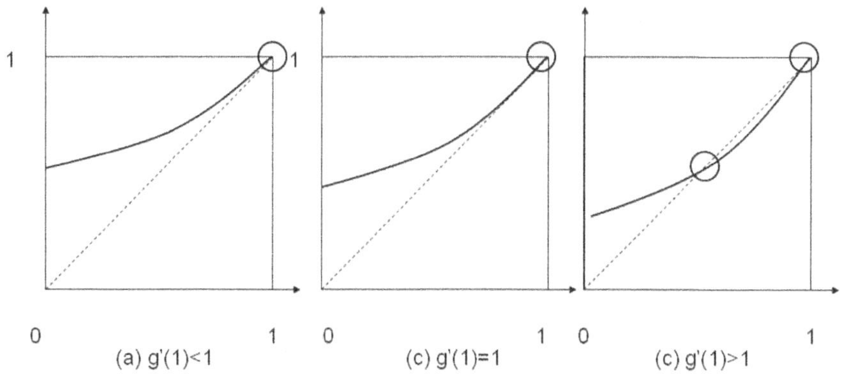

Fig. 13.3 Solutions of the self-consistent equation (Eq. (13.39)), **a** $g'(1) < 1$ **b** $g'(1) = 1$ and **c** $g'(1) >$. Below the critical point $g'_c(1) = 1$, we obtain solution (**a**). We refer to this phase as the one peak phase. In contrast, above the critical point, we obtain two solutions. One of them is stable, and one is unstable (**c**)

We consider the transition point from an impact analysis. The impact is the expected value of the number of offspring events, $v_\infty = \sum_{k=1}^{\infty}(M_0/L_0)^k = M_0/L_0 \cdot 1/(1 - M_0/L_0)$ for the SE-NBD and Hawkes processes determined using Eq. (13.21), where we assume that $\hat{d}_1 = 1$ and $\hat{d}_s = 0, s = 2, 3, \ldots$. Hence, we can confirm that the transition point is $M_0/L_0 = 1$. At this limit, the survival probability is 1.

13.4.2 Coefficients of the Variables

In this subsection, we calculate the coefficients of the variables (CV). $CV(\tilde{N}_T)$ is the coefficients of CV defined in Chap. 2. For the branching process, we obtain

$$\mathrm{E}(\tilde{N}_T) = \frac{\theta_0 T}{1 - \nu}, \tag{13.40}$$

and

$$\mathrm{V}(\tilde{N}_T) = \frac{\theta_0 T}{(1 - \nu)^3}. \tag{13.41}$$

We show the proof of them in Sect. 13.9 Appendix C. Therefore, we can calculate CV,

$$\mathrm{CV}(\tilde{N}_T) = \frac{1}{(1 - \nu)T}. \tag{13.42}$$

In the limit $T \to \infty$, Eq. (13.42) converges to 0 when $\nu < 1$. In contrast, when $\nu = 1$, Eq. (13.42) converges to $\neq 0$. Therefore, it is the phase transition between self-averaging process and non-self-averaging process.

13.5 Parameter Estimation for Default Data

In this section, we present the application of parameter estimation to the quarterly default data. S&P default data from 1981 to 2020 were used in this study. The dataset included 13 sectors (S&P 2020). We classified the 13 sectors as follows (Table 13.1). We considered the interaction as a network among the aforementioned sectors.

We estimated the parameters using the Bayes' formula:

$$P(K_0, M_0, L_0, r|k_1, \ldots, k_T) = (\frac{P(k_T|K_0, M_0, L_0, r)}{)} P(k_T) \cdots \frac{P(k_1|K_0, M_0, L_0, r)}{P(k_1)} \times f(K_0, M_0, L_0, r), \quad (13.43)$$

where $f(K_0, M_0, L_0, r)$ is the prior distribution. We used a uniform distribution for the prior distribution. We applied the maximum a posteriori (MAP) estimation described in Eq. (13.43). If we use the NBD for the distribution P, the process is the parameter estimation of the discrete MD-SE-NBD process introduced in Sect. 13.3. If we use the Poisson distribution instead of the NBD, the process is the parameter estimation for the discrete MD-Hawkes process. In this case, the parameters were M_0, L_0, and r (Table 13.1).

We applied the following five models: (1) multidimensional SE-NBD model (MD-SE-NBD), (2) multidimensional Hawkes model (MD-Hawkes), (3) SE-NBD model, (4) Hawkes model, and (5) NBD model. Models (3), (4), and (5) exclude interactions among sectors. Model (5) does not include self-excitation. Note that we can fit the model independently as sector models because of the definition of models (1) MD–SE-NBD and (2) MD-Hawkes. The models are summarized in Table 13.2. The AIC values of the models are given in Table 13.3. The parameters of models (1), (2), (3),

Table 13.1 Summary of 13 sectors

Number	Sector
1	Forest and building products/home builders
2	Consumer/service sector
3	Energy and natural resources
4	Financial institutions (FI)
5	Healthcare/chemical
6	Hi-tech/computers/office equipment
7	Insurance
8	Leisure time/media
9	Aerospace/automotive/capital goods/metal
10	Real estate
11	Healthcare/chemical
12	Telecommunications (TC)
13	Utility

Table 13.2 Summary of the models

Property	(1) MD-SE-NBD	(2) MD-Hawkes	(3) SE-NBD	(4) Hawkes	(5) NBD
Variance of the intensity function	○		○		○
Interaction among sectors	○	○			
Self exciting	○	○	○	○	

(4), and (5) are listed in Tables 13.4, 13.5, and 13.6. Here, S_{ij} denotes the interaction among sectors, as defined in Eq. (13.17).

From Table 13.3, we can confirm that (1) MD-SE-NBD has the lowest AIC among all sectors except real estate. For the real-estate sector (2) MD-Hawkes has the lowest AIC. In fact, we can confirm the large K_0 of the real-estate sector with models (1) MD-SE-NBD and (3) SE-NBD from Tables 13.4 and 13.6. A large K_0 indicates that the variance of the intensity function is small, similar to the Hawkes model. The Hawkes model has a zero-variance intensity function. In other words, except for the real-estate sector, these processes have a large variance in the intensity function.

Comparing the AIC of (1) MD-SE-NBD and (3) SE-NBD, we can confirm the effect of the interaction owing to other sectors based on the smaller AIC of (1) MD-SE-NBD in all sectors. In (3) SE-NBD, there is no interaction among the sectors. This can also be confirmed by comparing the AIC of the (2) MD-Hawkes and (4) Hawkes models because the MD-Hawkes model has a smaller AIC.

Comparing the AIC of (3) SE-NBD and (5) NBD, we can confirm the effect of self-excitation because of the smaller AIC of (3) SE-NBD in all sectors. In (5) NBD, there is no self-excitation.

Next, we confirm the interaction among the sectors using (1) MD-SE-NBD and (2) MD-NBD models. The interactions are represented by the matrix S, defined in Eq. (13.17). The interactions between the sectors are depicted in Fig. 13.4a and b using the parameters defined in Tables 13.4, 13.5, and 13.6. In the Hawkes process, these interactions are referred to Hawkes graphs (Embrechts et al. 2017). However, it is difficult to confirm the interactions between the upstream and downstream sectors of these networks.

To define the upstream and downstream sectors, we calculate the impacts $\mathbf{v}_{\infty}^{(i)}$ for (1) MD-SE-NBD and (2) MD-Hawkes, depicted in Figs. 13.5 and 13.6, respectively. The figures show the contagion of a shock in each sector. The impact $\mathbf{v}_{\infty}^{(i)}$ denotes the total increase in defaults due to the shock in the i-th sector. We show $\mathbf{v}^{(i)}$ for (1) MD-SE-NBD and (2) MD-Hawkes in Fig. 13.7a and b. The real-estate and FI sectors were upstream in both models. If we apply the hybrid model, we can obtain almost the same result as with the MD-SE-NBD model. Note that in the multiple sectors, the most affected sector is the real-estate sector, shown in Figs. 13.5 and 13.6. The shock in the real-estate sector affects the FI sector. Through the FI sector, the shock spreads to all sectors and returns to the real-estate sector. This is the cycle

Table 13.3 AIC for the discrete SE-NBD, discrete Hawkes, discrete NBD, and discrete Hawkes processes

		(1)	(2)	(3)	(4)	(5)
		MD-SE-NBD	MD-Hawkes	SE-NBD	Hawkes	NBD
Number	Model	AIC	AIC	AIC	AIC	AIC
1	Building	369.76	377.56	386.65	403.04	420.21
2	Consumer	616.63	626.16	627.20	647.18	710.24
3	Energy	498.82	552.08	504.77	570.07	625.28
4	FI	402.54	408.21	414.26	427.76	476.45
5	Health	387.40	390.22	389.28	395.27	420.75
6	Hi-tech	278.24	276.42	290.52	289.49	311.62
7	Insurance	253.37	261.49	256.48	267.26	257.72
8	Leisure	543.29	570.96	559.42	608.11	625.08
9	Metal	556.09	582.94	566.31	609.69	621.96
10	Real estate	124.39	122.34	133.37	131.37	166.12
11	TC	327.39	331.41	345.65	359.99	426.96
12	Transport	359.57	357.73	378.34	377.81	405.68
13	Utility	252.26	266.78	270.09	313.79	285.89

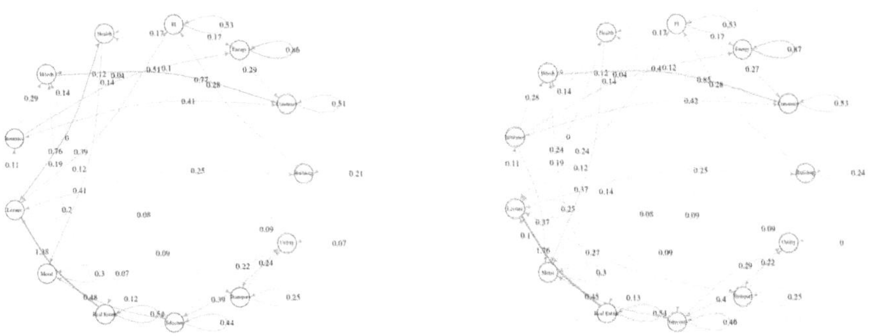

(a) (b)

Fig. 13.4 Interaction among the sectors: **a** MD-SE-NBD and **b** MD-Hawkes models

between the real-estate, FI, and other sectors. The cycle plays the role of an amplifier of the shock, as depicted in Fig. 13.8. This is consistent with the narrative of the Great Recession, 2009 because the origin is the shock of the price decrease in the real-estate sector in the US.

Table 13.4 Parameter estimation for (1) discrete MD-SE-NBD

Number	Model	K_0	M_0/K_0	S_{ij}
1	Building	0.41	0.82	$S_{11} = 0.21$, $S_{18} = 0.25$
2	Consumer	1.12	0.78	$S_{22} = 0.51$, $S_{24} = 0.29$, $S_{26} = 0.77$, $S_{27} = 0.41$
3	Energy	0.00	0.00	$S_{33} = 0.86$, $S_{34} = 0.17$, $S_{37} = 0.51$
4	FI	0.00	0.38	$S_{44} = 0.53$, $S_{41} = 0.28$, $S_{48} = 0.14$
5	Health	0.48	0.04	$S_{55} = 0.17$, $S_{52} = 0.10$, $S_{58} = 0.14$
6	Hi-tech	1.75	0.05	$S_{66} = 0.12$, $S_{63} = 0.04$, $S_{67} = 0.29$, $S_{610} = 0.19$, $S_{611} = 0.12$
7	Insurance	0.30	0.47	$S_{77} = 0.00$, $S_{78} = 0.11$
8	Leisure	0.35	0.82	$S_{88} = 0.20$, $S_{84} = 0.39$, $S_{85} = 0.76$, $S_{810} = 1.38$
9	Metal	0.74	0.78	$S_{99} = 0.30$, $S_{95} = 0.41$, $S_{911} = 0.48$
10	Real estate	23.84	0.00	$S_{1010} = 0.54$, $S_{104} = 0.08$
11	TC	0.00	0.38	$S_{1111} = 0.44$, $S_{118} = 0.07$, $S_{119} = 0.12$, $S_{1113} = 0.39$
12	Transport	4.45	0.04	$S_{1212} = 0.25$, $S_{122} = 0.09$, $S_{128} = 0.09$
13	Utility	0.02	0.73	$S_{1313} = 0.07$, $S_{1311} = 0.22$, $S_{1312} = 0.24$

13.6 Summary

In this chapter, we applied a multidimensional self-exciting negative binomial distribution (MD-SE-NBD) process to default portfolios with 13 sectors. The SE-NBD process is a Poisson process with a gamma-distributed intensity function. Using the multidimensional SE-NBD process, we estimated the interactions between these sectors as a network. We compared these results with those obtained using the multidimensional Hawkes (MD-Hawkes) process, which is a Poisson process with a zero-variance intensity function. By applying the empirical data, we created networks of contagious defaults among sectors. We confirmed that the upstream of contagious defaults is the real-estate and financial-institution (FI) sectors. These conclusions are consistent with our assessment of the default contagions.

Table 13.5 Parameter estimation for (2) discrete MD-Hawkes

Number	Model	M_0	S_{ij}
1	Building	0.16	$S_{11} = 0.24$, $S_{18} = 0.25$
2	Consumer	0.45	$S_{22} = 0.53$, $S_{24} = 0.27$, $S_{26} = 0.85$, $S_{27} = 0.42$
3	Energy	0.00	$S_{33} = 0.86$, $S_{34} = 0.17$, $S_{37} = 0.39$
4	FI	0.00	$S_{44} = 0.53$, $S_{41} = 0.28$, $S_{48} = 0.14$
5	Health	0.12	$S_{55} = 0.17$, $S_{52} = 0.12$, $S_{58} = 0.14$
6	Hi-tech	0.09	$S_{66} = 0.12$, $S_{63} = 0.04$, $S_{67} = 0.29$, $S_{610} = 0.19$, $S_{611} = 0.12$
7	Insurance	0.14	$S_{77} = 0.00$, $S_{78} = 0.11$
8	Leisure	0.32	$S_{88} = 0.25$, $S_{84} = 0.24$, $S_{85} = 0.24$, $S_{89} = 0.10$, $S_{810} = 1.76$, $S_{812} = 0.27$
9	Metal	0.36	$S_{99} = 0.30$, $S_{94} = 0.14$, $S_{95} = 0.37$, $S_{97} = 0.37$, $S_{911} = 0.45$
10	Real estate	0.00	$S_{1010} = 0.54$, $S_{104} = 0.08$
11	TC	0.00	$S_{1111} = 0.46$, $S_{114} = 0.09$, $S_{119} = 0.13$, $S_{1113} = 0.40$
12	Transport	0.19	$S_{1212} = 0.25$, $S_{122} = 0.09$, $S_{128} = 0.09$
13	Utility	0.00	$S_{1313} = 0.00$, $S_{1311} = 0.29$, $S_{1312} = 0.22$

Table 13.6 MAP estimation of the parameters for (3) discrete SE-NBD, (4) discrete Hawkes, and (5) NBD processes

		(3) SE-NBD			(4) Hawkes		(5) NBD	
Number	Model	K_0	M_0/K_0	S_{ii}	M_0	S_{ii}	K_0	M_0/K_0
1	Building	0.58	0.65	0.56	0.33	0.62	0.69	0.79
2	Consumer	1.06	0.66	0.75	0.62	0.78	1.33	0.49
3	Energy	0.22	1.40	0.88	0.27	0.90	0.38	16
4	Financial Institution	0.37	0.56	0.82	0.22	0.81	0.78	0.72
5	Health	0.48	0.41	0.81	0.17	0.83	0.98	1.17
6	Hi-tech	1.63	0.13	0.57	0.21	0.57	1.32	2.70
7	Insurance	0.34	0.52	0.53	0.23	0.37	0.67	1.86
8	Leisure	0.41	1.11	0.78	0.43	0.79	0.87	0.43
9	Metal	0.62	0.99	0.69	0.55	0.73	0.89	0.45
10	Real estate	2213709.45	0.00	0.78	0.05	0.78	0.48	2.70
11	Telecommunication	0.28	0.59	0.84	0.17	0.83	0.34	0.35
12	Transport	2.84	0.15	0.49	0.42	0.48	2.22	2.70
13	Utility	0.19	1.07	0.56	0.16	0.65	0.31	0.69

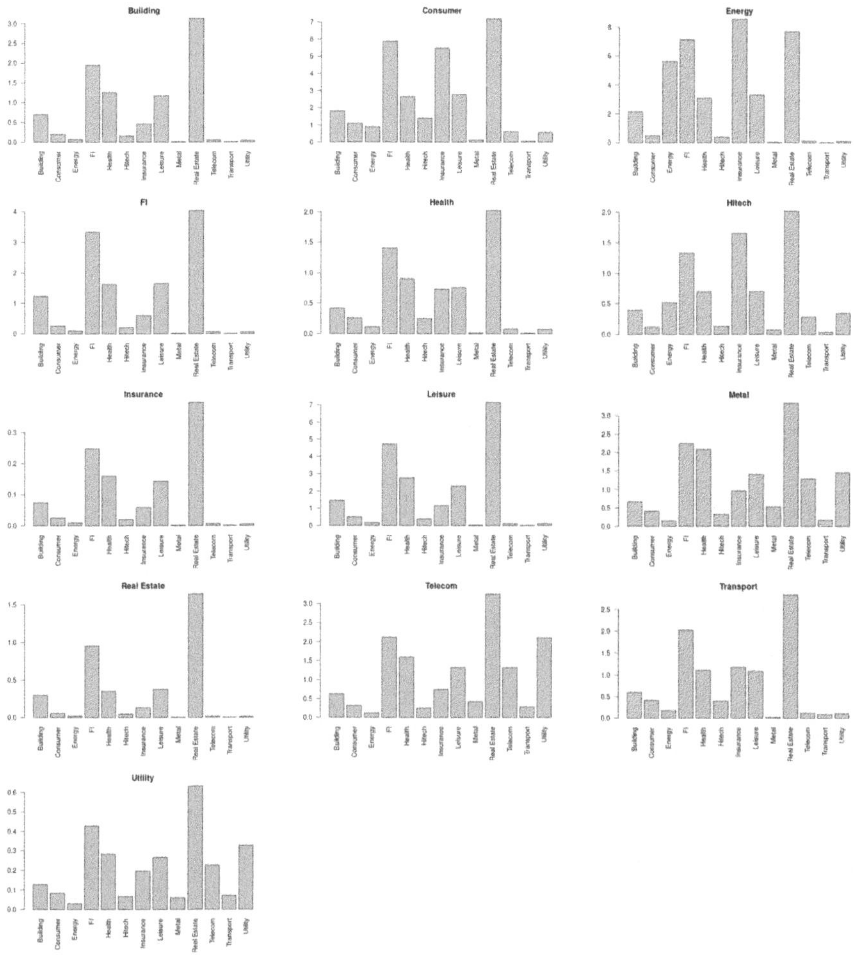

Fig. 13.5 Impact of each sector $\mathbf{v}_{\infty}^{(i)}$ obtained using (1) MD-SE-NBD

13.7 Appendix A Proof of Eq. (13.21)

In this section, we present the calculation of Eq. (13.21). We consider an event at $t = 0$ and the number of affected events. We consider the sequence a_t, which represents the effect at t and $v_t = \sum_t a_t$. The sequence is $a_0 = 1$ and

$$a_{t+1} = \frac{M_0}{L_0} \sum_{s=0}^{t} \hat{d}_{t-s} a_s = \frac{M_0}{L_0} \sum_{s=0}^{t} r^{t-s} a_s. \tag{13.44}$$

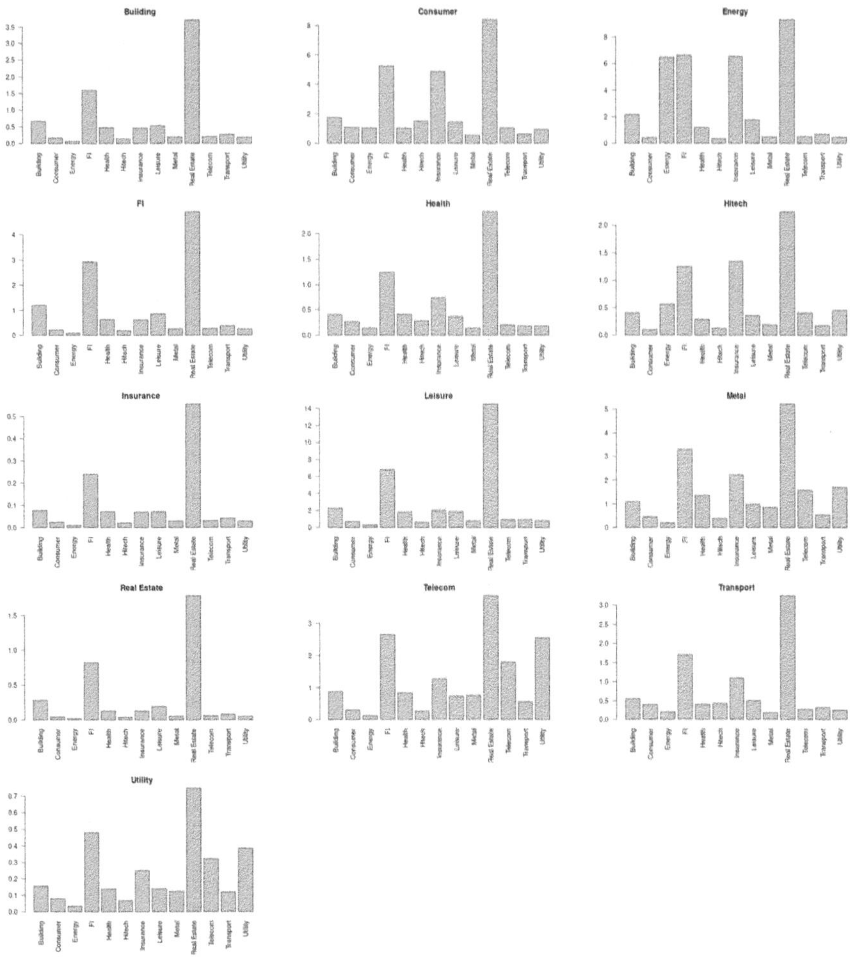

Fig. 13.6 Impact of each sector $\mathbf{v}_{\infty}^{(i)}$ obtained using (2) MD-Hawkes

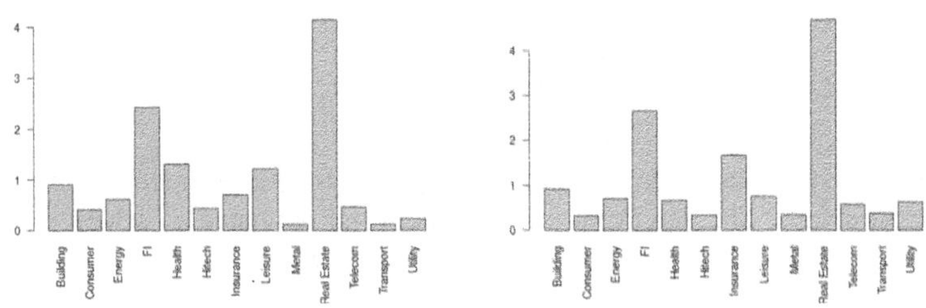

(a) (b)

Fig. 13.7 Impact $\mathbf{v}^{(i)}$ of the (**a**) MD-SE-NBD and (**b**) MD-Hawkes models

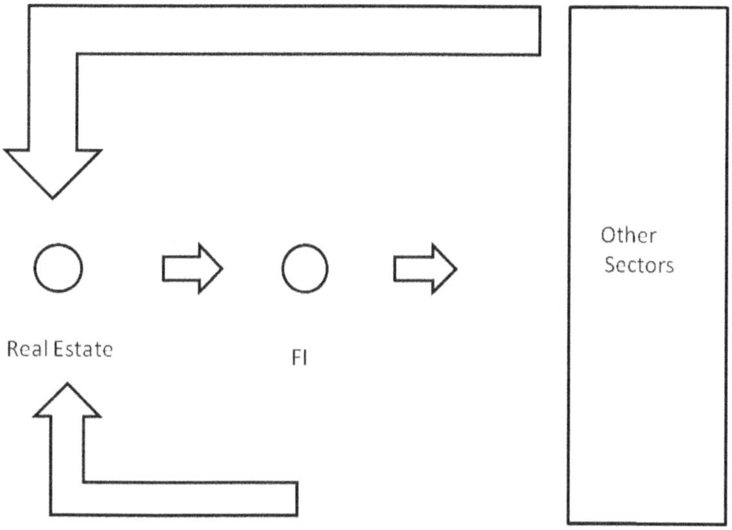

Fig. 13.8 Image of the contagion of shock among sectors. It plays the role of an amplifier of the shock

After solving this recurrence formula, we obtain

$$a_{t+1} = \frac{M_0}{L_0}\left(\frac{M_0}{L_0} + r\right)^t. \tag{13.45}$$

Subsequently, the impact is,

$$v_t = \frac{M_0}{L_0} \frac{1 - (r + \frac{M_0}{L_0})^t}{1 - r - \frac{M_0}{L_0}}. \tag{13.46}$$

In the limit $t \to \infty$, we can obtain Eq. (13.21).

Here, the scaled impact is

$$\frac{v_t}{t} = \frac{M_0}{L_0} \frac{1 - (r + \frac{M_0}{L_0})^t}{1 - r - \frac{M_0}{L_0}} \frac{1}{t}. \tag{13.47}$$

When $M_0/L_0(1-r) = M_0\hat{T}/L_0 < 1$, the scaled impact is 0. In contrast, in the case $M_0/L_0(1-r) = M_0\hat{T}/L_0 \to 1$, the scaled impact is not 0. This means that the impact does not decay and it represents the transition from the self-averaging process to non-self-averaging process. When the NBD process, $d_i = 0$, the process has no phase transition and becomes self-averaging process.

13.8 Appendix B Proof of Eq. (13.35)

In this section, we consider the correlation functions of the MD-SE-MBD process.

The covariance density function is defined as

$$Cov[N_t^{(i)}, N_{t'}^{(j)}] = E[N_t^{(i)}, N_{t'}^{(j)}] - E[N_t^{(i)}]E[N_{t'}^{(j)}]$$
$$= E[N_t^{(i)}, N_{t'}^{(j)}] - \bar{\upsilon}_i\bar{\upsilon}_j\Delta^2 = C_{ij}^*(t,t')\Delta^2, \tag{13.48}$$

where $C_{(ij)}^*(t,t')$ denotes the covariance density function.

We can obtain for $t = t'$,

$$E[N_t^{(i)}, N_{t'}^{(i)}] = E[(N_t^{(i)})^2] = (1+\omega_i')\bar{\upsilon}_i\Delta. \tag{13.49}$$

and

$$E[N_t^{(i)}, N_{t'}^{(j)}] = \bar{\upsilon}_i\bar{\upsilon}_j\Delta^2, \tag{13.50}$$

because the processes in lines i and j at the same time are independent. Note that the difference between the SE-NBD and Hawkes processes is only the correlation at the same time $t = t'$. Hence, the covariance density function can be decomposed as

$$C_{(ij)}^*(t,t') = (1+\omega_i')\bar{\upsilon}_i\delta_{ij}\delta(t-t') + C_{(ij)}(t,t'), \tag{13.51}$$

where $C_{ij}(t,t')$ is the correlation function, δ_{ij} is the Kronecker's delta, and $\delta(x)$ is the delta function.

If we set $\tau = t' - t$, we can rewrite Eq. (13.48) using Eq. (13.51),

$$E[N_t^{(i)}, N_{t+\tau}^{(j)}] = [C_{ij}(\tau) + \bar{\upsilon}_i\bar{\upsilon}_j + (1+\omega_i')\bar{\upsilon}_i\delta_{ij}\delta(\tau)]\Delta^2. \tag{13.52}$$

Here, we define $g_{ij}(x) = \tilde{\omega}_{ij}\hat{d}_t^{(i)}(x)$ and set $\tau > 0$.

We can calculate the following correlations:

$$E[N_t^{(i)}N_{t+\tau}^{(k)}] = [C_{ik}(\tau) + \bar{\upsilon}_i\bar{\upsilon}_k]\Delta^2$$
$$= \theta_0^{(k)}\bar{\upsilon}_i\Delta^2 + \sum_{j=1}^{D}\lim_{\Delta\to 0}\sum_{s=0}^{\infty} g_{kj}(s\Delta)E[N_t^{(i)}N_{t+\tau-(s+1)\Delta}^{(j)}]\Delta,$$
$$= \theta_0^{(k)}\bar{\upsilon}_i\Delta^2 + \sum_{j=1}^{D}\lim_{\Delta\to 0}\sum_{s=0}^{\infty} g_{kj}(s\Delta)$$
$$\cdot[C_{ij}(\tau-(s+1)\Delta) + \bar{\upsilon}_i\bar{\upsilon}_j + (\omega_i'+1)\bar{\upsilon}_i\delta_{ij}\delta(\tau-(s+1)\Delta)]\Delta^3,$$
$$= \theta_0^{(k)}\bar{\upsilon}_i\Delta^2 + \Delta^2\sum_{j=1}^{D}\int_0^{\infty} g_{kj}(w)$$
$$\cdot[C_{ij}(\tau-w) + \bar{\upsilon}_i\bar{\upsilon}_j + (\omega_i'+1)\bar{\upsilon}_i\delta_{ij}\delta(\tau-w)]dw,$$

$$= \theta_0^{(k)}\bar{v}_i\Delta^2 + \Delta^2(\omega_i' + 1)\bar{v}_i g_{ii}(\tau)$$
$$+\Delta^2 \sum_{j=1}^{D} \int_0^\infty g_{kj}(w)[C_{ij}(\tau - w) + \bar{v}_i\bar{v}_j)]dw,$$
$$= \bar{v}_i\bar{v}_k\Delta^2 + \Delta^2(\omega_i' + 1)\bar{v}_i g_{ii}(\tau) + \Delta^2 \sum_{j=1}^{D} \int_0^\infty g_{kj}(w)C_{ij}(\tau - w)dw$$

where we used the mean field approximation defined in Eq. (13.11),

$$\bar{v}_k = \theta_0^{(k)} + \sum_{j=1}^{D} \int_0^\infty g_{kj}(w)\bar{v}_j dw. \tag{13.53}$$

We can obtain the integral equation for $\tau > 0$ as

$$C_{ik}(\tau) = (\omega_i' + 1)\bar{v}_i g_{ii}(\tau) + \sum_{j=1}^{D} \int_0^\infty g_{kj}(w)C_{ij}(\tau - w)dw. \tag{13.54}$$

This is the integral equation for the correlation functions, Eq. (13.35). The difference between the SE-NBD and Hawkes processes is only the first term, which signifies the effect of the correlation at the same time.

We can rewrite Eq. (13.54) as

$$C_{ik}(\tau) = (\omega_i' + 1)\bar{v}_i g_{ii}(\tau) + \sum_{j=1}^{D} \int_0^\tau g_{kj}(w)C_{ij}(\tau - w)dw$$
$$+ \sum_{j=1}^{D} \int_0^\infty g_{kj}(\tau + w)C_{ji}(w)dw, \tag{13.55}$$

where we use the relation $C_{ij}(\tau) = C_{ji}(-\tau)$.

In the single-line case Eq. (13.55) becomes

$$C(\tau) = (\omega' + 1)\bar{v}g(\tau) + \int_0^\tau g(w)C(\tau - w)dw + \int_0^\infty g(\tau + w)C(w)dw. \tag{13.56}$$

Here, we set the kernel function $g(\tau) = ab\exp(-b\tau)$. We can solve this equation using the Laplace transformations,

$$C(\tau) = \frac{ab(\omega' + 1)\bar{v}}{2(1 - a)} \exp[-b(1 - a)\tau]. \tag{13.57}$$

Based on these results, the difference between the SE-NBD and Hawkes processes is only the intensity of the autocorrelation function. In the case of $\omega' = 0$, it corresponds to the correlation function of the Hawkes process.

13.9 Appendix C Proof of Eqs. (13.40) and (13.41)

Here, we consider the time series $\{N1.N_2, \ldots, N_T\}$ and the probability of $P(\{N1, N_2, \ldots, N_T\})$. Here we use the symbol, $\{N1, N_2, \ldots, N_i\} = \mathrm{N}_{1:i}$. The probability generating function is

$$\begin{aligned} F_T(s) &= \sum_{N_1}\sum_{N_2}\cdots\sum_{N_T} s^{N_1+N_2+\cdots+N_T} P(\{N1.N_2, \ldots, N_T\}), \\ &= \sum_{N_1}\sum_{N_2}\cdots\sum_{N_T} s^{N_1+N_2+\cdots+N_T} P(\mathrm{N}_{1:\mathrm{T}}), \\ &= \sum_{N_1}\sum_{N_2}\cdots\sum_{N_{T-1}} s^{N_1+N_2+\cdots+N_T} P(N_T|N_{T-1})P(\mathrm{N}_{1:\mathrm{T}-1}), \\ &= \sum_{N_1}\sum_{N_2}\cdots\sum_{N_{T-1}} s^{N_1+N_2+\cdots+N_{T-2}}[sf(s)]^{N_{T-1}} P(\mathrm{N}_{1:\mathrm{T}-1}), \end{aligned} \tag{13.58}$$

where $f(s) = e^{\nu(s-1)}$ which is the probability generating function of the Poisson distribution. We obtain repeating this process,

$$\begin{aligned} F_T(s) &= \sum h_T(s)^{N_1} P(N_1), \\ &= \exp[\theta_0 T(h_T(s) - 1)], \end{aligned} \tag{13.59}$$

where

$$h_1(s) = s, \tag{13.60}$$

and

$$h_T(s) = sf(h_{T-1}(s)). \tag{13.61}$$

In the limit $T \to \infty$,

$$F_\infty(s) = \sum h_\infty(s)^{N_1} P(N_1), \tag{13.62}$$

and

$$h_\infty(s) = s\exp[\nu(h_\infty(s) - 1)]. \tag{13.63}$$

We obtain using Eqs. (13.60), (13.61), and (13.63),

$$
\begin{aligned}
h_{\infty}(1) &= 1, \\
h'_{\infty}(1) &= \frac{1}{1-\nu} \\
h''_{\infty}(1) &= \frac{1}{(1-\nu)^3} - \frac{1}{1-\nu}
\end{aligned}
\tag{13.64}
$$

In the same way, we obtain

$$
\begin{aligned}
F'_{\infty} &= \frac{\nu T}{1-\mu}, \\
F''_{\infty} &= \frac{\theta_0 T}{(1-\mu)^3} - \frac{\theta_0 T}{1-\nu} + (\frac{\theta_0 T}{1-\nu})^2,
\end{aligned}
\tag{13.65}
$$

We calculate

$$
\begin{aligned}
\mathrm{E}[\tilde{N}_{\infty}] &= F'_{\infty}(1) = \frac{\theta_0 T}{1-\nu}, \\
\mathrm{V}[\tilde{N}_{\infty}] &= F''_{\infty}(1) + F'_{\infty}(1) - [F7_{\infty}(1)]^2 = \frac{\theta_0 T}{(1-\nu)^3}.
\end{aligned}
\tag{13.66}
$$

References

2020 Annual Global Corporate Default Study and Rating Transitions (S&P Global Ratings, 2021).

Embrechts P and Krichner M, Hawkes graphs, *Theory Probab Appl***62(1)** 163 (2017) Affine point processes and portfolio credit risk *SIAM J. Fin Math* **1** 642–665 (2010).

Florescu I, Mariani M C, Stanley H E, and Viens F G (Eds.) *Handbook of High-Frequency Trading and Modeling in Finance*, John Wiley & Sons (2016).

Hawkes A G, Spectra of some self-exciting and mutually-exciting point processes, *Biometrika* **58** 83–90 (1971).

Hisakado M, Hattori K, and Mori S, Multi-Dimensional self-exciting NBD process and Default portfolios *The Review of Socionetwork Strategies* **16(2)** 493–512 (2022).

Langville A N and Meyer C D, *Google's PageRank and Beyond*, Princeton University Press (2006).

Schönbucher P J, *Credit Derivatives Pricing Models: Models, Pricing, and Implementation* John Wiley & Sons, Ltd. (2003).

Chapter 14
Merton Model and Its Poisson Process

14.1 Introduction

We studied a Bayesian estimation method using the Merton model in Chap. 11. Under normal circumstances, the Merton model uses the correlation of asset price movements (asset correlation) to estimate the PD and the correlation. A Monte Carlo simulation is an appropriate tool to estimate the parameters, except under the limit of large homogeneous portfolios (Schönbucher 2003). In this case, the distribution becomes a Vasicek distribution that can be calculated analytically (Vasicek 2002).

In our previous chapters, we discussed parameter estimation using a beta binomial distribution with the default correlation and considered a multi-term case with a temporal correlation. In the double scaling limit, we can obtain the SE-NBD process and the Hawkes process. In this chapter, we consider the limit of the Merton model process. In this study, we discuss the phase transition when using the Merton model and compare this model to the SE-NBD and Hawkes models.

The remainder of this paper is organized as follows. In Sect. 14.3, we introduce the stochastic process of multi-term Merton model and obtain the Poisson process in the limit. In Sect. 14.4, we discuss the impact analysis. In Sect. 14.5, we discuss the super-normal phase transition of the convergence of the variance. Finally, a summary is presented in Sect. 5.

14.2 Poisson Process and Multi-term Merton Model

14.2.1 Multi-term Merton Model

In this section, we review the time series of a stochastic process using the Merton model (Merton 1974). We take the limit of the process and obtain the Poisson process with log-normal intensity function.

M. Hisakado, *Urn Models and Their Applications in Finance*,
https://doi.org/10.1007/978-981-96-3825-3_14

Normal random variables, S_t, are hidden variables that explain the status of economics, and S_t affects all obligors in the t-th term. Through S_t temporal correlation affects the obligors. To introduce the temporal correlation from different terms, let $\{S_t, 1 \le t \le T\}$ be the time series of the stochastic variables of the correlated normal distribution with the following correlation matrix:

$$\Sigma_A \equiv \begin{pmatrix} 1 & \hat{d}_1 & \hat{d}_2 & \cdots & \hat{d}_{T-1} \\ \hat{d}_1 & 1 & \hat{d}_1 & \ddots & \vdots \\ \ddots & \ddots & \ddots & \ddots & \ddots \\ \vdots & \ddots & \ddots & \ddots & \hat{d}_1 \\ \hat{d}_{T-1} & \cdots & \hat{d}_2 & \hat{d}_1 & 1 \end{pmatrix}, \tag{14.1}$$

where $(S_1, \ldots, S_T)^T \sim \mathrm{N}_T(0, \Sigma_A)$.

In this chapter, we consider two cases of temporal correlation: exponential decay, $\hat{d}i = r^i, 0 \le r \le 1$, and power decay, $\hat{d}i = 1/(i+1)^{-\gamma}, \gamma \ge 0$. Here we use r for the exponential decay of the Merton model. The exponential decay corresponds to a short memory and the power decay corresponds to intermediate and long memories (Florescu et al. 2016). In the case of exponential decay, we can write

$$S_{t+1} = rS_t + \sqrt{1-r^2}\xi_{t+1}, \tag{14.2}$$

where ξ_t is i.i.d. and the correlation between S_t and S_{t+1} is r. The first term corresponds to the temporal correlation decay.

Without loss of generality, we assume that the number of obligors in the t-th term is constant and denotes it as N. The asset correlation, ρ_A, is the parameter that describes the correlation between the value of the assets of the obligors in the same term. We consider the i-th asset value, $\hat{U}_{it}$, at term t, to be

$$\hat{U}_{it} = \sqrt{\rho_A} S_t + \sqrt{1-\rho_A}\epsilon_{it}, \tag{14.3}$$

where $\epsilon_{it} \sim \mathrm{N}(0, 1)$ is i.i.d. The first term is the effect of economics and the second term is the effect of the individual obligor. Using this formulation, the equal-time correlation of U_{it} is ρ_A.

The discrete dynamics of the process are described by

$$X_i(t) = 1_{\hat{U}_{it} \le Y}, \tag{14.4}$$

where Y is the threshold and $1 \le i \le N$. When $X_i(t) = 1(0)$, the i-th obligor in the t-th term is default (non-default). Equation (14.4) corresponds to the conditional default probability for $S_t = S$ as

$$G(y) \equiv \mathrm{P}(X_t = 1 | S_t = S) = \Phi\left(\frac{Y - \sqrt{\rho_A} S}{\sqrt{1-\rho_A}}\right), \tag{14.5}$$

where $\Phi(x)$ is the standard normal cumulative distribution, $G(S_t)$ is the distribution of the default probability during the t-th term in the portfolio, and the average PD is $p' = \Phi(Y)$, which is the long-term average of PD. It is the original process.

14.2.2 Vasicek Distribution

In this subsection, we consider the ratio of defaulted obligors in the portfolio for the large N_1 limit. The ratio is defined as

$$X = \frac{k_1}{N_1}, \tag{14.6}$$

in the limit $N_1 \to \infty$. Here, we study only one term. Here we set $S_1 = y$. We calculate the distribution,

$$\begin{aligned} P[X \leq x] &= \int_{-\infty}^{\infty} P[X \leq x | S = y]\phi(y)dy \\ &= \int_{-\infty}^{\infty} P[X = G(y) \leq x | S = y]\phi(y)dy \\ &= \int_{-y^*}^{\infty} \phi(y)dy = \Phi(y^*), \end{aligned} \tag{14.7}$$

where $\phi(y)$ is the normal probability distribution. We obtain

$$G(y^*) = \Phi\left(\frac{Y - \sqrt{\rho_A}y^*}{\sqrt{1-\rho_A}}\right) = x. \tag{14.8}$$

Therefore, using Eq. (14.7) we obtain

$$y^* = \frac{\sqrt{1-\rho_A}\Phi^{-1}(x) - Y}{\sqrt{\rho_A}}. \tag{14.9}$$

The distribution of ratio of loss is

$$F(x) = P[X \leq x] = \Phi\left(\frac{\sqrt{1-\rho_A}\Phi^{-1}(x) - Y}{\sqrt{\rho_A}}\right). \tag{14.10}$$

It is Vasicek distribution (Vasicek 2002).

When we change from $\Phi^{-1}(x)$ to the logistic function,

$$\Phi(x) \approx \frac{1}{1 + e^{-\beta x}}, \tag{14.11}$$

$\beta \sim 1.3$, we obtain the Johnson SB distribution,

$$F(x) = P[X \leq x] = \Phi\left(\frac{\sqrt{1-\rho_A}}{\sqrt{\rho_A}}\frac{1}{\beta}\log\frac{x}{1-x} - \frac{1}{\sqrt{\rho_A}}Y\right). \tag{14.12}$$

14.2.3 Limits of the Process

Next, we consider the limit of this process. We set $S_t = y$ here. The non-conditional distribution of the number of defaults is

$$\begin{aligned} P[N_t = k_t] &= \int_{-\infty}^{\infty} \frac{N!}{k_t!(N-k_t)!} G(y)^{k_t}(1-G(y))^{N-k_t}\phi(y)dy, \\ &= \int_{-\infty}^{\infty} \frac{N!}{k_t!(N-k_t)!}\Phi\left(\frac{\Phi^{-1}(p') - \sqrt{\rho_A}y}{\sqrt{1-\rho_A}}\right)^{k_t} \\ &\quad (1-\Phi\left(\frac{\Phi^{-1}(p') - \sqrt{\rho_A}y}{\sqrt{1-\rho_A}}\right))^{N-k_t} \times \phi(y)dy, \end{aligned} \tag{14.13}$$

where $\phi(y)$ is the normal probability distribution, $1/\sqrt{2\pi}e^{-y^2/2}$. Here, we use an approximation for the normal distribution,

$$\Phi(x) \approx \frac{1}{1+e^{-\beta x}}, \tag{14.14}$$

logistic function and $\beta \sim 1.3$. Using this approximation, we obtain

$$G(y) = \Phi\left(\frac{\Phi^{-1}(p') - \sqrt{\rho_A}y}{\sqrt{1-\rho_A}}\right) \equiv p \sim \frac{1}{1 + ((1-p')/p')^{1/\sqrt{1-\rho_A}} e^{\beta\sqrt{\rho_A}y/\sqrt{1-\rho_A}}}, \tag{14.15}$$

where we use the relation,

$$Y = \Phi^{-1}(p') \approx \frac{1}{\beta}\log\frac{p'}{1-p'}. \tag{14.16}$$

In the case $p = G(y) << 1$, we obtain,

$$G(y) = p'^{1/\sqrt{1-\rho_A}} e^{-\frac{\sqrt{\rho_A}}{\sqrt{1-\rho_A}}\beta y}. \tag{14.17}$$

Here, we take the limit $p', p \to 0$ and $N \to \infty$ with the condition, fixed $\lambda_0 = Np'^{1/\sqrt{1-\rho_A}}$. Note that the relation p and p' is Eq. (14.15) We define the number of defaults at term t, $\lambda(S_t) = \lambda(t)$ as

$$\lambda(y) = Np'^{1/\sqrt{1-\rho_A}} e^{-\frac{\sqrt{\rho_A}}{\sqrt{1-\rho_A}}\beta y} = \lambda_0 e^{-\frac{\sqrt{\rho_A}}{\sqrt{1-\rho_A}}\beta y} \tag{14.18}$$

The expected value of $\lambda(y)$ is

$$\bar{\lambda} = \int_{-\infty}^{\infty} \lambda(y)\phi(y)dy = \lambda_0 e^{\frac{\rho_A}{2(1-\rho_A)}\beta^2} = \lambda_0 e^{\alpha^2/2}, \tag{14.19}$$

where $\alpha = \frac{\sqrt{\rho_A}}{\sqrt{1-\rho_A}}\beta$.

Here, we change the variable from y to $\hat{y} = -y$. The standard normal distribution has symmetry in y, so we can change the variable without loss of generality. We obtain

$$\lambda(\hat{y}) = \lambda_0 e^{\frac{\sqrt{\rho_A}}{\sqrt{1-\rho_A}}\beta\hat{y}} = \lambda_0 e^{\alpha\hat{y}}. \tag{14.20}$$

In the exponential decay case, we obtain another representation

$$\lambda(\hat{y}_{t+1}) = \lambda_0 e^{\alpha\hat{y}_{t+1}} = \lambda_0^{1-r} e^{\sqrt{1-r^2}\alpha\hat{\xi}_{t+1}} \lambda(\hat{y}_t)^r, \tag{14.21}$$

using Eq. (14.2) and $\hat{\xi}_t = -\xi_t$.

In this limit, we can calculate Eq. (14.13) as

$$\begin{aligned} P[N_t = k_t] &\sim \int_{-\infty}^{\infty} \frac{\lambda(\hat{y})^{k_t} e^{-\lambda(\hat{y})}}{k_t!}\phi(\hat{y})d\hat{y} \\ &= \int_{0}^{\infty} \frac{\lambda(\hat{y})^{k_t} e^{-\lambda(\hat{y})}}{k_t!} f(\lambda)d\lambda, \end{aligned} \tag{14.22}$$

where

$$f(\lambda) = \phi(\hat{y})\frac{d\hat{y}}{d\lambda} = \frac{1}{\sqrt{2\pi}\alpha\lambda} e^{-\frac{(\log\lambda - \log\lambda_0)^2}{2\alpha^2}} \tag{14.23}$$

Then, the distribution of the intensity function is the log-normal distribution. The expected value is $\bar{\lambda} = \lambda_0 e^{\alpha^2/2}$ and the variance is $\bar{V} = \bar{\lambda}^2(e^{\alpha^2} - 1)$. Note that the expected value is a function of ρ_A in this process. In contrast, in the original process, the expected value of the probability of defaults does not depend on the function of ρ_A.

In the limit $\rho_A \to 0$, $\bar{\lambda} = \lambda_0$ and $\bar{V} = 0$, it is the Poisson process. In the limit $\rho_A \to 1$, the intensity function is $\alpha \to \infty$ and $f(\lambda) \sim 1/\lambda$ where $\bar{\lambda} \to \infty$ and $V \to \infty$.

We show an image of the relationship among the variables of this process in Fig. 14.1(a). For the comparison, we show that of the SE-NBD and Hawkes processes in Fig. 14.1(b). **The difference between this process and SE-NBD is the intensity function. The intensity function of this process is the gamma distribution and that of SE-NBD is the log normal distribution.** In this process S_t is independent from the historical data of λ_t and k_t. The temporal correlation is presented by the

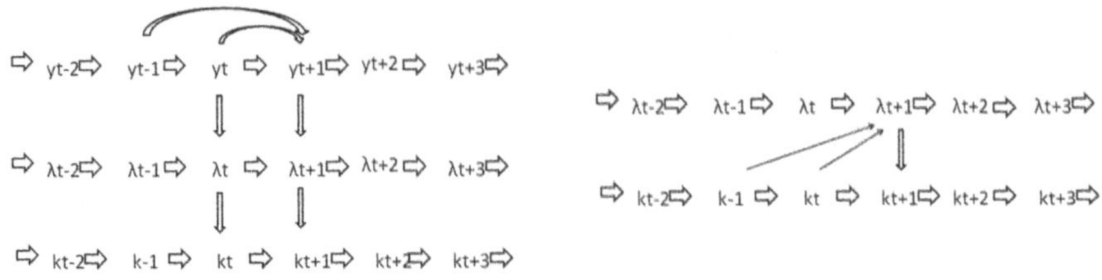

(a) (b)

Fig. 14.1 The relationship among the variables of (**a**) this process and (**b**) SE-NBD and Hawkes process

time series of S_t and the hidden process. y affects the number of defaults through the intensity function λ_t. On the other hand, in the NBD and Hawkes processes, the feedback exists between the intensity function λ_t and number of defaults k_t.

14.3 Impact Analysis

Next, we consider the impact analysis. This is the study of the effect of the shock of the noise δ at term t. When the shock is added in $\hat{y}_t$, we consider the additional expected value of the intensity function from $t+1$ to ∞.

14.3.1 Exponential Decay Case

First, we consider the exponential decay model $\hat{d}_i = r^i, r \leq 1$. We consider the ratio of the impact δ added at the time $t \to \infty$,

$$\frac{\bar{\lambda}_{\delta\infty}}{\bar{\lambda}_{\infty}} = \mathrm{E}\left[\log \frac{\lambda_0 e^{\alpha\delta+\alpha\hat{y}_t}}{\lambda_0 e^{\alpha\hat{y}_t}} \frac{\lambda_0 e^{\alpha\delta r+\alpha\hat{y}_{t+1}}}{\lambda_0 e^{\alpha\hat{y}_{t+1}}} \frac{\lambda_0 e^{\alpha\delta r^2+\alpha\hat{y}_{t+2}}}{\lambda_0 e^{\alpha\hat{y}_{t+2}}} \cdots\right] = \frac{\alpha\delta}{1-r}. \tag{14.24}$$

Then, in this case, the impact is finite.

14.3.2 Power Decay Case

Next, we consider the case, $\hat{d}i = 1/(i+1)^{\gamma}, i = 1, 2, \ldots$, where $\gamma \geq 0$ is the power index. We consider the ratio for the impact δ added,

$$\frac{\bar{\lambda}_{\delta\infty}}{\bar{\lambda}_{\infty}} = \mathrm{E}[\log \frac{\lambda_0 e^{\alpha\delta+\alpha\hat{y}_t}}{\lambda_0 e^{\alpha\hat{y}_t}} \frac{\lambda_0 e^{\alpha\delta/2^{\gamma}+\alpha\hat{y}_{t+1}}}{\lambda_0 e^{\alpha\hat{y}_{t+1}}} \frac{\lambda_0 e^{\alpha\delta/3^{\gamma}+\alpha\hat{y}_{t+2}}}{\lambda_0 e^{\alpha\hat{y}_{t+2}}} \cdots] \tag{14.25}$$

When $\gamma > 1$, we obtain

$$\frac{\bar{\lambda}_{\delta\infty}}{\bar{\lambda}_{\infty}} < \alpha\delta(1 + \frac{1}{\gamma - 1}). \tag{14.26}$$

Then, the impact is finite. When $\gamma = 1$, we can obtain

$$\frac{\bar{\lambda}_{\delta\infty}}{\bar{\lambda}_{\infty}} \sim \alpha\delta \log T. \tag{14.27}$$

Then, the impact is infinite. When $\gamma < 1$, we obtain

$$\frac{\bar{\lambda}_{\delta\infty}}{\bar{\lambda}_{\infty}} \sim \alpha\delta T^{1-\gamma}. \tag{14.28}$$

Then, the impact is infinite.

In summary, the impact is finite when $\gamma > 1$ and is infinite when $\gamma \leq 1$. Even when the impact is infinite, the average number of defaults converges because the process does not have an absorption process. The Hawkes process and SE-NBD process are absorption processes. Hence, there exists an absorption transition, which is a type of the non-equilibrium transition. In the next section, we consider the phase transition of this process and its impact.

14.4 Super-Normal Phase Transition

We are interested in the effect of temporal correlation. Here, we consider the variance of λ_T, $V(\lambda_T)$ the variance of λ of the term T is given by

$$\mathrm{V}(\Sigma^T \lambda(\lambda_T = \hat{y}_i)) = T\bar{V} + 2\bar{V}\sum_{i=1}^{T-1} \hat{d}_i(T - i).$$

The second term is from the temporal correlation. We consider the behavior of the second term in limit $T \to \infty$.

14.4.1 Exponential Decay

In this subsection, we study $V(\lambda_T)$ for the exponential decay model $\hat{d}_i = r^i, r \leq 1$:

$$V(\lambda_T) \simeq \bar{V}T + +2\bar{V}\sum_{i=1}^{T-1} r^i(T-i).$$

The first term on the right-hand side (RHS) behaves as $\propto T$, thus, this is normal diffusion. For $r \neq 1$, the second term is

$$2\bar{V}[T\frac{1-r^{T-1}}{1-r} + \frac{(T-1)r^{T-1}(1-r)-(1-r^{T-1})r}{(1-r)^2}] \propto T$$

and it also normal diffusion. We conclude that as the number of data samples increases, variance increases with T. When $r = 1$, there is no temporal correlation decay and all obligors are correlated ρ_A. Hence, there is no phase transition for $r < 1$.

14.4.2 Power Decay

In this subsection, we consider the power decay case $\hat{d}i = 1/(i+1)^\gamma, i = 1, 2, \ldots,$ where $\gamma \geq 0$ is the power index. The power correlation affects the number of defaults over long periods of time. Ranges $\gamma \leq 1$ and $\gamma > 1$ are called long memory and intermediate memory, respectively (Florescu et al. 2016). In contrast, exponential decay affects short periods of time and is called short memory. The asymptotic behavior of $V(\lambda_T))$ is given as:

$$\mathrm{V}(\lambda_T) \simeq \bar{V}T + 2\bar{V}\sum_{i=1}^{T-1}(i+1)^{-\gamma}(T-i).$$

$\gamma > 1$ Case

We obtain

$$\begin{aligned}\mathrm{V}(\lambda_T) &\simeq \bar{V}T + +2\bar{V}\sum_{i=1}^{T-1}(T-i)/(i+1)^\gamma \\ &\simeq \bar{V}T + +2\bar{V}T^{-\gamma+2}/(\gamma-1).\end{aligned} \tag{14.29}$$

The first term is normal diffusion and the second term is the $T^{-\gamma+2}$, where $\gamma > 1$. It is slower than normal diffusion. Hence, the significant terms are the first term, which is the normal diffusion.

$\gamma = 1$ Case

$\mathrm{V}(\lambda_T)$ behaves as

$$\mathrm{V}(\lambda_T) \simeq \bar{V}T + 2\bar{V}\sum_{i=1}^{T-1}(T-i)/(i+1). \tag{14.30}$$

The RHS of Eq. (14.30) is evaluated as

$$\mathrm{V}(\lambda_T) \simeq \bar{V}T + +2\bar{V}T\log T - T + 2] \sim T\log T. \tag{14.31}$$

In conclusion, $\mathrm{V}(\lambda_T)$ behaves asymptotically as

$$\mathrm{V}(\lambda_T) \sim T\log T \tag{14.32}$$

and becomes an anomalous faster diffusion.

$\gamma < 1$ Case

$\mathrm{V}(\lambda_T)$ is calculated as

$$\mathrm{V}(\lambda_T) \simeq \bar{V}T + 2\bar{V}[\frac{1}{(1-\gamma)(2-\gamma)]}T^{-\gamma+2} \sim T^{-\gamma+2}. \tag{14.33}$$

Thus, we can conclude that $\mathrm{V}(\lambda_T)$ behaves as

$$\mathrm{V}(\lambda_T) \sim T^{-\gamma+2}, \tag{14.34}$$

which also exhibits an anomalous faster diffusion.

In conclusion, a phase transition occurs when the temporal correlation decays according to the power law. This is the same as the original process, which is before the limit. When the power index, γ, is less than one, the variance, $\mathrm{Z}(\lambda_T)$, anomalous diffusion. Conversely, when the power index, γ, is greater than one, it is a normal diffusion. This phase transition is called a super-normal phase transition which is the transition between long memory and intermediate memory. In this study, when the power index is less than one, the estimator does not converge to the steady state.

14.5 Summary

This study adopted the Merton model. We have shown that the Merton model becomes the Poisson process with the log-normal distributed intensity function in the limit. We discuss the relation between this model and the Hawkes and SE-NBD processed. We apply this model to the default portfolios. In addition, we have shown why the beta distribution and Vasicek model is similar. The difference depends on the difference between the log-normal distribution and the gamma distribution.

When the power index, γ, is larger than one, the estimator distribution of the PD converged normally. When the power index was less than or equal to one, the diffusion is anomalous fast diffusion. This is the super phase. This phase transition is called the super-normal transition. In the case of exponential decay, there was no phase transition.

References

Florescu I, Mariani M C, Stanley H E, and Viens F G (Eds.) *Handbook of High-Frequency Trading and Modeling in Finance*, John Wiley & Sons (2016).

Merton R C, On the Pricing of Corporate Debt: The Risk Structure of Interest Rates, J. Fin. **29(2)**, 449–470 (1974). The difference depends on the difference between the log-normal distribution and the gamma distribution.

Schönbucher P J, *Credit Derivatives Pricing Models: Models, Pricing, and Implementation* John Wiley & Sons, Ltd. (2003).

Vasicek O A, The Distribution of Loan Portfolio Value, *Risk* **15(12)**, 160 (2002).

Appendix A
Appendix: Discontinuous Phase Transition and Gap

Introduction

In this Appendix, we analyze a discontinuous phase transition and the gap at critical point. We use a sequential voting model to discuss the phase transition. In this analysis, we can confirm the "voter" and his /her referred "voters". In this model, public perception is represented by r votes selected from previous votes. The relation between voters is a network. There are two types of voters: herders and independents.

In this model, there is a transition between a state in which most voters make the correct choices and a state in which most voters are incorrect. This is an absorption phase transition that belongs to a non-equilibrium transition (Hinrichsen 2000). In the symmetric case, we demonstrate that the phase transition is continuous and the universality class is the same as the nonlinear Pólya model. The network affects only the super-normal transition point. In contrast, in the asymmetric case, the phase transition is a discontinuous phase transition. The gap in the discontinuous phase transition depends on the network.

The remainder of this chapter is organized as follows. In Sect. A.2, we introduce networks with the parameter ω. In Sect. A.3, we introduce our voting model mathematically and define two types of voters: independents and herders. In Sect. A.4, we combine the voting model with the networks and consider the phase transition. In Sect. A.5, we describe the relation to the elephant random walk. In Sect. A.6, numerical simulations are performed to confirm the phase transition. Finally, the conclusions are presented in Sect. A.7.

Review of Network

In this section, we first review how to create voter networks, as described in Chap. 8. The network means that the voter selects referred voters to decide their opinion. The voter corresponds to the node. When a voter joins the network, in the first step, the

M. Hisakado, *Urn Models and Their Applications in Finance*,
https://doi.org/10.1007/978-981-96-3825-3

voter selects r voters to refer the information. This relation corresponds to the edge. In the second step, the voter decides his/her opinion using the information. When a voter is an independent, he/she does not use this information. We consider the case wherein a voter selects r voters based on popularity. This process is sequential, as in the BA model. When voter i joins at time $t = i$, site i selects r voters for connections, as shown in Fig. A.1. We show that the incoming arrows correspond to the seed popularity that all voters have when they join. The seed popularity is r, the number of in-degrees.

We denote the number of incoming and outgoing edges of node i as k_i^{IN} and k_i^{OUT}, respectively. The degree of node i is given by $k_i = k_i^{IN} + k_i^{OUT}$. The popularity l_i of node i is defined as

$$l_i = k_i^{IN} + \omega \cdot k_i^{OUT}. \tag{A.1}$$

In the BA model, weight 1 is for the in-degree and out-degree. We set the weight ω for the out-degree. The latest popularity is used for the evolution of the network and is updated with time. The probability that node $i < t$ is selected by node t is

$$P(\text{node } i \text{ is selected by node } t) = \frac{l_i}{\sum_{s=1}^{t-1} l_s}. \tag{A.2}$$

As popularity increases, the probability that the site is selected increases.

We can extend this to the negative ω. In this case, negative feedback can be observed. Note that if $l_i < 0$, we set $l_i = 0$ when $\omega < 0$, in the case where $-r/\omega$ is indivisible. Popularity decreases as the number of incoming links of the voter increases. After the t-th voter joins, the total number of l_j after the t-th voter joins is $\sum_j l_j = (\omega + 1)r(t - r + 1)$, where $t \geq r$. l_i represents the popularity of voter i. Using the parameter ω, we can represent several networks using the same process.

In this network, a voter joins the network one by one. This corresponds to sequential voting, which is explained in the next section. Using this system voters can refer without fail because a voter creates the network and decides the opinion at the same time. When we can separate the creation of the network and sequential voting, the ordering of the votes may be the problem generally because the referred voter has not decided the opinion yet. If we assume that the voters can refer to previous votes without fail, the order of the voting does not affect our conclusions.

When $\omega = 1$ and $\omega = 0$, the network becomes the BA model and random graph, respectively. This is because when $\omega = 0$, all voters are selected with the same probability. In this case, the older node has many outgoing arrows because the lifetime of the older nodes is longer than that of the new node. Hence, the probability that a node is selected in life increases with age. We consider the range of negative ω to be $-1 \leq \omega < 0$. The maximum number of selections was $\lceil(-r/\omega)\rceil$ where $\lceil x \rceil$ is the ceiling function. This means that there is a maximum hub size of $-1 \leq \omega < 0$. There is no maximum hub size in $0 \leq \omega$. As the size of the hub increases, ω increases owing to positive feedback.

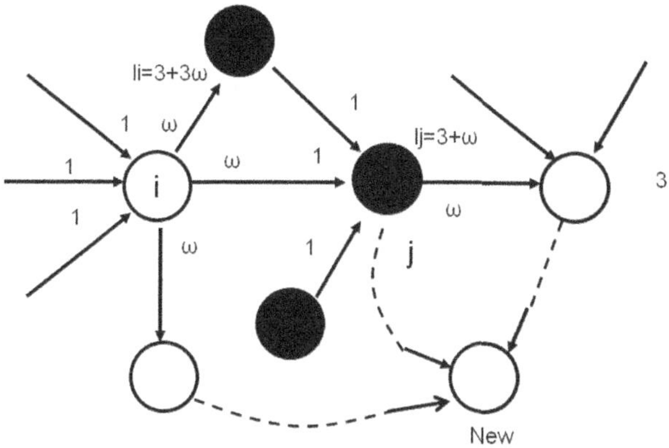

Fig. A.1 Sample graph with $r = 3$. Voter i selects three voters and is selected by three voters. The arrow is from the selected voters to selecting voter. The weight of the outgoing arrow is ω and incoming arrow is 1. The popularity is set to each voter and the some of the weights of incoming arrows and outgoing arrows. Hence, voter j obtains information from voter i. The popularity for voter i is $l_i = 3 + 3\omega$ and that for voter j is $l_j = 3 + \omega$. The color of the node shows the voting of the voter. The site is white, as he/she voted for candidate C_0

Voting Model

In this section, we introduce the voting model. It is the process of deciding an opinion. We model the voting behavior of two candidates, C_0 and C_1, at time t, and C_0 and C_1 have $c_0(t)$ and $c_1(t)$ votes, respectively. At each time step, one voter votes for one candidate, indicating that voting is sequential. Hence, at time t, the t-th voter votes, and the total number of votes is t. Voters are allowed to view them. r previous votes for each candidate are selected as public perception, where r is a constant. Voter votes simultaneously when they join the network. Therefore, the herder obtains the information from voters who have already voted. It is the process of creating the network. In the next step, the voter decides his/her opinion using information. The edge shows the flow of the information. The direction is from the information provider (selected voter) to the information user (selecting voter).

In our model, we assume an infinite number of voters of each of the two types: independents and herders. Independent votes for candidates C_0 and C_1 have probabilities $1 - q$ and q, respectively. Their votes are independent from others' votes; that is, their votes are based on their fundamental values. We assume $q \geq 1/2$. Independent does not refer to other voters but is referred by other voters. Hereafter, we set C_1 as the correct candidate, and the voting ratio to candidate C_1 is the correct ratio used to consider voting performance.

Herders' votes are based on the number of previous r votes. Note that the voter does not necessarily refer to the latest r votes. We consider previous r votes to refer to those that are selected from the voters' network. Therefore, at time t, r previous votes are the number of votes for C_0 and C_1, represented by $c_0^r(t)$ and $c_1^r(t)$. Hence,

$c_0^r(t) + c_1^r(t) = r$ holds. If $r > t$, voters can view t previous votes for each candidate. In the limit $r \to \infty$, voters can view all previous votes. We define the number of previous votes for C_0 and C_1 as $c_0^\infty(t) \equiv c_0(t)$ and $c_1^\infty(t) \equiv c_1(t)$. In the real world, the number of references r depends on the voters; however, we considered r constant in this study.

We consider the Bernoulli random variables, $X(t)$ which corresponds to the vote at time t. $X(t) = 1(0)$ is the vote to candidate 1(0) at t. Then, we obtain

$$c_1^r(t) = \sum_{i=1}^{t} X(i), \tag{A.3}$$

and

$$c_0^r(t) = \sum_{i=1}^{t} (1 - X(i)). \tag{A.4}$$

The herder is considered a digital herder. We define $c(t)_1^r/r = 1 - c(t)_0^r/r = Z_r(t)$. A digital herder's behavior is defined by the function $f(z) = \theta(Z_r - 1/2)$, where $\theta(Z)$ is a Heaviside function, with the threshold of 1/2, corresponding to the majority decision. In Watts (2002), a model whose threshold is different from 1/2 that was discussed. The model corresponds to voting without independent voters. We define $P_{h1}^r(t)$ as the probability that the herder would vote for C_1 at time t,

$$P_{h1}^r(t) = \left\{ \begin{array}{ll} 1 & ; c_1^r(t) > r/2, \\ 1/2 & ; c_1^r(t) = r/2, \\ 0 & : c_1^r(t) < r/2. \end{array} \right\} \tag{A.5}$$

Independents and herders appear randomly and vote. We set the ratio of independents to herders as $(1 - p)/p$. In this study, we focus mainly on the upper limit of t, which refers to voting by an infinite number of voters.

Voting Model on Network

In this section, we combine the network and voting models discussed in the previous two sections. We consider that a voter can view r previous votes. In Section II, we define how to select previous voters as references. The influence of reference voters is represented as a voting model in a network. The problem lies in how the network affects the voting model. In this study, we analyzed the cases by comparing models based on networks created using parameter ω. The network includes a random graph, the BA model case, and extended lattice, which correspond to $\omega = 1, 0, -1$, respectively. In Fig. A.2, we illustrate each of these three cases for $r = 2$. A white (black) dot indicates a voter who voted for candidate $C_0(C_1)$. The two arrows point toward a dot, indicating that a voter refers to two other voters when $r = 2$. In the

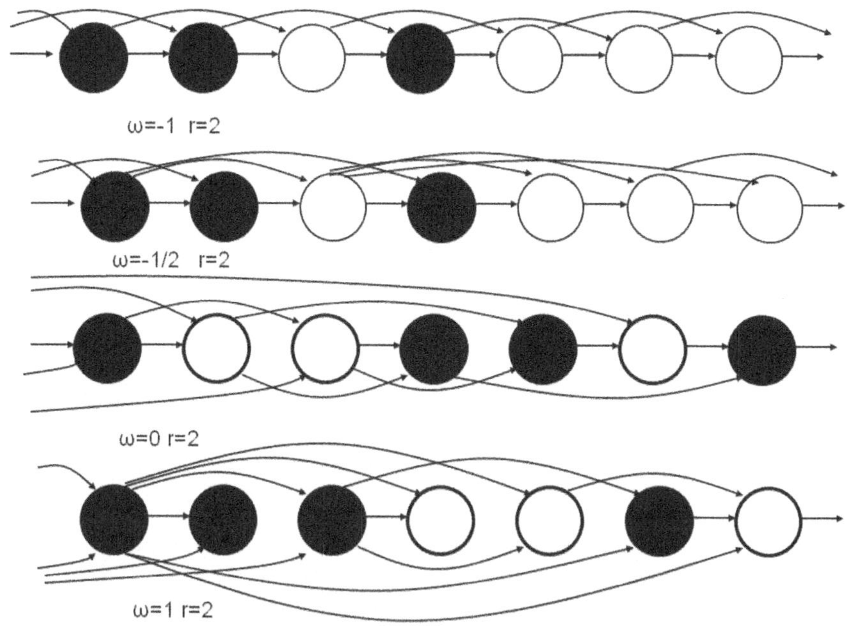

Fig. A.2 Representation of graphs. An extended 1D lattice, Fermion-like graph, random graph, and BA model, for $\omega = -1, -1/2, 0, 1$, $r = 2$, respectively. The white (black) dot is a voter who voted for candidate $C_0(C_1)$. The two arrows pointing toward a dot represent a voter who refers to two voters when $r = 2$. In the case of an extended 1D lattice, a voter refers to the last two voters. In the case of a random graph, a voter refers to two previous voters who are selected randomly. In the case of the BA model, a voter refers to two previous voters who are selected by popularity. Hence, there are voters who play the role of a hub in the BA model. The first voter in the network graph is a hub that influences many other voters

case of a 1D extended lattice, the voter refers to the latest two voters. In the case of a random graph, a voter refers to two previous voters who are selected randomly. In the BA model, a voter refers to two previous voters selected through the voter's popularity network. A positive ω network has the characteristics of a scale-free network with hubs. However, in the negative ω network, the variance in the number of connections in each node is small.

We define $P_1^r(t)$ as the probability that the t-th voter would vote for C_1 using Eq. (A.5),

$$P_1^r(t) = \left\{ \begin{array}{ll} p + (1-p)q & ; c_1^r(t) > r/2, \\ p/2 + (1-p)q & ; c_1^r(t) = r/2, \\ (1-p)q & : c_1^r(t) < r/2. \end{array} \right\} \tag{A.6}$$

It is the sum of the voting of herders and independents.

In the scaling limit, $t = c_0(t) + c_1(t) = c_0^\infty + c_1^\infty \to \infty$, we define

$$Z(t) = \frac{c_1(t)}{t} \Longrightarrow Z_\infty. \tag{A.7}$$

$Z(t)$ is the ratio of voters who vote for C_1 at t.

Here, we define $\pi(Z)$ as the majority probability of the binomial distributions of Z, in other words, the probability that $c_1^r(t) > 1/2$. When r is odd, we have

$$\pi(Z) = \sum_{g=\frac{r+1}{2}}^{r} \binom{r}{g} Z^g (1-Z)^{r-g}. \tag{A.8}$$

π can be calculated as follows:

$$\pi(Z) = \frac{(2n+1)!}{(n!)^2} \int_0^Z x^n (1-x)^n dx = \frac{1}{B(n+1, n+1)} \int_0^Z x^n (1-x)^n dx. \tag{A.9}$$

Equation (A.9) can be applied when the referred voters overlap. In fact, the referred voters do not select an overlap. However, in this study, we used this approximation to investigate the large t limit.

Each popularity has a color in the voting model. Color depends on whether the voter voted for candidate C_1 or C_0. In Fig. A.1, voters who voted for $C_1(C_0)$ are represented by black (white) circles.

We define the total popularity of voters who vote for candidate $C_1(C_0)$ at t as $g_1(t)(g_0(t))$. In the scaling limit, $g_0(t) + g_1(t) = (1+\omega)r(t-r+1) \to \infty$, we define

$$\frac{g_1(t)}{(1+\omega)r(t-r+1)} = \hat{Z}(t) \Longrightarrow \hat{Z}_\infty. \tag{A.10}$$

$\hat{Z}(t)$ is the ratio of popularity for C_1 at t.

We can denote the evolution of popularity as

$$\begin{aligned} g_1(t) = \hat{k} &\to \hat{k} + i : \\ r/2 \le j \le r, i = \omega j + r, P_{\hat{k},t}(i) &= {}_rC_j \hat{Z}^j (1-\hat{Z})^{r-j}[(1-p)q + p], \\ 0 \le j < r/2, i = \omega j + r, P_{\hat{k},t}(i) &= {}_rC_j \hat{Z}^j (1-\hat{Z})^{r-j}(1-p)q, \\ r/2 \le j \le r, i = \omega j, P_{\hat{k},t}(i) &= {}_rC_j \hat{Z}^j (1-\hat{Z})^{r-j}(1-p)(1-q), \\ 0 \le j < r/2, i = \omega j, P_{\hat{k},t}(i) &= {}_rC_j \hat{Z}^j (1-\hat{Z})^{r-j}[(1-p)(1-q) + p], \end{aligned} \tag{A.11}$$

where $P_{\hat{k},t}(i)$ is the probability of the transition from $\hat{k}$ to $\hat{k}+i$ at t.

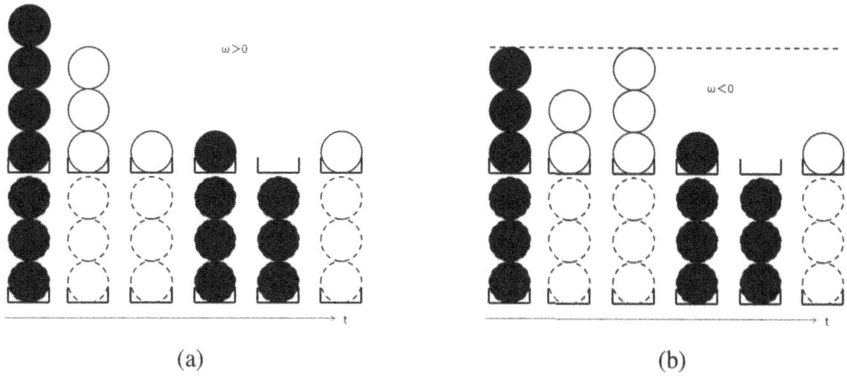

Fig. A.3 **a** $\omega > 0$ **b** $\omega < 0$ cases. The horizontal axis represents to time t and the new joined node. The dotted (real) balls correspond to the incoming(outgoing) arrows. The colors of balls in a urn are the same

Relation to the Urn Model

In this subsection, we consider this voting model on a network using the urn problem. The urn corresponds to the node, and outgoing and incoming links correspond to the balls. We show the image in Fig. A.3. In the urn, the balls are same colors, white or red. It corresponds to the opinion of the voter.

One urn joins as time passes. When the i-th urn enters the network at time $t = i$, r balls are taken out from all urns with a probability that is proportional to the popularity. When the urn joins, r balls which corresponds to the outgoing link are in the urn. The weighted number of balls corresponds to popularity. The color of balls in the new urn is decided by Eq. (A.6). As the number of votes we counts the white and red balls which corresponds to candidates, C_0 and C_1.

The new node (urn) connects to the selected nodes (urns) by incoming and outgoing links. When the mode (urn) is selected, the outgoing links (balls) are added to it. This affects ω to popularity of the urn. In the negative ω case, there is a cap on the number of balls in the urn. In contrast, positive ω case, there is no cap. We show the image in Fig. A.3. In the limit $\omega = -1$, the network becomes lattice and the urn model was discussed in Chapter 6.

Phase Transition

Here, we consider self-consistent equations for popularity at a large t limit,

$$(1+\omega)r\hat{Z}_{\infty} = \sum_{j=1}^{r} P_{\hat{k},t}(i)\cdot i = r(1-p)q + rp\pi(\hat{Z}_{\infty}) + r\omega\hat{Z}_{\infty}. \qquad \text{(A.12)}$$

Hence, we obtain

$$\hat{Z}_\infty = (1-p)q + p\pi(\hat{Z}_\infty). \tag{A.13}$$

This is a self-consistent equation that does not depend on ω. The self-consistent equation for the voting ratio Z_∞ is also

$$Z_\infty = (1-p)q + p\pi(Z_\infty). \tag{A.14}$$

We define a new variable $\hat{\Delta}_t$ such that

$$\hat{\Delta}_t = g_1(t) - r(t-r+1) = \frac{1}{2}\{g_1(t) - g_0(t)\}. \tag{A.15}$$

For convenience, we change the notation from $\hat{k}$ to $\hat{\Delta}_t$. Therefore, $\hat{\Delta}_t$ holds within $\{-r(t-r+1), r(t-r+1)\}$. Given $\hat{\Delta}_t = \hat{u}$, we obtain a random walk model:

$$\begin{aligned}
&\hat{\Delta} = \hat{u} \to \hat{u} + i: \\
&r/2 \le j \le r, i = (r + (2j - r)\omega/(1+\omega), P_{\hat{u},t}(i) = {}_rC_j \hat{Z}^j (1-\hat{Z})^{r-j}[(1-p)q + p], \\
&0 \le j < r/2, i = (r + (2j - r)\omega/(1+\omega), P_{\hat{u},t}(i) = {}_rC_j \hat{Z}^j (1-\hat{Z})^{r-j}(1-p)q, \\
&r/2 \le j \le r, i = (r + (2j - r)\omega/(1+\omega), P_{\hat{u},t}(i) = {}_rC_j \hat{Z}^j (1-\hat{Z})^{r-j}(1-p)(1-q), \\
&0 \le j < r/2, i = (r + (2j - r)\omega/(1+\omega), P_{\hat{u},t}(i) = {}_rC_j \hat{Z}^j (1-\hat{Z})^{r-j}[(1-p)(1-q) + p],
\end{aligned} \tag{A.16}$$

where $\hat{Z} = \hat{k}/(2r(t-r+1)) = \hat{u}/(2r(t-r+1)) + 1/2$ and $P_{\hat{u},t}(i)$ is the probability of the transition from $\hat{u}$ to $\hat{u}+i$ at t.

We now consider the continuous limit $\epsilon \to 0$

$$\hat{X}_\tau = \epsilon \hat{\Delta}_{[\tau/\epsilon]}, \tag{A.17}$$

where $\tau = t\epsilon$. Approaching the continuous limit, we obtain the following stochastic partial differential equation:

$$\begin{aligned}
\mathrm{d}\hat{X}_\tau = &\left[\frac{2r(1-p)q}{1+\omega} - \frac{r}{1+\omega} + \frac{\omega}{(1+\omega)} \frac{\hat{X}_\tau}{(\tau - r + 1)}\right. \\
&\left.+ \frac{2rp}{1+\omega} \frac{(2n+1)!}{(n!)^2} \int_0^{\frac{1}{2} + \frac{\hat{X}_\tau}{2r(\tau - r+1)}} x^n (1-x)^n dx\right] \mathrm{d}\tau + \sqrt{\epsilon}\mathrm{d}B,
\end{aligned} \tag{A.18}$$

where $\mathrm{d}B$ is the Wiener process. For $r = 1$, the equation becomes:

$$\mathrm{d}\hat{X}_\tau = \left[\frac{(1-p)(2q-1)}{1+\omega} + \frac{p+\omega}{1+\omega} \frac{\hat{X}_\tau}{\tau}\right] \mathrm{d}\tau + \sqrt{\epsilon}\mathrm{d}B. \tag{A.19}$$

The relation between the voting ratio for C_1 and $\hat{X}_\infty$ is

$$\frac{\hat{X}_{\infty}}{2r(\tau - r + 1)} = \hat{Z}_{\infty} - \frac{1}{2}. \quad \text{(A.20)}$$

We assume that the stationary solution is

$$\hat{X}_{\infty} = r\bar{\hat{v}}\tau + r(1-p)(2q-1)\tau, \quad \text{(A.21)}$$

where $\bar{\hat{v}}$ denotes a constant. As Eq. (A.20) and $0 \leq \hat{Z} \leq 1$, we obtain

$$-1 \leq \bar{\hat{v}} + (1-p)(2q-1) \leq 1. \quad \text{(A.22)}$$

Substituting Eq. (A.21) into Eq. (A.18), we obtain

$$\bar{\hat{v}} = -p + \frac{2p \cdot (2n+1)!}{(n!)^2} \int_0^{\frac{1}{2} + \frac{(1-p)(2q-1)}{2} + \frac{\bar{\hat{v}}}{2}} x^n (1-x)^n dx. \quad \text{(A.23)}$$

This equation is self-consistent equation. Equation (A.23) does not depend on parameter ω. Hence, the transition point is the same for all $\omega > -1$.

When $p \leq p_c$, the phase is referred to as the peak phase, and the solution is $\bar{\hat{v}}_0$. Equation (A.23) admits three solutions for $p > p_c$: When $p > p_c$, the upper and lower solutions are stable; in contrast, the intermediate solution is unstable. The two stable solutions correspond to good $\bar{\hat{v}}_+$ and bad equilibria $\bar{\hat{v}}_-$, respectively, and the distribution becomes the sum of the two Dirac measures. This is a two-peak phase.

The phase transition point, p_c is common in all models. If $r = 2n + 1 \geq 3$, then a phase transition occurs in the range $0 \leq p \leq 1$. If a voter obtains information from either one or two voters, there is no phase transition.

Super-Normal Transition

We expand $\hat{X}_\tau$ around solution $\bar{v} = r\bar{\hat{v}}\tau + r(1-p)(2q-1)\tau$,

$$\hat{X}_\tau = r\bar{\hat{v}}\tau + r(1-p)(2q-1)\tau + r\hat{W}_\tau. \quad \text{(A.24)}$$

We set $\hat{X}_\tau \gg \hat{W}_\tau$. This result indicates that $\tau \gg 1$. We rewrite Eq. (A.18), using Eq. (A.24) and obtain the following:

$$\mathrm{d}\hat{W}_\tau = [\frac{\omega + pA}{1+\omega}]\frac{\hat{W}_\tau}{\tau}\mathrm{d}\tau + \sqrt{\epsilon}\mathrm{d}B, \quad \text{(A.25)}$$

where

$$A = \frac{(2n+1)!}{(n!)^2 \cdot 2^{2n}}(1 - \{\bar{\hat{v}} + (1-p)(2q-1)\}^2)^n. \quad \text{(A.26)}$$

The trend in this solution is $(\omega + pA)/(1 + \omega)$ where: there is a transition in the trend at the $(\omega + pA)/(1 + \omega) = 1/2$. We set the solution to this equation as p_{vc}. When $(\omega + pA)/(1 + \omega) > 1/2$, the convergence speed is slower than that in the normal case. However, when $(\omega + pA)/(1 + \omega) < 1/2$, the convergence speed is slower than that in the normal case. We call this transition a super-normal transition, and its transition point is p_{vc}.

When $\omega \geq 1$, $(\omega + pA)/(1 + \omega) \geq 1/2$ because $A \geq 0$. Thus, there is no super-normal transition at $\omega \geq 1$. We observed a super-normal transition at $1 > \omega > -1$.

In summary, the convergence of $Z(t)$ is

$$V(Z(t)) \propto \begin{cases} t^{-1} & p < p_{vc}, \\ t^{(2pA-2)/(1+\omega)} & p > p_{vc}, \\ \frac{\log(t)}{t} & p = p_{vc}, \end{cases} \tag{A.27}$$

where $V(Z(t))$ is the variance of $Z(t)$.

Symmetric Case, q = 1/2

Here, we consider a symmetric model $q = 1/2$. When $r \geq 3$, there are two stable solutions and one unstable solution $\hat{v} = 0$ above p_c. The vote ratio for C_1 is a good or bad equilibrium. In one sequence, Z is taken as $\bar{\hat{v}}/2 + 1/2$ in the case of good equilibrium or as $-\bar{\hat{v}}/2 + 1/2$ in the case of bad equilibrium where $\bar{\hat{v}}$ is the solution of Eq. (A.23). This indicates a two-peak phase, and the critical point is $p_c = \frac{(n!)^2}{(2n+1)!}2^{2n} = 1/A$, where the gradient of the RHS of Eq. (A.23) at $\bar{\hat{v}} = 0$ is 1. For $r = 3 (n = 1)$, $p_c = 2/3$, and $r = 5 (n = 2)$, $p_c = 8/15$. As r increases, p_c moves toward 0. In the large limit, $r \to \infty$, p_c becomes 0. This is consistent with the case where herders obtain information from all previous voters. The discussion above does not depend on ω. In the limit $\omega = -1$, there is no phase transition in the lattice case.

We consider the super-normal transition in the symmetric case $q = 1/2$ by considering the case $r = 2n + 1 \geq 3$. In this case, we observe a phase transition. If $r \leq 2$, no phase transition is observed.

In the one-peak phase $p \leq p_c$, the only solution is $\bar{\hat{v}} = 0$. p_c is the critical point of the phase transition. The critical point of convergence is $p_{vc} = (1 - \omega)/2A = \frac{(1-\omega)}{2} p_c$.

Above p_c, in the two-peak phase, we obtain two stable solutions that are not $\bar{\hat{v}} = 0$. At p_c, $\bar{\hat{v}}$ moves from 0 to one of the two stable solutions. In one voting sequence, the votes converge to one of these stable solutions. Here, $\bar{\hat{v}}$ is the solution of Eq. (A.23). In the case $r = 3$ we obtain $\bar{\hat{v}} = \pm\sqrt{(3p - 2)/p}$ and $p_c = 2/3$. The critical point of convergence in the two-peak phase was $p_{vc} = (5 + \omega)/6$.

In Fig. A.4, we show the phase diagrams for cases $q = 1/2$, $r = 3$, and $r = 1$. In the case $r = 1$, there is only a super-normal transition. The super-normal transition

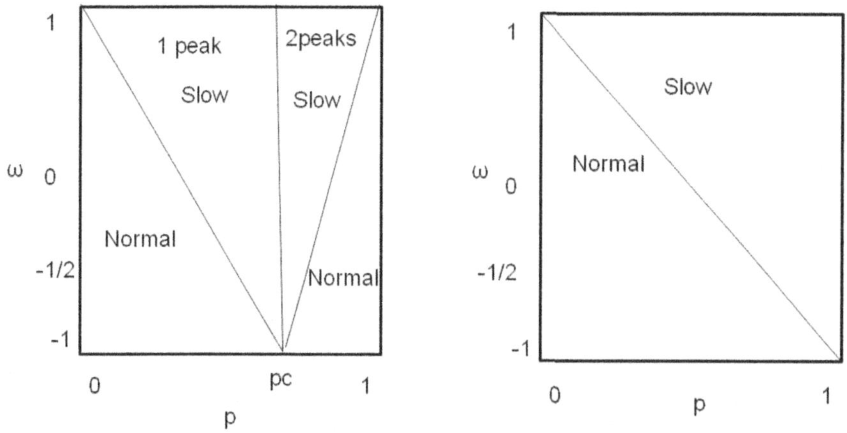

Fig. A.4 Phase diagram of the symmetric case ($q = 1/2$) for $r = 1$ and $r = 3$ cases. The left figure shows $r = 1$ and the right figure show $r = 3$. In the case of $r = 3$, p_c is 2/3. The horizontal axis is p, and the vertical axis is ω

disappears at $\omega = -1$. In the case $r = 3$, we can confirm two types of phase transitions, the phase transition and a super-normal transition. The super-normal transition disappears at $\omega = -1$. In the analog herder case, the phase diagram becomes the same as that in the $r = 1$ case. Analog herders vote for each candidate with probabilities proportional to the candidates' votes.

Next we consider the self-consistent equation, Eq. (A.23),

$$\bar{\hat{v}} = -p + \frac{2p \cdot (2n+1)!}{(n!)^2} \int_0^{\frac{1}{2}+\frac{\bar{\hat{v}}}{2}} x^n (1-x)^n dx = f(\bar{\hat{v}}). \qquad \text{(A.28)}$$

$f(x)$ is the odd function because of the symmetry and $f(0) = 0$ and $f'(0) = 1$ at the critical point $p = p_c$. Therefore, we can set

$$\begin{aligned} f(\bar{\hat{v}}) &= \frac{p \cdot (2n+1)!}{2^{2n}(n!)^2} \bar{\hat{v}} - \frac{p \cdot n \cdot (2n+1)!}{3 \cdot 2^{2n}(n!)^2} \bar{\hat{v}}^3 \cdots, \\ &= \frac{p}{p_c} \bar{\hat{v}} - \frac{n \cdot p}{3 p_c} \bar{\hat{v}}^3 + \cdots, \end{aligned} \qquad \text{(A.29)}$$

where $p_c = \frac{(n!)^2}{(2n+1)!} 2^{2n}$. We obtain

$$\bar{\hat{v}} = \pm \sqrt{\frac{p - p_c}{p_c} \frac{3}{n \cdot p}} \sim \pm |p - p_c|^{\beta}. \qquad \text{(A.30)}$$

Therefore, the critical exponent $\beta = 1/2$ when n is finite. It is same as the Tanh-type model discussed in Chap. 4 and ACO discussed in Chap. 8.

Asymmetric Case, $q \neq 1/2$

Next, we consider the asymmetric case. In this case, the phase transition is of the first order that exhibits a discontinuity at the critical point p_c. We investigate the correct ratio for this phase transition. To define the correct ratio, we assume that candidate 1(0) is the correct(incorrect) one. The correct ratio of independent voters is $q > 1/2$, and the correct ratio of the herder determines the total correct ratio. Using the correct ratio, we can confirm the voting performance. The phase transition has a negative effect on performance. In Fig. A.5, we show the image of the correct ratio for several ω. The horizontal axis represents, p and the vertical axis represents the correct ratio. We observe a gap in the correct ratio at transition point p_c, when $\omega > -1$ p_c does not depend on ω. When $\omega = -1$, no phase transition occurred.

In the limit $\omega \to \infty$, the trend of Eq. (A.25) is 1 for all Z. In this case, the hubs have the maximum number of links. At the critical point, p_c the correct ratio depends only on the initial voting. The two equilibria are $\bar{\hat{v}}_+$ and $\bar{\hat{v}}_-$. The probabilities of good and bad equilibria are given by α and β. The correct ratio is $Z_\infty = (\alpha\bar{\hat{v}}_+ + \beta\bar{\hat{v}}_-)/2 + 1$. In the case $\omega \to \infty$, we obtain $Z_\infty = (\alpha\bar{\hat{v}}_+ + \beta\bar{\hat{v}}_-)/2 + 1 = p/2 + (1-p)q$, which is the dotted bottom line in Fig. A.5. In general, for $\omega > -1$, we obtain

$$\lim_{p \to p_c} \bar{v} = 2\bar{\hat{v}} - 1 > (\alpha\bar{\hat{v}}_+ + \beta\bar{\hat{v}}_-)/2 + 1 \geq p/2 + (1-p)q. \tag{A.31}$$

This means that the the first vote decides the equilibrium. In other words, the initial impact decides the equilibrium.

In summary, in the one-peak phase, the correct ratio does not depend on ω. In contrast, in the two-peak phase, the correct ratio depends on ω. As ω increased, the gap at the transition point increases and the correct ratio decreases. In other words, the correct ratio of the herders decreases. In the lattice case $\omega = -1$, no phase transition occurs, and the correct ratio increases without the phase transition as p increases.. The effects of hubs are observed in the correct ratio during the two-peak phase. As the hub effects increase, the gap also increases.

Relation to Elephant Random Walk

In this Appendix, we consider the relationship between the voting model used in this study and the elephants random walk model over a network (Schütz et al. 2004). In the random graph case, and $r = 1$, the model corresponds to standard elephant random work. Here, the number of referred voters is r. In the elephant random walk,

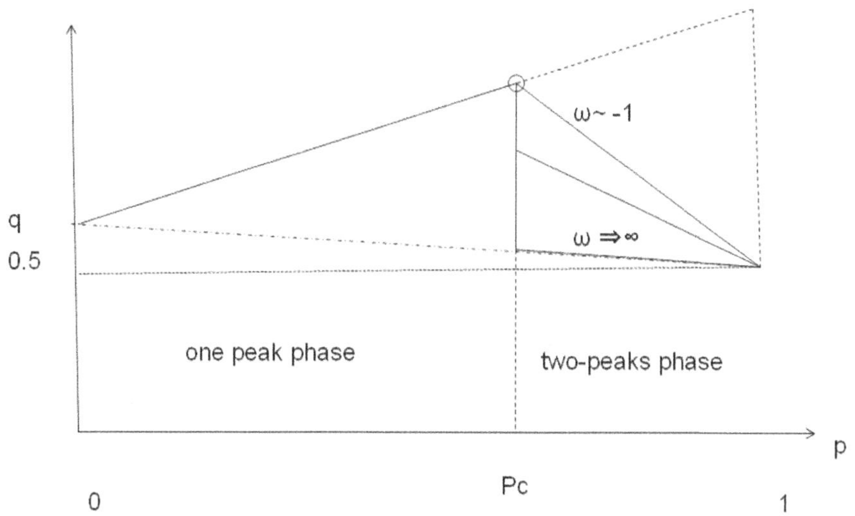

Fig. A.5 Image of expected correct ratio and first-order phase transition in the asymmetric case. The effects of the hubs are observed in the correct ratio in the two-peak phase. As the effects of the hub increase, the gap also increases. The numerical simulation of expected correct ratio is in Fig. A.8

ω is the parameter related to the memory. As ω increases old memory becomes important, and the old behavior is frequently remembered. This corresponds to that the size of hub becomes large.

Here, we set $p = 1$ and consider the case where all voters are herders in the voting model. We use $\hat{p}$ as the parameter instead of p in the voting model. We define $P_1^r(t)$ as the probability that the t-th voter would vote for C_1. We change Eq. (A.6) to

$$P_1^r(t) = \left\{ \begin{array}{ll} \hat{p} & : c_1^r(t) > r/2, \\ \hat{p}/2 & ; c_1^r(t) = r/2, \\ (1-\hat{p}) & : c_1^r(t) < r/2. \end{array} \right\} \tag{A.32}$$

With probability $\hat{p}$, the voter decides on the basis of the referred voters. That is, the difference between p and $\hat{p}$ is the introduction of noise. In the voting model, a voter behaves as a herder. In other words, the voter behaves in accordance with the information obtained. In contrast, in the elephant random walk model, the voter behaves against the referred information with probability $1 - \hat{p}$. In the case $\hat{p} < 1/2$, the probability that the voter behaves against the referred information is larger than the voter behaves according to the obtained information. In the case $r = 1$, the model becomes an elephant random walk model. When the analog herder case corresponds to the elephant random walk, the phase diagram becomes the same as that of the $r = 1$ digital herder case. A voter refers to only one previous voter. Analog herders vote for each candidate with probabilities proportional to the candidates' votes. The initial condition of the elephant random walk is $P_1^r(1) = \hat{q}$ and includes an asymmetric case. Here, we consider the symmetric case, $\hat{q} = 1/2$.

The evolution of connectivity can be written as,

$$
\begin{aligned}
& g_1(t) = \hat{k} \rightarrow \hat{k} + i : \\
r/2 \leq j \leq r, i = \omega j + r \quad & P_{\hat{k},t}(i) = {}_rC_j \hat{Z}^j (1 - \hat{Z})^{r-j} \hat{p}, \\
0 \leq j < r/2, i = \omega j + r \quad & P_{\hat{k},t}(i) = {}_rC_j \hat{Z}^j (1 - \hat{Z})^{r-j} (1 - \hat{p}), \\
r/2 \leq j \leq r, i = \omega j \quad & P_{\hat{k},t}(i) = {}_rC_j \hat{Z}^j (1 - \hat{Z})^{r-j} (1 - \hat{p}), \\
0 \leq j < r/2, i = \omega j \quad & P_{\hat{k},t}(i) = {}_rC_j \hat{Z}^j (1 - \hat{Z})^{r-j} \hat{p},
\end{aligned}
\tag{A.33}
$$

and $P_{\hat{k},t}(i)$ is the probability of the transition from $\hat{k}$ to $\hat{k} + i$ at t.

Approaching the continuous limit, as in Section IV, we can obtain the stochastic partial differential equation:

$$
\begin{aligned}
\mathrm{d}\hat{X}_\tau =& [\frac{\omega}{(1+\omega)} \frac{\hat{X}_\tau}{(\tau - r + 1)} \\
&+ \frac{r(2p-1)}{1+\omega} \{\frac{2 \cdot (2n+1)!}{(n!)^2} \int_0^{\frac{1}{2} + \frac{\hat{X}_\tau}{2r(\tau - r + 1)}} x^n (1-x)^n dx - 1\}]\mathrm{d}\tau + \sqrt{\epsilon}\mathrm{d}B.
\end{aligned}
\tag{A.34}
$$

When we set $2\hat{p} - 1 = p$, Eq. (A.34) becomes Eq. (A.18).

For $r = 1$, the equation becomes

$$
\mathrm{d}\hat{X}_\tau = [\frac{2p - 1 + \omega}{1 + \omega} \frac{\hat{X}_\tau}{\tau}]\mathrm{d}\tau + \sqrt{\epsilon}\mathrm{d}B. \tag{A.35}
$$

We assume that the stationary solution is

$$
\hat{X}_\infty = r\bar{\hat{v}}\tau + r(1-p)(2q-1)\tau, \tag{A.36}
$$

where $\bar{\hat{v}}$ denotes a constant. As Eq. (A.20) and $0 \leq \hat{Z} \leq 1$, we obtain

$$
-1 \leq \bar{\hat{v}} + (1-p)(2q-1) \leq 1. \tag{A.37}
$$

Substituting Eq. (A.36) into Eq. (A.34), we obtain

$$
\bar{\hat{v}} = (2p-1)\{\frac{2 \cdot (2n+1)!}{(n!)^2} \int_0^{\frac{1}{2} + \frac{\bar{\hat{v}}}{2}} x^n (1-x)^n dx - 1\}. \tag{A.38}
$$

This equation is self-consistent. The equation does not depend on parameter ω.

We expand $\hat{X}_\tau$ around solution $r\bar{\hat{v}}\tau + r(1-p)(2q-1)\tau$,

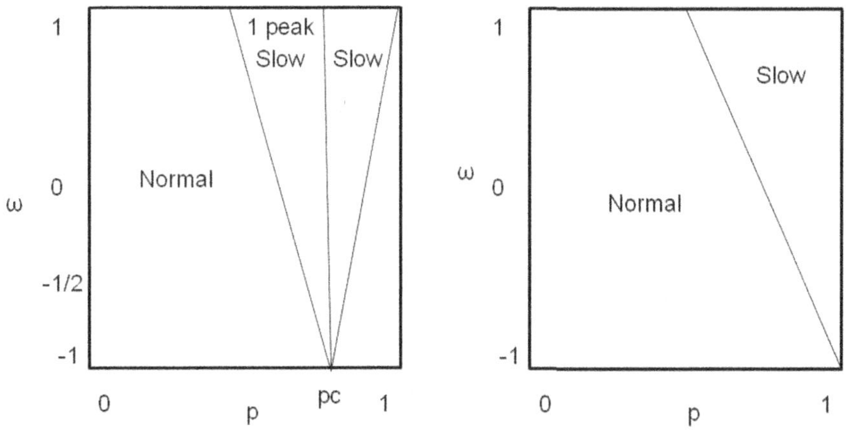

Fig. A.6 Phase diagram of elephant random walk for $r = 1$ and $r = 3$ cases in the symmetric case $q = 1/2$. The left figure shows $r = 1$ and the right figure show $r = 3$. In the case $r = 3$, p_c is 2/3

$$\hat{X}_\tau = r\bar{\hat{v}}\tau + r\hat{W}_\tau. \tag{A.39}$$

We set $\hat{X}_\tau \gg \hat{W}_\tau$. This result indicates that $\tau \gg 1$. We rewrite Eq. (A.34), using Eq. (A.39), and obtain the following:

$$\mathrm{d}\hat{W}_\tau = [\frac{\omega + (2p-1)A}{1+\omega}]\frac{\hat{W}_\tau}{\tau}\mathrm{d}\tau + \sqrt{\epsilon}\mathrm{d}B, \tag{A.40}$$

where

$$A = \frac{(2n+1)!}{(n!)^2 \cdot 2^{2n}}(1-\bar{\hat{v}}^2)^n. \tag{A.41}$$

The phase diagram of the elephant random walk is shown in Fig. A.6. We can map from Figs. A.6 to A.4, if we set $2\hat{p} - 1 = p$. The case $r = 1$ corresponds to the elephant random walk. We confirm a super-normal transition of the basic elephant random work at $\hat{p} = 3/4$ (Schütz et al. 2004).

Numerical Study of Phase Transitions

In this section, we present the results of numerical studies on the phase transitions of the voting model in networks. We adopt the model parameters (r, q) as $r = 3$ and $q \in \{0.5, 0.6\}$ and change the control parameter $p \in [0, 1]$. With regard to ω, we adopt $\omega \in \{-0.9, -2/3, -1/2, 0, 1, 2, 9\}$. We estimate the statistical quantities of the model using simple Monte Carlo sampling. Using the probabilistic rules of the network and voting model, we generated 10^2 networks and 10^4 sample sequences in

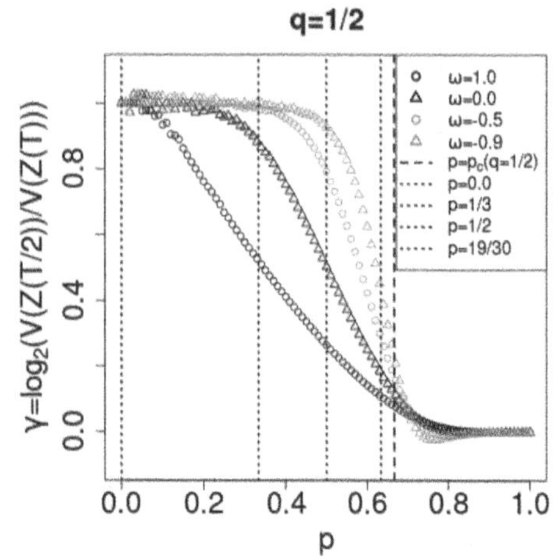

Fig. A.7 Plot γ versus p for $q = 0.5$, $r = 3$ and $T = 10^4$, $\omega = 1$(black○), 0.0(black △), −0.5(gray∘) and −0.9(gray△), where T is the time. The broken and dotted lines show the value of $p_c = 2/3$ and $p_{vc}(q, \omega)$ for $\omega \in \{1, 0.0, -0.5, -0.9\}$

each network. After estimating the statistical quantities by taking the average over the sample sequences, we calculated the average of the networks.

When $\omega = -1$, we adopt an extended lattice with $r = 3$, which is different from that of the network with $\omega = -1$ because of the initial condition of the network. We used a perfect network as the initial condition for this numerical simulation.

The probability that voter 1 was chosen by the next voters is zero because of the in-degrees is 0. In addition, the in-degrees and popularity of Voters 2 and 3 are 1 and 2, respectively. Because they are smaller than those of the voters for $t > 3$, they are chosen less by the next voters.

Symmetric Case, q = 1/2

In this subsection, we consider the symmetric case, $q = 1/2$. A phase transition occurred at $p = p_c = 2/3$. In the one-peak phase $p < p_c = 2/3$, $Z(t)$ converges to the unique solution to Eq. (A.14). There is also a super-normal transition at $p = p_{vc} = (1 - \omega)/3$. As Eq. (A.27), the power-law exponent γ of $V(Z(t)) \propto t^{-\gamma}$ is given by

$$\gamma = \begin{cases} 1 & p < p_{vc}, \\ (2 - 3p)/(1 + \omega) & p > p_{vc}, \end{cases}$$

where $V(Z(t))$ is the variance of $Z(t)$. At $p = p_{vc}$, $V(Z(t)) \propto \log t / t$. Here, $\gamma = 1$ is the normal phase and $0 < \gamma < 1$ is the super diffusion phase. We estimate γ using the following estimator: $\hat{\gamma} = \log_2 V(Z(T/2))/V(Z(T))$.

Figure A.7 plots γ versus p for $\omega \in \{1, 0, -0.5, -0.9\}$. The horizontal axis represents the ratio of herders p. The vertical axis represents the speed of convergence γ. The vertical broken line indicates the critical point, $p_c(1/2) = 2/3$. The other

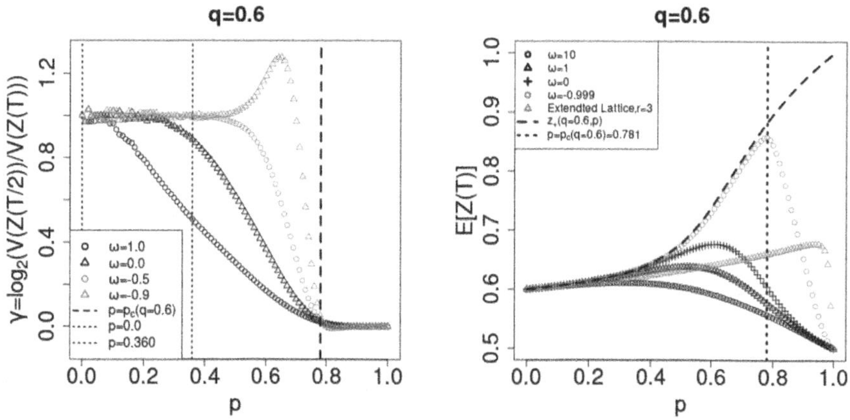

Fig. A.8 Plot γ versus p for $q = 0.6, r = 3$ and $T = 10^4$, $\omega = 1$(black$\circ$), 0.0(black$\triangle$), -0.5(gray$\circ$) and -0.9(gray$\triangledown$). The broken solid and dotted lines show the values of $p_c(q = 0.6) = 0.781$ and $p_{vc}(q = 0.6, \omega)$ for $\omega \in \{1, 0.0\}$. There is no super-phase for $\omega \in \{-0.5, -0.9\}$. Right: Plot of the $z_+(q = 0.6, p)$, the unique solution of Eq. (A.13) for $p < p_c$ and the larger solution of Eq. (A.13) for $p > p_c$ and the estimated expected value of $Z(T)$ versus p. $\omega = 10$(black$\bigcirc$), 1(black$\triangle$), 0(black+), -0.999(gray$\bigcirc$) and the extended lattice ($gray\triangle$). The thin dotted line indicates $p = p_c(0.6)$. In the extended lattice case, we confirm there is no phase transition as ($gray\triangle$)

vertical dotted lines show p_{vc} for $\omega \in \{1, 0, -0.5, -0.9\}$, respectively. The phase transition occurs at $p = p_c(1/2)$. For $p > p_c$, γ is almost zero, and the variance in $Z(t)$ becomes constant. There are two stable states. For $p > p_c$, $V(Z(t))$ does not decrease to zero. For $p < p_c$, a super-normal transition occurs. The positions of the critical points are vague; evidently, region p for the normal phase for $\omega = 1$ is extremely narrow, and it considers $p_{vc}(\omega = 1) = 0$. For $\omega = 0$, $p_{vc} = 1/3$ and a plateau region was observed for $p < 0.25$. As ω decreases from $\omega = 1 \to -0.9$, the plateau region widens. Near the critical point p_{vc}, γ is smaller than one, and estimating p_{vc} from the figure is difficult.

Asymmetric Case, q = 0.6

A phase transition occurred at $p = p_c(q = 0.6) = 0.781$. As in the symmetric case $Z(t)$ converges to a unique solution as Eq. (A.14) for the one-peak phases for $p < p_c$. The variances of $Z(t)$ and $V(Z(t))$ also decrease to zero in the limit. There is also a super-normal transition at $p = p_{vc}(0.6)$. When $\omega = 1(0)$, $p_{vc} = 0(0.360)$. When $\omega = -0.5, -0.9$, only the normal convergence phase exists.

The left figure of Fig. A.8 plots γ versus p for $\omega \in \{1, 0, -0.5, -0.9\}$. The horizontal axis represents the ratio of herders p. The vertical axis represents the speed of convergence γ. The vertical broken line indicates the critical point, $p_c(q = 0.6)$. The other vertical dotted lines show p_{vc} for $\omega \in \{1, 0\}$, respectively. It is clear that the phase transition at $p = p_c(q = 0.6)$. For $p > p_c$, γ is almost zero, and the variance in $Z(t)$ becomes constant. There are two stable states. For $p > p_c$, $V(Z(t))$ does not decrease to zero. For $p < p_c$, a super-normal transition occurs. The positions of the critical points are vague; evidently, region p for the normal phase for $\omega = 1$ is extremely narrow, and it considers $p_{vc}(\omega = 1) = 0$. For $\omega = 0$, $p_{vc} \simeq 0.36$, and a plateau region was observed for $p < 0.3$. As ω decreases from $\omega = 1 \rightarrow -0.9$, the plateau region widens. Near the critical point p_{vc}, γ is smaller than one, and estimating p_{vc} from the figure is difficult.

In the asymmetric case, the phase transition at $p_c(q), q > 1/2$ is discontinuous. The image on the right side of Fig. A.8 shows the average value of $Z(T)$, which is the correct value. The dashed curve z_+ in the right figure of Fig. A.8 corresponds to the upper slanting solid line and the connected dotted line in Fig. A.5. We confirm the gap increases as ω increases. Note that if we increase T, p_c becomes the thin dotted line. In addition to this network model, $\omega \in \{10, 1, 0, -0.999\}$, we present the results for the extended lattice case. Because there is only one stable state for $p < p_c(q)$, $Z(t)$ converges to the solution to Eq. (A.14).

We denote this solution as $z_+(q, p)$ in Fig. A.9. The figure shows that the convergence speed crucially depends on ω. When $\omega = 10$ and $T = 10^4$, as p increases from 0 to 1, the average value of $Z(T)$ deviates from $z_+(q = 0.6, p)$. As ω decreases to -1, the departure timing is delayed, and at $\omega = -0.999$, the average value coincided with $z_+(q, p)$ until $p_c(q)$.

For $p \geq p_c$, there are two stable solutions to Eq. (A.14), and we denote them as $z_+(q, p) = 2\hat{v}_+ - 1$ and $z_-(q, p) = 2\hat{v}_- - 1$. $z_+(q, p)$ is a larger solution, and the unique solution for $p < p_c(q)$ continuously becomes $z_+(q, p)$ for $p > p_c(q)$. In the case $p < p_c(q)$, the expected value is the weighted average of $z_+(q, p)$ and $z_-(q, p)$. In the limit $T \rightarrow \infty$, $E(Z(T))$ converges to $z_+(q, p)$. These results suggest that the discontinuity in $E(Z(T))$ in the limit $T \rightarrow \infty$ at $p = p_c$ depends on ω.

From the viewpoint of performance, network $\omega \sim -1$ exhibits the best performance among all networks. There was a phase transition; however, the gap was small. Hence, the highest correct ratio can be confirmed for $\omega \geq -1$. However, when there is a large hub, the gap increases and the performance worsens in the two-peak phase. In the lattice case $\omega = -1$, there is no phase transition; however, the performance is worse than that of the other networks.

Summary

We investigated the phase differences between the different networks as models for voter information. This is based on the assumption that voters obtain information from a network, including hubs. A voting model represents how public perceptions are conveyed to voters. Our voting model was constructed using two types of voters—herders and independents—and two candidates. Herders vote for the majority of

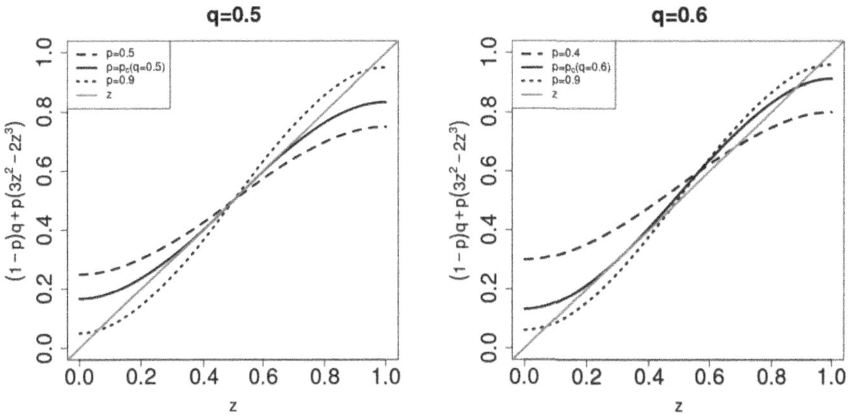

Fig. A.9 Solutions of Eq. (A.14) for case $r = 3$ and $q = 0.5, 0.6$. The left figure shows the symmetric case, and the right figure shows the asymmetric case. The horizontal axis is Z, and the vertical axis is the RHS of Eq. (A.14). In both figures, when $p < p_c$, there is only one solution. When $p > p_c$, there are three solutions. The right solutions are z_+ and the left solutions are z_-. z_+ and z_- are the stable solutions

candidates and obtain information related to previous votes from their networks. We examined the differences between the phases on which networks depend.

In $\omega > -1$, we observed two types of phase transitions: phase transition and super-normal transition. As the number of herders increases, the model features a phase transition beyond which a state in which most voters made the correct choice coexisted with one in which most of them were wrong. During this transition, the distribution of votes changes from the one-peak phase to the two-peak phase. This is an absorption phase transition that belongs to the non-equilibrium phase transition. The other transition is a super-normal transition in terms of convergence speed.

In the symmetric case, the phase transition is continuous transition. However, in the asymmetric case, thus is discontinuous. We confirmed that the transition point does not depend on ω for both symmetric and asymmetric cases. In the symmetric case, the phase transition does not depend on the network. **However, in the asymmetric case, the gap in the discontinuous phase transition depends on the network.** As ω increases, the gap increases. In other words, the hubs affect the gap in the discontinuous phase transition. In the two-peak phase, hubs affect the correct ratio. However, in the one-peak phase, the hubs do not affect the correct ratio.

The super-normal transition also depends on the network. As ω increases, the transition point p_c decreases. At $\omega = 1$, the transition disappears, and at $\omega \geq 1$, convergence is slow for any p. Hence, a phase transition is observed in $1 > \omega > -1$. This is effective for both symmetric and asymmetric cases.

For $\omega = -1$ which belongs to the lattice network, there was no phase transition. Hence, the correct ratio increases as p increases because there is no phase transition. The lattice case is summarized in Chap. 6.

Table A.1 Networks and universality class in the voting model

Model	All $r = t$	Networks $\omega \neq -1$	Lattice $\omega = -1$
Symmetry	$\beta = 1$	$\beta = 1/2$ (Mori et al. 2015)	Oscillation See Chap. 6
Asymmetry	$\beta = 1$ See Chap. 5	First kind phase transition	Oscillation See Chap. 6

From the viewpoint of performance, near the lattice case, $\omega \sim -1$ exhibits the best performance of the voting in all networks. As the hub size decreases, the performance improves.

Table A.1 summarizes the relations between networks and the phase transition. In this study, we considered networks corresponding to $\omega \neq -1$ and the lattice with $\omega = -1$. In the lattice case, there was no phase transition. In contrast, in networks with $\omega \neq -1$, we observe a phase transition. In the symmetric case, it is a continuous transition with critical exponent $\beta = 1/2$. The universality class is the same as that of the nonlinear Pólya model (Hill et al. 1980; Pemantle 1991, 2007). In the asymmetric case, the transition is discontinuous. When all voters are referred to , the model belongs to the universality class of the voter model (Hinrichsen 2000). In this case, the critical exponent is $\beta = 1$. This is also an absorption phase transition that belongs to the non-equilibrium phase transition.

References

Hill R, Lane D and Sudderth W, Exchangeable Urn Processes, *Ann. Probab* **8**, 214 (1980).

Hinrichsen H, Non equilibrium Critical Phenomena and Phase Transitions into Absorbing States, *Adv. Phys* **49** 815 (2000).

Mori S and Hisakado M, Correlation function for generalized Pólya urns: Finite-size scaling analysis, Phys. Rev. E., **92** 052112 (2015).

Mori S and Hisakado M, Finite-Size Scaling Analysis of Binary Stochastic Processes and Universality Classes of Information Cascade Phase Transition, *J. Phy. Soc. Jpn* **84** 054001 (2015).

Pemantle R, When are touchpoints limits for generalized Pólya urns? *Proc. Am. Math. Soc* **113** 235 (1991).

Pemantle R, A survey of random processes with reinforcement, *Prob Surv* **4** 1 (2007).

Schütz G M and Trimper S, Elephants can always remember: Exact long-range memory effects in a non-Markovian random walk, *Phys. Rev. E* **70**, 045101(R) (2004).

Watts D J, A simple model of global cascades on random networks, Proc. National Academy of Sciences, **99(9)** 5766 (2002).

Post Script

I joined a rating agency about 20 years ago and began researching credit risk. In my research, I was interested in similar distributions, such as the beta distribution and the Vasicek distribution. Both distributions exhibit flexibility and can represent continuous uniform distribution.

The Vasicek distribution is connected to the Merton model and is frequently used by financial institutions. This model has a hidden value as an economic indicator, allowing us to interpret the mean of the distribution in terms of finance. This is the economic motivation. Using this hidden value, the distribution represents the correlation. Conversely, the beta distribution is sometimes used for Collateralized Debt Obligations (CDOs) in certain financial institutions, although it lacks a solid foundation in finance. This distribution represents the correlation using the positive feedback. However, there is no economic motivation for this distribution. The methods to represent the correlation are different and it is difficult to compare them.

In the limit, we derive Poisson processes with a gamma distribution for the beta-binomial distribution and a log-normal distribution for the Vasicek distribution as the intensity function. The gamma and log-normal distributions share a similar shape, extending from 0 to (∞). Both models can be applied to financial time series, as discussed in Chaps. 13 and 14. From a mathematical standpoint, the self-exciting negative binomial distribution (SE-NBD) process or the Hawkes process is straightforward to apply because of the reproductive property of the gamma distribution. This is the conclusion of this book.

We hope that these tools will be useful in various fields, considering the properties of urn processes discussed in the first half of this book.

M. Hisakado, *Urn Models and Their Applications in Finance*,
https://doi.org/10.1007/978-981-96-3825-3

Name Index

M. Hisakado, *Urn Models and Their Applications in Finance*,
https://doi.org/10.1007/978-981-96-3825-3

Subject Index

M. Hisakado, *Urn Models and Their Applications in Finance*,
https://doi.org/10.1007/978-981-96-3825-3

The manufacturer's authorised representative in the EU is Springer Nature Customer Service Centre GmbH, Europaplatz 3, 69115 Heidelberg, Germany. If you have any concerns regarding our products, please contact ProductSafety@springernature.com

Printed and bound by CPI Group (UK) Ltd, Croydon, CR0 4YY
30/09/2026
02225296-0011